"Impeccably researched, Ruchira Talukdar's book is a timely contribution to the ongoing debates on coal in two most coal-dependent countries of the world: India and Australia. This book is far greater than the poisoning curse of coal; it offers a passionate and devastating critique of dirty coal and an enriching analysis of the resistances and different approaches to decarbonisation in these two countries. A fine book, immensely significant in shaping how the world will think about just transition."

Kuntala Lahiri-Dutt, *Professor in Resource, Environment and Development, The Australian National University*

"One of the provocative aspects of this book is the ultimate moral argument that combines human rights and land justice. This argument is quite significant. Human rights alone can fail to link with anti-colonial, resurgence, abolition, and other land-based social justice movements. Ruchira Talukdar makes a brilliant connection between human rights and land rights, and offers a solution to what are problematic rhetoric, policies, and proposals that privilege human rights against the deeper aspirations of many communities and populations suffering from injustice, and facing risks from climate change and climate change drivers. There are very few studies comparing northern and southern environmentalisms, placing historic and contemporary environmentalism in both contexts in cross-communication."

Kyle Whyte, *George Willis Pack Professor and University Diversity and Social Transformation Professor, University of Michigan*

"In this meticulous and in-depth investigation of anti-coal politics in Australia and India, Ruchira Talukdar shows us the power of subsistence communities in India and Indigenous peoples in Australia in slowing the pace of coal mine expansion and driving the shift towards renewable energy. While often on the frontline of coal extractivism – tied to human rights abuses, destruction of country and exposure to environmental pollutants – subsistence and Indigenous communities are also the front line of its resistance. At times this resistance lines up alongside the environment movement, and other times it does not. This book comprehensively draws out the continuities and discontinuities between these rights-based campaigns and the broader environment movement. In so doing, it exposes how national and global environment movements sometimes sideline and/or silence a rights-based agenda as they seek to meet their own goals.

By giving voice to the environmentalism of the poor in India and Indigenous rights in Australia, this book demonstrates why centring rights, including Indigenous rights, will be vital to achieve social justice and environmental responsibility in a decarbonised world. Everyone aspiring for a climate-just future should read this book. And those of us in the environment movement should definitely down tools long enough to read this book and let its message soak in; then pick our tools back up just a little more carefully."

Kristen Lyons, *Professor in Environment and Development Sociology, University of Queensland*

I0042097

Politics and Resistance of Coal in Australia and India

Since 2009, international climate activism has focused on stopping coal mining in solidarity with local and Indigenous struggles that are resisting coal mining. Based on ethnographic and historic research in Australia and India, this book compares the politics and resistance to coal in the two countries, particularly focusing on the time period between 2009 and 2018 and the case of the Carmichael coal mine in Queensland and the Mahan coal mine in central India.

This book shows differences and similarities in the political economy of coal and creates understanding about the significantly different imperatives and narratives of anti-coal environmentalism in Australia and India. Through the Stop Adani movement and its collaboration with the Wangan and Jagalingou Traditional Owners and farmers against coalmining in Queensland, and Greenpeace and forest-based communities resisting coalmining in Madhya Pradesh, Ruchira Talukdar explores not only anti-coal movement dynamics but also how these movements grapple with the violation of Indigenous land rights through coal extraction in both places. Drawing on differences and patterns in Australian and Indian anti-coal activisms, this book proposes a global outlook – an intersectional framework beyond the singularity of 'stopping coal' that can encapsulate visions for secure futures of communities on the frontlines of fossil fuel struggles – for climate activism. The conclusions help to decolonise climate activism as well as make it cognizant of global North–South contextual differences for effective solidarity.

The author's unique vantage point through experience in environmental activism over 20 years across Australia and India combined with research in both countries makes this book a crucial resource for scholars and practitioners in just transition, climate politics and environmental activism across the global North and South..

Ruchira Talukdar has worked in the environment movement in India and Australia, in Greenpeace, Australian Conservation Foundation and Friends of the Earth, for two decades. Her research and writing focusses on comparative aspects of climate justice between the global North and South, with specific reference to Australia and South Asia. Her PhD thesis compared the politics and resistance to coal in Australia and India. Ruchira co-founded Sapna South Asian Climate Solidarity, a climate justice project based out of Australia, for effective global North solidarity for just climate futures in the global South. She is based out of Melbourne and Calcutta.

Routledge Studies in Environmental Justice

This series is theoretically and geographically broad in scope, seeking to explore the emerging debates, controversies and practical solutions within Environmental Justice, from around the globe. It offers cutting-edge perspectives at both a local and global scale, engaging with topics such as climate justice, water governance, air pollution, waste management, environmental crime, and the various intersections of the field with related disciplines.

The *Routledge Studies in Environmental Justice* series welcomes submissions that combine strong academic theory with practical applications, and as such is relevant to a global readership of students, researchers, policy-makers, practitioners and activists.

Solar Technology and Global Environmental Justice
The Vision and the Reality
Andreas Roos

Environmental Justice in Early Victorian Literature
Adrian Tait

Guerrilla Ecologies
Green Capital, Nature, and the Politics of Catastrophe
John Maerhofer

Rectifying Climate Injustice
Reparations for Loss and Damage
Laura García-Portela

Environmental Justice in Nepal
Origins, Struggles, and Prospects
Edited by Jonathan K London, Jagannath Adhikari, and Thomas Robertson

For more information about this series, please visit: www.routledge.com/Routledge-Studies-in-Environmental-Justice/book-series/EJS

Politics and Resistance of Coal in Australia and India

Climate Justice Activism in the Global North and South

Ruchira Talukdar

Routledge
Taylor & Francis Group

LONDON AND NEW YORK

First published 2025
by Routledge
4 Park Square, Milton Park, Abingdon, Oxon OX14 4RN

and by Routledge
605 Third Avenue, New York, NY 10158

Routledge is an imprint of the Taylor & Francis Group, an informa business

British Library Cataloguing-in-Publication Data
A catalogue record for this book is available from the British Library

ISBN: 978-1-032-53124-3 (hbk)
ISBN: 978-1-032-53125-0 (pbk)
ISBN: 978-1-003-41041-6 (ebk)

DOI: 10.4324/9781003410416

Typeset in Times New Roman
by Newgen Publishing UK

Contents

Figures

Tables

Foreword

Climate change is now urgently upon us, but the strategies we use to address it must bring about justice. And that justice must meet the needs of the disadvantaged and exploited of the world, the global South as well as the global North. Ruchira Talukdar allows us to see how such alliances can be built to challenge climate change and to do it with justice as the outcome.

This book compares environmental conflicts in two locations, one in the global North and one in the global South, where coal mining is taking place. The global North example is the Galilee Basin in Queensland, Australia. The global South case study is in the Singrauli district of Madhya Pradesh in India. Since 2009, when the Copenhagen Climate Summit failed, environmental activists concerned about climate change have had to develop new tactics because it was clear that 'smoke-stack' mitigation of greenhouse gas emissions would not be enough. Instead, the only way to intervene in global warming would be to end the use of any fossil fuels – to keep coal in the ground. Yet, climate change per se was not the priority of many grassroots campaigners for environmental justice with whom environmental activists collaborated to counter coal mining, and it is this dilemma that the book explores by tracing these two case studies.

This moving book draws on deep knowledge of each place, which has allowed Ruchira to conduct her thorough research through careful observation and many interviews. She builds on a strong body of political ecology analysis to trace further the roles of governments, international activists and – most importantly – local, grassroots communities. This book allows us to see how central the concerns of local people must be in any campaign to build justice.

Australia and India are seldom compared in research, yet there is much in common in these two case studies although there are important differences as well. It is long overdue to have analysts consider these two countries together – they are linked to each other by their heavy dependence on coal, although this arises from different reasons. Australia relies on coal mining for export and for a long time in politics has denied much of the science that has identified climate change is occurring. India presents itself as committed to developing a strong renewable energy industry but claims that – as a country with a large proportion of its popu-lation living in poverty – its priority is rapid development for which it must remain dependent on coal, either imported or domestically produced. Now, with Australia's economic turn towards India, the value of research that takes an in-depth approach to helping understand the political economies of coal and energy, the state of democracy and how environmental movements can function in these contexts gives insights about how climate action and climate justice can be achieved, or not, in these places.

To build a transition to renewable energy which has justice at its core demands uncomfortable alliances. One of the many strengths of this book is that it demands that we recognise the challenges this poses for all involved. A focus in each case study is on Indigenous people who have faced immense threats in both countries. The priorities of local, Indigenous peoples in each place may be different to those of environmental activists because local communities face urgent threats around dispossession and damage to Country (in the Australian case) and loss of economically and spiritually essential old growth forest (in the Indian case). As we read this book, we come to understand the perspectives of these local Indigenous communities – it is clear that the immediate threats are those to local environments.

The need to meet the threat of coal mining has meant, however, that Indigenous people in the Galilee Basin have had to build alliances with the very grazier-settlers who have dispossessed them and denied them access to their Country. This has been no easier for the graziers and we have valuable glimpses of the unresolved tensions in this necessary alliance. In Singrauli, it has been the environmental activists who have found it most challenging to build the alliance – they have had to prioritise the damage to forests to empower local Adivasi communities to challenge coal mining which will contribute to greenhouse gas emissions. These case studies are each complex but so powerfully explored in this book that it brings the challenges and dilemmas to life, allowing us as readers to explore with the actors in each area the difficult decisions they have to make.

This book takes us all on an important journey, beyond our experiences, whether they are Indian or Australian, to consider the pathways open to us to find just solutions to the common problems we face. We learn the immense power of the coal corporations in influencing – and at times corrupting – governments in both places, to achieve short-term goals to keep the mining profits going for as long as possible. We come to know the urgency of the fears felt by local communities, bringing them together with unlikely allies to meet the rising threats they all face. And we recognise how environmental activists – at local and international levels – grapple with difficult strategic choices to fulfil their growing commitment to work with local communities to achieve a just and safe future.

While the research in this book focusses on the political economy of coal and its resistance, its findings can also hold lessons for an extractive 'clean' energy economy, effectively pointing to the ground where communities are fighting for climate and environmental justice, and sending forth narratives and assertions that can apply for large-scale renewable energy projects as well as the rush for critical minerals mining unfolding today.

So, this book is both challenging and exciting. It challenges us to see the scale of the threats still being faced in India and Australia. But it is powerfully exciting because it opens up insights into the courageous decisions being taken by local communities and environmental activists as they build strategies for justice.

– Heather Goodall

Heather Goodall is Professor Emerita of History and has published on Indigenous histories and environmental conflicts in Australia and intercolonial networks between Australia and India and around the Indian Ocean.

Acknowledgements

This book has been 20 years in the making. When I stepped into environmental activism in Greenpeace India in 2005, I was struck by the challenges experienced in telling stories on climate change from the ground in India for the global Greenpeace network. The messiness of everyday lives lived in proximity with nature in South Asia – a reality for the vast majority of the population – could not be rolled into neat stories with a singular advocacy for our campaigns. It is at that point that I read Ramachandra Guha's essay *Radical American Environmentalism and Wilderness Preservation: A Third World Critique* (1989) and Guha and Joan Martinez-Aliers' *Varieties of Environmentalism: Essays North and South* (1997). These two texts remain as my earliest influence for writing about environmentalism in the global North and South.

A decade of working in Australia-based environmental organisations offered a chance to see various northern environmentalisms from up close; numerous conversations with fellow activist about environmental movements in India further whetted my appetite to write on this topic. By the time I began my PhD research in 2017, Adani had become a byword for climate change in Australia. It felt opportune to ride the wave of interest that had grown in Australian environmentalism to learn about Indian environmental politics and environmental resistances so as to understand the Adani conglomerate's environmental legacy.

From my PhD supervisors Professors James Goodman and Devleena Ghosh at the University of Technology Sydney I received ample encouragement and freedom to design a layered research project that could address the multiple questions I wanted to ask. Attending Professor Heidi Norman's subject on the political history of Aboriginal Land Rights and realising striking parallels between settler colonial and postcolonial land struggles in Australia and India shaped one of the core investigations of the book on whether Indigenous land rights-based legislations have begun to redress historic colonial wrongs towards these communities. My PhD on the politics and resistance to coal in Australia and India found a stimulating home in UTS's Climate Justice Research Centre, with its commitment towards climate, land and Indigenous justice, and its emphasis on comparative research on climate and energy. Professor Kuntala Lahiri-Dutt at the Australian National

University, whose work on the 'Diverse Worlds of Coal in India' (2016) shaped my own research and writing, has been a mentor throughout.

I have been inspired by activists and communities that I spent time with during my research. In Australia, I particularly thank Anthony Esposito – a long-time ally for Indigenous justice – for the uncalculating gift of time spent talking about Green and Black politics. In India, the time spent with activist and researcher Priya Pillai – who has worked on Indigenous issues and land rights over two decades – helped me to appreciate the smaller details and moments in the life of an environmental resistance. I was struck by the clarity and optimism of the Wangan and Jagalingou youth leader who I met and spoke to for hours on several occasions in Brisbane, while she juggled writing university assignments with the responsibilities of fighting Australia's (potentially) largest coal mine on her people's Land. From the Mahan community, especially women, I received effusive warmth during months of visits. Their sense of pride in their place will always remain with me, as one village elder asked me in a dialect of the Hindi language '*Didi, aapko dehaat kaisa laga*?', meaning, sister, how do you like our Country?

I thank journalists Paranjoy Guha Thakurta and James Norman for reading final drafts of my chapters. I am grateful to friends in Sydney, Melbourne, Calcutta and Santiniketan who have been at hand during the completion of the dissertation and the book. To my parents in Calcutta and my brother in Atlanta, I owe gratitude for unquestioned support and patience. I dedicate this book to Sreedhar Ramamurthi, dear to many and gone too soon, a tireless advocate for communities struggling for land, livelihoods, recognition and dignity in India's mining landscapes.

Ruchira Talukdar, Kolkata, India

Acronyms

ALP	Australian Labor Party
BJP	Bharatiya Janata Party
CAG	Comptroller and Auditor General
CBA	Coal Bearing Area
CEA	Central Electricity Authority
CPR	Centre for Policy Research
CSE	Centre for Science and Environment
ENGOs	Environmental non-governmental organisations
EPBC	Environment Protection and Biodiversity Conservation Act 1999
FRA	Forest Rights Act 2006
GHG	Greenhouse gas
IEA	International Energy Agency
LNP	Liberal National Party (Queensland)
MCG	Mackay Conservation Group
MoEF	Ministry of Environment and Forests
MSS	Mahan Sangharsh Samiti
NAPCC	National Action Plan for Climate Change
NAPM	National Association of People's Movements
NDA	National Democratic Alliance (India)
NDC	Nationally determined contribution
NMP	National Minerals Policy
NTA	Native Title Act 1993
NTPC	National Thermal Power Corporation (Ltd)
PESA	Panchayat Extension to Schedule Areas Act 1996
UNFCCC	United Nations Framework Convention on Climate Change
UPA	United Progressive Alliance (India)
W&J	Wangan and Jagalingou (traditional owners from Central Queensland)

1 Introduction

A comparative ethnography of anti-coal activism in Australia and India

> We must be able to create even larger coalitions, to change the way activism is perceived, to express the same variety and diversity we see in our cities and territories...the leadership must come from below, from the community.
>
> Klein (2015)

The climate crisis requires the most drastic global reduction of the burning of coal, oil and gas, collectively called fossil fuels, which make the single biggest contribution towards exacerbating atmospheric carbon dioxide levels.

In response to the climate crisis, global climate activism aims to hasten the transition of economies towards renewable energy by also facilitating a transition away from fossil fuels. The 2009 Copenhagen Climate Summit failed to arrive at a definitive pathway to reduce global greenhouse gas (GHG) emissions. And this failure is regarded as a significant milestone that transformed the approach of global climate activism, towards directly stopping the extraction of coal, oil and gas. While activism's earlier 'end-of-pipe' or 'smokestack' approach had centred on demands for the reduction of emissions from the burning of fossil fuels, its post-2009 politics aimed to leave fossil fuels in the ground.

Community struggles at sites of fossil fuel extraction have been an ongoing feature in both the global North and the South. However, such conflicts might not explicitly challenge coal as a fossil fuel, but for the deleterious effects of its extraction on the lands, livelihoods and cultures of those communities (Roy and Schaffartzik 2021). But after Copenhagen, climate activists began widely politicising coal extraction as the main source of carbon emissions, adding a global scale of relevance to local struggles against coal. Since 2009 climate groups in the global North and international activist networks have increasingly politicised fossil fuel extraction through direct disruption at extraction sites and along transport corridors, targeting institutions that fund fossil fuel projects, and mass demonstrations against fossil fuel projects amongst other tactics.

Fossil fuels are inextricably linked to economic development. Targeting coal, oil and gas extractions brought national political economies of energy-development under the purview of environmentalism. It also brought environmental activism

DOI: 10.4324/9781003410416-1

into direct conflict with issues of energy security and the national interest in fossil fuel-centred economies.

Six years after Copenhagen, the Paris Climate Summit set emission reduction targets to limit global warming to within two degrees. It is well understood that meeting the goals of the 2015 Paris Agreement requires a rapid phasing out of fossil fuels, and for those fossil fuels not yet extracted to be left in the ground.[1] In the case of coal which is the biggest contributor to global emissions, being able to meet the Paris targets implied that all existing coal reserves must be phased out within the next decades, and no new coal mines developed (Steffen 2015).

The Paris targets deepened the contradictions between the economic imperatives of fossil fuel-producing countries and the global imperative of addressing climate change. Nations dependent on the production of fossil fuels have experienced economic conundrums owing to a simultaneous increase in globalised trade of fossil fuels during the time period that the world has also tried to forge a climate change agreement (Eckersley 2009). Governments in coal-producing countries around the globe have followed contradictory policy tracks of agreeing to emissions reductions while simultaneously expanding coal production. The entrenched power of the fossil fuel sector has often hijacked the ability of governments to move away from fossil fuel extractive projects, despite global market withdrawals and the decreasing costs of renewables (Healey et al. 2019).

This conflicting political economic backdrop set the framework within which the post-Copenhagen environmentalisms of challenging fossil fuel extraction played out. This book focuses on resistance to the extraction of coal in the global North and South, through an investigation of two anti-coal movement case studies, one in Australia and another in India. In *This Changes Everything*, activist Naomi Klein (2014) proposes a vision and politics of solidarity by climate activists with grassroots and local resistance to fossil fuel extraction for climate activism's new approach. This book investigates how this new approach plays out in the contexts of Australia and India and reflects on the similarities and differences between their contexts, politics, and narratives. It is based on ethnographic field research of an anti-coal movement in Australia, and one in India, roughly during this globally relevant time period for climate action through stopping coal extraction between the 2009 Copenhagen summit and the 2015 Paris summit.

Australia and India stand out in the global coal economy as two countries still continuing to expand coal production even after global coal demands have declined, and the converging timeline of the Paris Agreement in 2015 has made it imperative to rapidly phase out coal production. As an industrialised country, Australia is considered as belonging to the global North, and as a developing country, India to the global South. Under the United Nations' framework for climate change negotiations (UNFCCC), the terms also indicate how countries politicise their climate responsibilities, with developing countries like India holding developed countries as historically responsible. The North–South economic divide between the two countries creates a deep unevenness in the imperatives and politics of coal-led growth.

A North–South divide is also reflected in their environmentalisms and can be partly attributed to their economic divide. A characteristically Northern approach to nature conservation is known to persist in Australian environmentalism, and a contrasting Southern focus on the livelihoods of ecosystem-dependent communities is known to dominate Indian environmentalism. Through an evaluation of the movement cases in Australia and India, this book seeks to understand whether the new approach creates possibilities for common ground across this divide in environmentalism between the North and South, and what that common ground looks like.

Section 1.1 introduces the two case studies, the research purpose and the questions. The Australian and Indian case studies focus on a contested coal mining site each and a concomitant anti-coal movement. I also situate the case studies within the broader political, economic and ecological contexts in their respective countries. Section 1.2 describes this book's ethnographic research structure, methods, and the conceptual framework I have used in analysing the Australian and Indian cases. The final Section 1.3 offers an outline of the book's chapters.

1.1 Research case studies and questions

The need for a rapid phase-out of fossil fuels has put coal at the centre of difficult questions that need to be asked about the relationship between climate action and development (Edwards 2019). But owing to a considerable difference in the imperative for economic development between the industrialised and industrialising economies of the world, the need to phase out coal has placed uneven challenges on the coal-led countries of the North versus the South.

Australia and India represent a global North–South unevenness in their developmental imperatives. Despite both regions being considerably affected by climate change, the scale of the internal challenges they face – balancing growth that is coal-oriented while addressing climate impacts from coal burning on their own populations – varies significantly due to the North–South divide. Both countries have continued to deepen their respective internal contradictions between development and climate action by increasing coal extraction. They have done so despite an increase in their renewable energy production. They stood out as two of the top four countries – the others being China and Indonesia – proceeding with major new coal mining developments since the Paris Agreement in 2015 (Climate Analytics 2019).[2]

Internationally, both countries have continued to extend characteristically Northern and Southern positions on expanding coal extraction as an extension of their climate politics: Australia on grounds of its relatively small net emissions on a global scale (as over 80% of its extracted coal is exported and does not count towards domestic emissions because it is burnt in overseas thermal plants), and India from the perspective of equity due to an extremely low per capita emissions, and the historic right to develop as a postcolonial nation. These contexts expose various contradictions in the political economies of coal in Australia and India and help to explain the significance of the respective anti-coal activisms that this book presents, against the reality of exacerbating climate change.

Another North–South difference that is central to the discussion in this book pertains to the history of environmental movements in both countries – specifically, within Australia's dominant legacy of nature conservation and India's movements for environmental justice for nature-dependent subsistence communities. These legacies and historical narratives pertain to movements from the 1970s and 1980s in both geographies that I discuss in Chapter 2.

The North–South divide in environmentalism between Australia and India brings into question what critical similarities and differences of narratives and politics are likely to be reflected in their anti-coal resistances. It raises the possibility of considering their respective transformations from historic legacies, due to sweeping socio-economic changes and increased environmental conflicts from the mid-1990s in both countries and how such alterations reflect in today's anti-coal movement narratives. Finally, a historic divide in Australian and Indian environmentalisms raises the question of whether a common ground can be achieved across Northern and Southern environmentalism today through a common anti-coal focus.

The following subsections lay out the Indian and Australian contexts along the points raised above, introduce the two cases of coal mining and anti-coal activism, and introduce the research questions I asked as an investigation of environmentalisms.

The Indian context

India is one of the fastest-growing major economies and coal supplies 56% of its electricity (Central Electricity Authority 2019). India has also parallelly developed a large renewable industry. In its Nationally Determined Contribution (NDC) submission for the Paris climate summit, India made a substantial commitment to source 40% of its electricity from renewables by 2030, by aiming to have 175 gigawatts of installed renewables capacity by 2022, and 500 by 2030 (Varadhan 2019). However, at Paris, India also stated the centrality of coal in its energy mix into the future; with economic growth anticipated to involve expanded coal extraction from domestic reserves and private mining from offshore reserves, although the aim remained to reduce coal imports for energy security (Rosewarne 2016).

In its NDC in Paris, India indicated its present-day developmental priority was to provide electricity to the nearly 300 million poor who still live without power (Government of India 2014, p. 4). Electricity use is strongly linked to development in the Human Development Index. India has close to 18% of the world's population, but also high social inequality and a significant poor population. Consequently (even though it is now the world's third highest GHG emitter), owing to the negligible carbon footprint of the poor, its per capita emissions have remained one of the lowest in the world. As the most abundant fuel, coal has played a central role in India's post-independence nation-building and has become synonymous with the national interest (Lahiri-Dutt 2016). Thermal power is strongly linked to poverty reduction and constitutes the moral basis for India's coal-led development.

India is also one of the region's most vulnerable countries to climate change, due to its disproportionate burden on the poor. Predicted impacts include displacements driven by sea level rise and coastal erosion (Hazra et al. 2002), increasing frequency and duration of heat stress (Somanathan et al. 2017), impacts of monsoon variability on agriculture on which 65% of the population relies (Pai et al. 2017; Roxy et al. 2015; Roxy et al. 2017), and risks to water supplies (Adve 2019).

While emphasising its historic disenfranchisement from and its present right to growth, India paid insufficient attention to its own vulnerabilities from climate change during the initial phase of international climate dialogues (Raghunandan 2019). Successive Indian governments have articulated 'climate justice' in terms of India needing the carbon space to grow as a postcolonial nation (Goodman 2016). However, its rising emissions from the mid-1990s brought pressure from large industrialised nations, particularly the United States, to acknowledge its contribution to future emissions (Vihma 2011).

The National Action Plan for Climate Change (NAPCC), adopted in 2008, linked developmental and climate objectives through a 'co-benefits' approach, aiming to 'increase the living standards of a vast majority' to reduce their 'vulnerability to the impacts of climate change' while simultaneously making this development path 'environmentally sustainable' (Government of India 2008, p. 2). Paradoxically, coal-led development was seen as central to the co-benefits approach to mitigate the effects of climate change on the poor.

India's coal-led development has been responsible for the land dispossession and loss of livelihoods of India's indigenous Adivasi ('native dweller') and peasant communities. From the mid-1990s, India started both to privatise and expand coal and thermal power production to sustain a high growth rate of 8% of GDP. Given India's inequality and highly uneven development, the change in the role of the State to that of a broker for private corporations – to acquire land for private coal mining – challenged the already fraught idea of public interest in India's growth (Levien 2011). Even without the challenge of climate change, the pursuit of coal-led industrialisation through neoliberal economic measures has exacerbated land dispossessions and livelihood disruptions for vulnerable communities and escalated ecological conflicts. Currently, India has the highest number of environmental conflicts in the world (Environment Justice Atlas 2016).

The Australian context

The alignment of the Australian economy with major emerging economies in Asia, to supply vast demands for minerals and fuels including coal, has been a primary pillar of Australia's economic policies. Due to this continuing and growing resource demand, Australia has steadily increased its coal production and exports since the beginning of the recent minerals boom from the mid-1990s and is now one of the world's largest coal exporters.

Australia has protected and prioritised its resource exports-driven economic pathway over climate action through the 'no regrets' approach, stipulating that 'Australia should not implement measures that would have net adverse economic impacts nationally or on Australia's trade competitiveness, in the absence of similar action by major GHG-producing countries' (Commonwealth of Australia 1992). Australia has also argued against emissions reductions as a principle, stating that 'we only put in 1.4% of the [net global] emissions' (Prime Minister John Howard, quoted in Bulkeley 2000, p. 725).

A well-coordinated campaign of climate change denial, operating across politics and Australia's mining sector, has consistently discredited climate science and the need for emissions reduction while strengthening the interests of coal and gas extraction (Baer 2016). This combination of policy approaches and political factors has also resulted in Australia having one of the highest per capita emissions in the world; Australia's 1.4% of net global GHG emissions is produced by a mere 0.3% of the global population. The emissions intensity comes from high reliance on coal-fired energy which supplies 75% of the country's electricity. Australia lacked a policy pathway to decarbonise its domestic energy at the time of the Paris climate summit (Climate Analytics 2019).

Paradoxically, Australia is one of the world's most climate-vulnerable industrialised nations. It is the driest inhabited continent on earth and is already experiencing increased temperatures, increased severity of droughts and heatwaves, increasing frequency and intensity of bushfires, and reduced patterns of rainfall in the dry interior regions. These results are expected to increase over time and affect crucial ecosystems, such as the Great Barrier Reef on which the tourism industry depends, and risk climate-exposed export sectors such as agriculture (CSIRO 2016). The demonstrable effects of climate change in Australia have worsened alongside the intensifying minerals boom, which has seen an unprecedented scale of resource-extraction, including coal.

The scale of Australia's minerals boom opened up regions previously unexplored by the coal and coal seam gas (CSG) industries to commercial development. This generated widespread social discontent, particularly where coal and CSG extraction encroached on prime agricultural lands. Australia has vast reserves of coal, which if extracted and burnt, can substantially alter the global climate. Therefore, it is considered the global frontline to keep fossil fuels in the ground (Rosewarne et al. 2014). The minerals boom generated various tensions and imperatives for mobilisation against coal and CSG mining, such as, for environmentalists, between the rampant scale of Australia's coal mining and the global climate imperative of phasing out coal reserves, and for farmers, through governments prioritising coal mining while marginalising their needs. Reflexively, climate change has emerged as an embodiment of various concerns over coal mining in Australia (Duus 2013).

Case studies

The Singrauli region in the central Indian state of Madhya Pradesh and the Galilee Basin region in the Central Queensland region in Australia, where the prospect of new coal mines sparked anti-coal protest movements, serve as the comparative cases for this book.

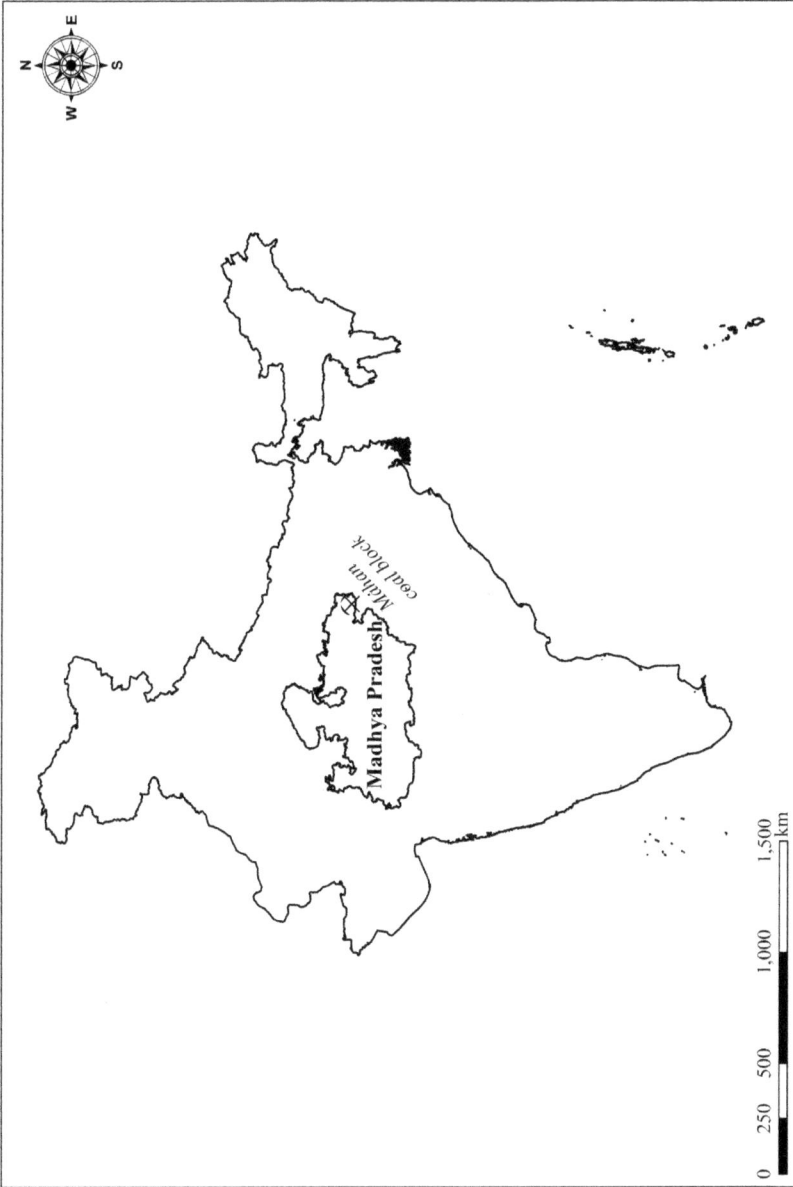

Figure 1.1 India with the location of the Mahan coal block.

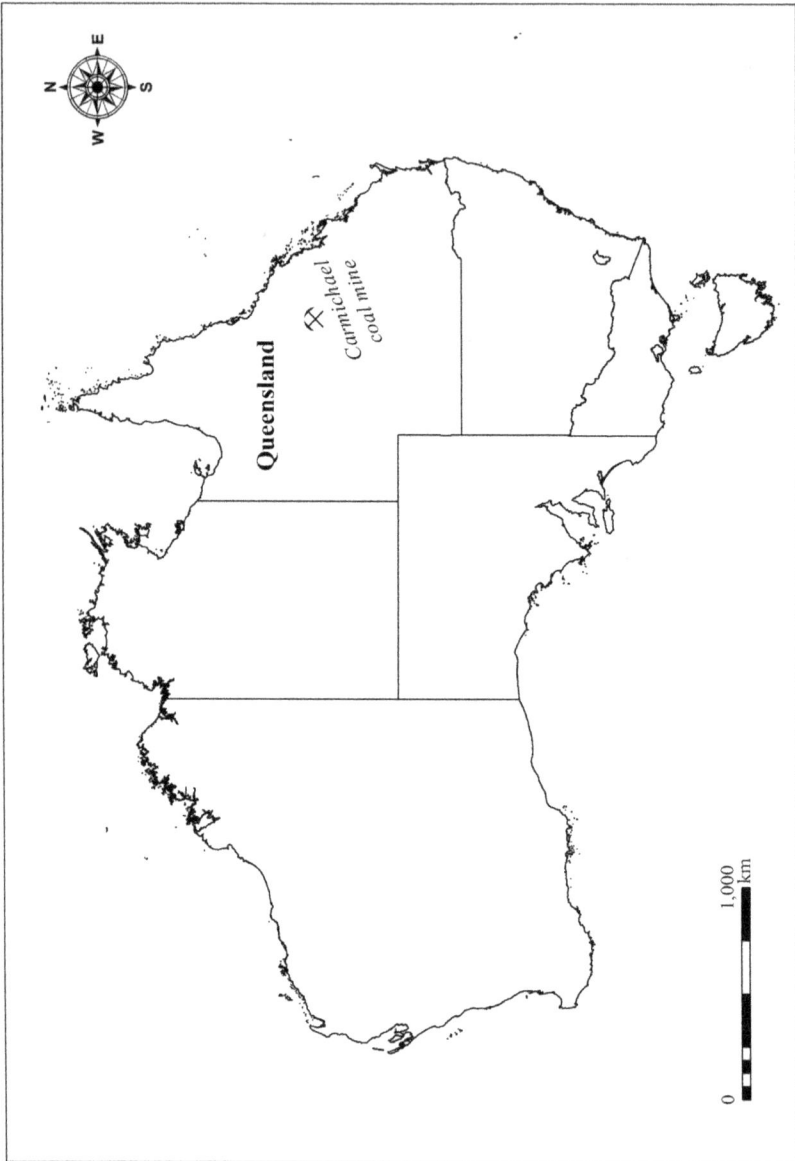

Figure 1.2 Australia with the location of the Carmichael coal mine.

In the Singrauli region in central India, an alliance between the Indian arm of the international environmental organisation Greenpeace and a local mobilisation of 11 forest-dependent village communities succeeded in their struggle to stop coal mining in the last remaining tracts of old-growth Sal tree forests in South Asia (Talukdar 2018a). Although Singrauli is one of India's oldest and most extensively mined coal regions, new coal developments are now spreading into the residual intact forests on the fringes of its coal fields, threatening wildlife connectivity and the livelihoods of subsistence-based and Adivasi communities.

The 1000-hectare open-cut coal mine in the Mahan forests, jointly owned by Essar Energy and Hindalco Ltd., was first proposed in 2009. It would have jeopardised the livelihoods of 54 forest-dependent villages (Padel 2016). The coal mine became the centre of an Indian government corruption scandal on coal mine allocation known as 'Coalgate'. The Indian government eventually cancelled the coal mine allocation in 2015, following the directive of the Supreme Court of India, based on a corruption investigation. Yet the same government targeted Greenpeace India, as part of a nationwide crackdown on non-governmental organisations (NGOs) working on human rights issues and exposing the ecological and social impacts of industrial projects. Greenpeace was singled out for its anti-coal campaign. Its bank accounts were frozen, and its capacity to operate its campaigns was severely constricted (Talukdar 2018b).

In Australia, the Carmichael coal mine, owned and operated by the Adani Group, was initially slated to be Australia's largest coal mine. However, it has become the strategic and symbolic focus of a multi-pronged resistance, including the environment movement under the 'Stop Adani' campaign, local farmers and the Indigenous traditional owners under the 'Adani, No means No!' land rights campaign. It is located in the Galilee Basin, Australia's largest and previously untapped coal reservoir, where nine mega coal mines were proposed at the peak of Australia's minerals boom in 2011.

Coal extracted and burnt at full capacity from all the proposed mines would make Galilee the seventh biggest emitter in the world, with the rest of Australia occupying 14th place (Greenpeace Australia 2012). It risked depleting water reserves from Australia's largest underground reservoir the Great Artesian Basin, it has affected the land rights of the Wangan and Jagalingou (W&J) traditional owners, and threatened to damage to the Great Barrier Reef through port expansion and increased coal traffic (Environmental Law Australia 2016). The coal mine received environmental clearances a year before the Paris climate summit. It faced a concerted civil society resistance for eight years but finally began operations (albeit in a drastically reduced capacity) in 2019, due to ongoing political support and despite its financial unviability (Talukdar 2019).

Research aims and questions

This research brings together and cuts across a few overlapping fields. Against the backdrop of global environmental activism's new strategic focus on stopping coal extraction, it attempts to understand how this approach interacts with other local oppositions around lands, livelihoods and shared natural resources affected by coal mining. Effectively, it aims to understand what collective meaning emerges from today's resistance to coal, and what politics and vision it generates for environmental activism.

Given the similar timescale of Australia's minerals boom and India's neoliberalised economic growth, which caused widespread ecological transformations and escalated land and natural resources related socio-environmental tensions, the book first takes a broader view of the profile of current environmental resistances, within which the specific anti-coal resistances are situated, in both countries. It attempts to understand how neoliberal economic forces, encouraged by the actions of governments, accelerated ecological destruction and triggered new environmental resistance. It endeavours to understand how the two anti-coal resistances contribute to the broader picture of contemporary environmental activism, in Australia and India respectively, and what critical new dimensions they add to these landscapes.

The book studies anti-coal activism against the contrary policies and politics of governments that promote coal production despite aggravating climate change and other destructive ecological effects, and against increasing social disaffection. This approach brings into consideration the role of the State, and the relationship of the State to the various actors and affected communities who are resisting coal projects, including environmental activists. The case studies attempt to draw out how the increasing disaffection of communities against governments is shaping new solidarities and alliances against coal mining.

Finally, the comparative purpose of this book is to articulate the critical similarities and differences between the Australian and Indian environmental movements. In Chapter 2, I discuss the distinctions between Australia's nature-conservation-focussed Northern environmentalism, and India's livelihood-centric movements of subsistence Adivasi and peasant communities; between eco-centric and human-centric concepts of environmental justice; between the notion of nature as an untrammelled wilderness, and nature as the place where people 'live, work and play'; and broadly between the rich and poor societies from which they arise.

Historically, these environmentalisms have lacked a shared ground. The fixed ontology of Australia's radical eco-centric environmentalism made collaborations challenging, with a majority of Indigenous–green interactions in Australia between the 1970s and 1990s indicating a fundamental mismatch of visions (Dodson 1997). The scientific conservation model of nature preservation has had a deleterious effect when exported to populous landscapes in the global South such as India, where Adivasi communities have been driven off their ancestral lands for the creation of national parks (Guha and Martinez-Alier 1997).

The universal nature of the climate problem and the need for global action for climate justice make it imperative to know the critical differences between environmental movements arising from disparate Northern and Southern contexts today and to understand what common ground can look like. At the same time, global movements driven out of the North run the risk of interpreting Southern contexts through lenses that can simplify and polarise the experience of Southern marginalised communities. In particular, transplanting Northern environmental discourses to the South has hampered a contextualised understanding of the relation between poverty and environmental justice to emerge (Lawhon 2013). Western environmental justice research and activism need to contextualise Southern environmental actors fully to avoid misunderstanding, misinformation and misguided action (Williams and Mawdsley 2006).

The strategic focus of international environmental activism on 'keeping coal in the ground' via an emphasis on anti-coal resistances, and more generally the global fight against all fossil fuel extraction, generates a set of normative and empirical drivers to attempt bridging the conceptual divide between Northern and Southern environmentalisms. A comparison of anti-coal resistances across the global North–South divide, one that offers a detailed engagement with all contexts and participants, constitutes a necessary step in that direction. Through its focus on the common challenge of stopping coal extraction in two disparate geographies, this book aims to uncover the common ideologies and politics that can bridge environmentalism's North–South divide. It aims to suggest a narrative of anti-coal activism that can be globally representative.

Research questions

Approaching the two anti-coal movements from the perspective of the historical differences in the politics and visions of their respective environmentalisms, this research asked four interconnected questions:

In both cases, how have the discourses and politics of environmentalism transformed from the previous era?
With regard to the respective anti-coal activisms, what has been the State and civil society dialectic, and how has coal extraction been countered?
What are the discourses, tactics and relations of the respective anti-coal activisms? What is their significance for environmentalism and its context in Australia and India, respectively?
What are the similarities and differences between the anti-coal resistances and their contexts in Australia and India? What outlook for global environmentalism emerges from their comparison?

1.2 Research approach, methods, materials and structure

This book is based on fieldwork and ethnographic research in Australia and India undertaken for my PhD. My primary tools for the ethnographic research were participant observations at multiple locations and events, daily field notes recorded in Mahan and the Galilee Basin, and semi-structured interviews.

As a part of my Indian fieldwork between 2017 and 2018, I conducted participant observations at the Greenpeace campaign office in New Delhi and at Mahan in Singrauli, Madhya Pradesh (the site where the coal mine had been proposed), and

22 semi-structured interviews which included Greenpeace staff, local movement leaders in Singrauli, and other civil society actors based in New Delhi..

My Australian fieldwork spanned over three years between 2016 and 2018 during the most active phase of the resistance to the Carmichael coal mine. I followed the actions of the Stop Adani movement – the environmental movement in the collective resistance to the Carmichael coal mine – in Sydney and Melbourne where it has very large volunteer bases, and in Central and North Queensland in the cities of Mackay and Townsville, respectively. I also travelled to the Galilee Basin.

I conducted 24 semi-structured interviews which included representatives from national as well as Queensland-based regional environmental non-governmental organisations (ENGOs), the farmers' advocacy network against the coal mine, and the W&J traditional owners' campaign against the Carmichael coal mine. I attended Stop Adani protest gatherings across Melbourne, Sydney and Brisbane between 2017 and 2018.

All interviewees have been de-notified in the book, and in the case of the local respondents from Mahan, their names changed, for privacy. Designations attributed to interviewees quoted in the Indian and Australian chapters are current as of the time of the interviews. The secondary materials for the research included news articles, including in the Hindi language for local news articles on Mahan in India, legal documents, ENGO reports, and expert analysis reports. Campaign materials included ENGO press releases and feature articles, campaign videos, pamphlets, and banners.

In India, a web-based community radio channel called Radio Sangharsh meaning Radio Resistance set up by Greenpeace, proved a chief resource to study the local movement at Mahan. It contained testimonials and accounts from local movement members on key movement moments and the interference of the local administration and company officials in their activities. Another source of information at Mahan was the diary entries of one of the movement leaders that he shared with me. The entries captured his personal experience of key moments in the movement between 2012 and 2014.

There are many ways in which environmentalisms can be compared. I now discuss the themes that emerged from my fieldwork and how they shaped the structure of my comparison and the choice of political ecology as the theoretical approach for this book.

Themes and comparability of the Australian and Indian cases

From India, three interconnected themes emerged from the interviews with the three respondent groups. The first, from the interviews with non-Greenpeace and non-Mahan respondents (other civil society actors), was around how neoliberal economic development since the mid-1990s had affected the environment and transformed environmental activism in India, and the relevance of the Mahan movement for the current landscape of environmental resistance.

The second, from the Greenpeace respondents, was around the formation of the Mahan campaign as a part of the new global climate activist strategy to 'keep coal in the ground'. And contextualising the global approach to the reality of a massive expansion in coal mining in India, particularly in the thickly forested central region. The third from the leaders of the local movement at Mahan consisted of biographical accounts of their motivations to join and fight against the coal mine, emphasising their dependence on the forests for their livelihoods.

The themes from the Indian interviews suggested taking a political–ecological approach in order to examine the socio-environmental impacts of India's post-liberalisation growth.

And against the broader tableau of environmental resistance in the neoliberal era, to analyse the significance of the Mahan movement, and the community's assertion of their newly found forest rights under the *Forest Rights Act 2006* (FRA) to reject coal mining.[3] Threatened livelihoods, linked to altered social and environmental conditions, are central to political ecological studies (Bryant and Bailey 1997). As a theoretical approach, political ecology addresses the underlying drivers of ecological change and develops ethical solutions. Political ecology assumes that the human impact of environmental change is unevenly distributed, with the poor and marginalised groups facing its impacts disproportionately (Watts 1983).

I trace the social and ecological impacts of both a massive increase and a privatisation of coal mining and thermal power generation since the mid-1990s. I analyse how a new language of rights over forests and lands is changing the discourse of the environmentalism of the poor today, and what role the neoliberal Indian State plays in producing environmental injustice towards subsistence-based communities. The build up of the Greenpeace and the Mahan community's anti-coal movement, and the subsequent government hostility towards Greenpeace, is projected against the special political treatment given to coal in India's development, and the nexus between the State and the coal-mining corporations. The significance of the Mahan movement, particularly the local community's rejection of coal mining based on an assertion of forest rights, is assessed against the Singrauli region's history of land dispossession and livelihood destruction from decades of intensive coal mining and thermal power generation. The Indian case study Chapters 3, 4 and 5 are structured around these lines of analysis.

In Australia, the focus of the interviews with the national ENGO representatives was on understanding how the minerals boom has transformed environmentalism, tracing the emergence of the Stop Adani movement along a continuum of environmental resistance since the mining boom, and understanding how its tools and tactics differed from earlier movements. For Queensland ENGO representatives, the interviews focussed on the environmental impact of the coal mining boom in Central Queensland, on the weakening of environmental regulations, and the challenges faced by local and state-level groups. For the farmers against the Galilee coal mines, the interviews focussed on their grievances and their motivation to advocate against coal mining. For the W&J representatives, the interviews focussed on the grounds for their rejection of the Carmichael coal mine, the stages of the 'No means No' campaign, and their motivation and reasons to fight.

Based on the themes emerging from the interviews, in Chapter 6 I locate the rise of new forms of environmental activism in response to a combination of Australia's lack of action on climate change and the environmental and social effects of an unprecedented scale of coal extraction and export from mineral-rich regions from the mid-1990s. The emergence of the Stop Adani movement, its tactics and tools, is analysed in Chapter 7 against the context of Queensland's coal export boom and Australia's continuing climate inaction after the Copenhagen summit, as a new wave of environmental activism strategically focussed on 'keeping coal in the ground'.

Farmers' activism against coal and CSG mining is analysed as due to the cumulative impacts of the minerals boom – the extensive scale of mining, its effects on land and water, and its encroachment on fertile agricultural lands – and growing grievances against the State's structural marginalisation of agriculture in favour of mining. The land rights campaign of the W&J is understood within the context of the shifting relations between the State and Indigenous groups during the resource boom, based on the State's role in promoting and encouraging extractive projects. The Australian case study discussed in Chapters 6, 7 and 8 is structured around these lines of analysis. Chapter 8 analyses the significance of the three anti-coal mining activisms – that of environmentalists, farmers and the W&J traditional owners – against the political and economic context of coal in Central Queensland, Australia's largest coal-exporting region.

As respondents on both sides referred to a similar time frame of two decades, beginning in the mid-1990s, as critical for understanding the considerable ecological destruction and social discontent that formed the backdrop to today's environmental activisms, this similarity helped to structure the cases as a longitudinal study of environmentalism. Another significant similarity is a central focus on the role of the State. The two case studies are designed to understand how the changing role of States and the changing political economies – under the effects of neoliberalism in India and the minerals boom in Australia – changed environmental activisms by intensifying the disaffection of impacted communities. The historic and continuing reality of the Indian State's role in deepening environmental injustices for the Indian poor and the Australian State's role in escalating injustices for Indigenous communities are somewhat directly comparable.

The focus on the State is another strong relevance of the political–ecological approach for this book. I now discuss further points of relevance of the political–ecological theoretical approach for this book.

Political ecology as a theoretical framework

This book brings together literature from multiple, intersecting fields. The two case studies combine historical relations, antagonisms and contexts of various environmentalisms in both Australia and India, as well as the political, social and ecological realities of coal extraction and its effects. With its place-based focus and consideration of social–ecological processes, historic power struggles and inequalities, the field of political ecology amalgamates the analysis of socio-ecological problems with the political economy. Despite relevant criticism regarding its primarily local focus that I discuss in Chapter 2, political ecology is relevant for a substantive and critical comparison of the two cases and it therefore forms the major theoretical approach for this book.

Political ecological studies have been primarily based in the global South, where they have traced the struggles of cultivators and hunter-gatherers due to the enclosure of commons (Bryant 1998; Peluso 1992), the continuation of colonial legacies through State-organised scientific forestry (Guha 1989; Jewitt 1995), and the role of gender in the construction of scientific knowledge, distribution of environmental rights and responsibilities, and grassroots activism (Agarwal 1997; Carney 1996; Joekes 1995; Rocheleau et al. 1996). India's earliest political ecological work *Economy of Permanence* (1945) by Gandhian economist J. C. Kumarappa discussed what constitutes a non-violent socio-economic order. Although generally lacking a global focus, some texts (see Ghai and Vivian 1992; Friedmann and Rangan 1993; Schroeder and Neumann 1995; Peet and Watts 1996) have probed the national and international significance of micro-political ecological struggles.

More recently, texts on ecological conflicts in industrialised societies and particularly North America – such as the transformation of wetlands in Minnesota and Illinois (Robertson 2000, 2004) and forests (Prudham 2003) and feral lands (McCarthy 2001, 2002) – have begun establishing First World Political Ecology as a field of study. They have begun to diversify the narrative and focus of political ecology by including notions of marginality, land management and place of nature emerging from Northern and particularly settler colonial contexts (Wilson 1999). Scholars have argued that this emerging area needs to critically reflect on what constitutes 'a context of, and for political ecology', on the relationship between political ecology and spatiality, in the First World (Wainwright 2005, p. 1034); that research in settler colonial societies needs to include a careful analysis of colonial practices (Braun 2002).

Guha and Martinez-Alier's *Varieties of Environmentalism: Essays North and South* (1997) provided a systematic critique of the supposed dominance of Northern wilderness-centric environmentalism, by highlighting ecological struggles of various marginalised communities across several societies and historical periods, in South America and South Asia. It took a multidisciplinary approach, combining anthropology, economics, sociology and ecology, to construct 'an alternative and sometimes oppositional framework from the conventional wisdom of Northern social sciences' (p. 14).

The cases delineated the specific geographies, material conditions and communities from which the environmental conflicts arose. It underscored that the theorisation of nature and hence the construction of environmentalism is grounded in what Harvey calls the 'materialities of place, space and environment' (1996, p. 44). Both the concept of varieties of environmentalism and the approach of delineating various cases and their contexts have played a central role in shaping and articulating this book's research on environmental activisms.

Despite historic limitations around its field of visions, owing to its central focus on place, the political ecological framework is an effective approach for studying multi-scalar contestations centred on sites of coal extraction in the two case studies in this book. Political ecology's focus on place helps expose the layers of histories, complexities in human–nature relations, and the relational dialectics between different interest groups over shared resources; these aspects constitute central lines of inquiry in the Australian and Indian cases.

With its strategic focus now on stopping coal projects, environmental activism's actions are inextricably tied with coal's infrastructure and systems – mines, ports, railroads and thermal power plants – as well as financial institutions that are funding coal projects, towards which it directs a variety of disruptive tactics (Brown and Spiegel 2019). These actions have the potential to impact the national and international political economy of coal. Other grievances brought by non-environmentalist actors, which relate to natural resource or land conflicts, represent a more immediate and material risk from the coal projects. How these different sets of interests against coal extraction intertwine and strengthen each other in the collective anti-coal resistance can offer a new environmental understanding as the Australian and Indian cases will demonstrate.

The case studies will demonstrate that this book contributes to research on cross-contextual comparisons of environmental movements in three significant ways. Using the political ecological approach, it investigates relations between environmentalist and non-environmentalist constituents in both the cases and compares the contextual differences in the nature of relations between the two cases. Through an investigation of the process and characteristics of the collective politics and narratives emerging from the Australian and Indian movements, it contributes to research on current environmentalisms across the North–South divide.

The various scales of resistance demonstrated by the two movement cases – the local, the national and the international – offer perspectives on the scalar relevance of micro-struggles against coal in the era of climate change. These three research directions contribute to the field of political ecology by extending the framework to areas where existing studies have paid insufficient attention – such as relations between local actors and NGOs, addressing contradictory and multiple claims around sites of conflict, and the multi-scalar relevance of micro-struggles particularly from the global South. In Chapter 2, I discuss gaps in existing political–ecological literature and how this research addresses them.

Next, I discuss the challenges faced in structuring a study in comparative environmentalism across the global North–South divide and how the design of this research reconciled these differences.

Challenges and opportunities in a North–South comparison of environmentalism

Between the Mahan forests and the Galilee Basin, a difference can be observed regarding what is vitally at stake across environmental justice movements in the settler colonial Northern and postcolonial Southern regions. The most significant resistance to the coal mine in India came from the project site itself, from forest-dependent subsistence-based communities living primarily outside the capitalist market-system. The proposed coal mine threatened to displace them from their homes and lands, and disrupt their forest-dependent livelihoods. This immediate risk added a critical urgency to their struggle for human rights, which is characteristic of Southern environmental conflicts. In Australia, however, the bulk of the resistance came from urban-based volunteers and supporters of ENGOs, a primarily tertiary-educated segment of Australian society with left-leaning and progressive political views and a high level of concern for green issues and climate change (see Chapter 2).

The W&J have ancestral ties to the land and the waters in the Galilee Basin, and in that respect come closest in comparison to the Adivasi people in Mahan. However, the spatial difference between Mahan and the Galilee – between significant human settlements dependent on subsistence livelihoods, and sparse habitations and industrial-scale agricultural operations of the Basin's farmers – remains. The cattle properties in the Galilee are massive in scale – it took us up to 45 minutes to drive through one – creating a sense of vastness and remoteness amongst the community living around the fringes of the proposed mines.

Literature from the field of comparative environmentalism acknowledges the disparities in the political contexts of industrialised and industrialising nations, even in the case of democracies (for example, see O'Neill 2012). In fact, while acknowledging the unavoidable disparity of cases, the North–South comparative framework for environmentalism actually seeks to highlight such critical contrasts, to demonstrate the diversity in social responses to environmental problems across countries being compared (O'Neill 2012). The North–South frame is one amongst various frames and debates comparing environmentalisms around the world (Kousis et al. 2008).

Although the conflict over the Mahan coal mine and the anti-coal activism of Greenpeace and the Mahan community became a national issue, the movement lacked multiple environmental campaigning organisations, unlike the case in Australia. Compared to Australia, where the entire national environmental movement transformed into a politics of stopping coal outright, in India the interviews verified what is commonly understood amongst civil society actors, that ENGOs and advocacy groups avoid taking a direct approach towards the 'sectoral targeting of coal'.

Further, climate change did not become an issue for the mass mobilisation of subsistence communities in India. These contrasts indicate two very different societies, who constitute the majority of environmental actors in those societies, and what their imperatives are. They also highlight the very different political challenges for environmental organisations and peoples' movements, as well as the very different framings of environmental conflicts, in the South (Haynes 1999).

However, the presence of Greenpeace in India, an international ENGO whose global campaign strategies are driven out of the global North, added a strongly comparable dimension to the two movement cases. Doyle (2005) suggests that one way in which the North–South gap can be bridged is through campaign-based studies. Greenpeace's anti-coal campaign in the Mahan forests is a reflection of the new global strategy for climate activism through politicising coal mining and directly intervening to stop coal mining projects. I selected the anti-coal mining movement in Mahan as the Indian case based on this factor. A second factor for the particular selection of these two movement cases for comparison was the multiple scales of contention against coal that they generated, from the local to the national and international as the case study chapters will demonstrate.

A third factor was the presence of Indigenous resistances to the coal projects. The disparity in North–South comparative environmentalism research can also be bridged through a comparison of global grassroots uprisings that face similar challenges (see Peluso and Watts 2001; Taylor 1995). More specifically, this research's comparison of Indigenous contestations to coal mining, between a post-colonial Southern context and a settler colonial Northern context, contributes to a global understanding of Indigenous resistance to fossil fuels in the climate era. Although the W&J formed solidarities with First Nations' struggles against fossil fuel extraction in Canada and the United States, no connections were forged with Adivasi coal-mining conflicts in India related to the Adani Group, the Indian conglomerate that is also the proponent of the Carmichael coal mine in Queensland. Tracing their similarities and disparities through research is even more significant given the North–South divide that can act as a barrier for them to join in solidarity.

Finally, there is a growing interest within the field of comparative environmental movements to expand the unit of analysis to include a variety of State and non-State actors (and their interactions) working in an area (Balsiger 2007). The relational politics between the various anti-coal constituents in the collective movement in Australia, and between Greenpeace and the Mahan community in India, contributes to this aspect of research in comparative environmentalism.

Next, in the final subsection, I discuss how I approached the ethnographic research for this book.

My research practice

This book was born out of my PhD research and dissertation that I completed in 2021. But why I chose to work on this topic has a longer context going back to when I first forayed into environmental campaigning in Greenpeace India in 2005. While in my daily work of communicating stories of our activism, I would take Greenpeace's 'global' environmental campaign frames and contextualise them to the Indian context, I wondered if and how global environmental activism could make space for stories to emerge from the ground up from the global South. It was evident to me how much of the context from a place like India was left out in projecting 'neat' stories on singular campaign issues to a global audience; that in the absence of us talking about intersecting vulnerabilities faced by community members daily, we were not building the global audience's understanding of the intricacies of environmental activism in India.

Working in environmental campaigning in Australia since 2008, first at Greenpeace and then the country's oldest ENGO Australian Conservation Foundation, I had a chance to observe the inner workings of a Northern environmentalism over a period of time. The intention of being able to write for fellow activists and climate justice researchers from the global North to demystify their gaze towards the global South and fully contextualise the ground there grew stronger during this period of time. When the Australian climate movement began campaigning against the Adani Group's Carmichael coal mine in 2014, they looked to India to understand the conglomerate's environmental track record in India, and to understand the landscape of environmental politics and resistance. To me, it felt like a strong moment to engage Australian researchers and campaigners in exercises to fully contextualise the ground in a global South location like India, to formulate narratives and approaches for effective solidarity with struggles there.

During my PhD candidature between 2016 and 2021, I ran workshops for Australian ENGOs and various chapters of the national Stop Adani movement to help them understand the contrasting context of India's national coal politics, and the imperatives and challenges of its anti-coal movements, with the aim of building transnational solidarity.

My PhD fieldwork took an ethnographic approach. Ethnography is a qualitative research method that can be described as embedded participant observation, or fieldwork that involves an immersion in the research site and its practices (Plows 2007). It approaches social movements from the perspective of the people involved (Bouma and Atkinson 1995). Therefore, it helps to identify how social movement actors who have been 'framed out' of formal policy discussions are framing issues in their own terms (Plows 2007). Although the original anthropological practice of ethnography involved living within the communities being studied, today (especially across other fields within the social sciences) ethnography is practised with a focus on 'what happens in a particular work locale or social institution when it is in operation' (Hammersley 2006, p. 5). This is how I applied ethnography to my fieldwork and research.

For this research, apart from observing sites and groups across two different countries, I also had to follow the flow of campaign events and the movement of actors at various locations where meetings and protests were taking place. Even in the Indian case, although the bulk of the resistance came from the actual project site in the Mahan forests, it would not have sufficed to observe the movement at Mahan alone. To observe and understand, and then be able to describe and analyse all the dimensions and scales of the Indian movement, I needed to spend time with urban-based environmental activists, primarily in the Greenpeace office. The range of sites that needed to be observed for the Australian movement was more extensive.

Trust and access are key issues in social movements, and gatekeeping can often occur in ethnographic settings. Insider research can prove to be a methodological bonus in such settings (Plows 2007). This was definitely the case during the field trips to Mahan in India. The local movement leaders opened up to me and discussed details of village-level politics, caste and land-related disputes, and recounted their personal experiences of the struggle, once they accepted me as 'one of them' (meaning the Greenpeace team which the local movement was working in close alliance with).

This was also the case during the Queensland league of my Australian field trip, where local environmental campaigners, farmers and W&J Traditional owners, located very close to the site of Carmichael coal mine, and feeling the hostility of the local pro-coal politics, would tend to be cautious of 'outsiders' finding out about their campaign strategies. Particularly, talking about grassroots struggles against coal mining in India in small sessions-based-interactions during multiple visits to Queensland helped me establish my trust amongst local movement members. Conversations about the struggles of Indigenous communities in India against coal mines were especially central to the relationship of trust I was able to establish with W&J campaign spokespersons.

The ethnographic research approach is conducive to studying the forms of resistance that Naomi Klein envisages as critical – a globally linked network where climate activists are standing in solidarity with and sharing the struggles of various local resistances to fossil fuel extraction – to break the global power of the fossil fuel industry. In *This Changes Everything,* Klein (2014) offers an account of the global phenomenon of Blockadia, which is 'not a specific location on a map but rather a roving transnational conflict zone…wherever extractive projects are attempting to dig and drill' (p. 295). Blockadia is characterised as a broad and disparate movement that is motivated by multiple grievances (also see Brown and Spiegel 2017; Martinez-Alier et al. 2016). Blockadia is however not a new concept; the idea originated from a peaceful uprising in the Niger Delta against the oil corporation Shell after oil spills destroyed the lands of the Ogoni and Ijaw peoples (EJAtlas 2014). It subsequently spread to other parts of the world (Martinez-Alier 2023).

As an activist-turned insider researcher, operating within the broader vision of 'Blockadia', I saw my role as that of interweaving understandings across the contextual differences between Australia and India, and highlighting their similarities, to conceptualise a common ground on which intersectional global North–South solidarity against coal mining can be built. Doing the work of tying the threads connected back with my original intent of writing this book for fellow activists and climate justice researchers about how to understand and create space for stories and contexts from the global South. This is how I enmeshed my own experience as an environmentalist in India and Australia into comparative research on the politics of and resistance to coal in these two democracies whose actions have a significant bearing on the world's climate.

Finally, because by training I am grounded in a global South perspective of climate justice, I likely bring an alternate perspective to North–South comparisons of environmentalism, as compared to the majority of the research undertaken in this field, which has been done from the global North. This alternate outlook has possibly allowed me to emphasise and compare elements in this research that might have traditionally been deprioritised.

1.3 Book chapters and literature

This book combines a few different analytical approaches. Overall, it compares two present-day environmental resistances in the North and the South, specifically against coal extraction. It approaches this comparison by analysing the transformation of the respective environmentalisms through the actions of the State and increased mining of coal (and other resources) since the mid-1990s. It dissects the two anti-coal resistances; what imperatives and events shaped their formation, and what political impact they had at various levels.

It evaluates the relations between different sets of actors in the respective movements, and how they generate a collective narrative and shared politics. Based on this, the book begins with an outline of theoretical debates on Northern versus Southern environmentalisms, followed by three chapters each on the Indian and Australian cases, and closes with an analysis of the comparative themes across those cases. Due to the multiple analytical approaches, the book uses assorted literature that informs various discussions. Each case study chapter is also specifically structured around a set of primary research findings. I now summarise the texts and research presented in the chapters and draw an outline of all the chapters.

In this chapter I have introduced the research topic and questions, the research approach and structure, and discussed my own ethnographic practice. I have also briefly discussed some political ecological texts that are relevant to the theoretical approach of this book. Chapter 2 analyses the historic divide in Northern and Southern environmentalisms, and the possibilities for common ground between them in the present era of climate change through literature. It contains a summary analysis of what constitutes Indian environmentalism of the poor, and the Australian conservation-focussed environmentalisms – their politics, challenges, and dominant criticisms. It sets the historic context of the respective environmentalisms against which the new environmental movements are evaluated.

Chapters 3, 4 and 5 constitute the Indian case study. Chapter 3 lays out the broader context of the political economy of development in India, and traces today's environmentalism of the poor as discontents emerging from within this milieu. The transformation of the Indian State under neoliberalism, and how this increased the grievances of livelihoods-based communities, is analysed through discussions of literature on the postcolonial Indian State and neoliberalism in the Indian context. Today's environmentalism of the poor is analysed by a discussion of some of the most significant peoples' struggles for livelihood and land in the last two decades.

Chapter 4 begins by discussing the centrality of coal in the Indian political economy, and the changing character of coal-development under neoliberalism. It locates the build up of the political conflict over the Mahan coal mine within this paradigm of coal-led development, particularly against the State–corporate nexus in coal. It traces the formation of the Greenpeace and local movement against illegal approvals and other controversial high-level government actions on the project. It analyses the government's suppression of Greenpeace for its anti-coal activism. Instead of mass assertions for climate action, after the crackdown, the government's assertion of coal was challenged by parts of civil society as a risk to democracy in India.

Chapter 5 traces the formation of the local anti-coal movement through the perspective of its local leaders, with a central focus on the significance of legal forest rights in the lives of indigenous forest-dwelling communities. It analyses the significance of the anti-coal uprising in India's energy capital, the Singrauli district.

Chapters 6, 7 and 8 comprise the three chapters of the Australian case study.

Chapter 6 lays out the broader context of the economic and political transformations brought by the minerals boom that began in the mid-1990s, and traces the
emergence of a new anti-coal approach of environmentalism, the rise of farmers'
protests against coal and coal seam gas, and the participation of Indigenous native
title groups in collective resistances against coal-mining projects, in response. It
discusses how the actions of the State were changed by the scale of the globalised
resource boom through literatures on the resource curse and neoliberalising political economy. It also discusses the changing priorities of the State with regard to
Indigenous land reforms. Today's environmental movements are analysed through
a discussion of some of the most prominent anti-coal resistances, involving
collaborations with farmers and Indigenous groups, that emerged in response to
the minerals boom in Australia's dominant coal-mining regions.

Chapter 7 begins by discussing the importance of coal extraction and export
to the Australian economy, and how the massive scale of coal-mining operations
during the resource boom started changing the economic and ecological balance
of coal regions, and changed the behaviour of governments. It traces the development of the Carmichael coal mine project through a series of special favours from
both state and federal governments. It traces the development of the environmental
activism against this State–corporate nexus, and analyses its impact on the national
politics on the issue of the Carmichael coal mine. It traces the farmers' opposition
to the coal mine through their growing discontent at the allocation of free and
unlimited water resources to the Galilee Basin coal mines. It traces the build up of
the W&J's land rights resistance as a challenge to Australia's native title system
as a response to having faced coercion by the mining corporation, and within the
Native title system, to consent to coal mining on their lands.

Chapter 8 traces the three different streams of resistance against the Carmichael
coal mine that emerged locally – the local arm of the Stop Adani environmental
movement, the local farmers' protests, and the W&J's land resistance. It analyses
their relations, and what their collective resistance means for the settler colonial
past and historical political relations in the region.

Based on these, Chapter 9, the Analysis, the final chapter, discusses similarities
and critical differences between the Indian and Australian case studies. It draws
multiple points of comparison – between the concept of climate justice and the use
of climate change issues for mobilisation, between the nature of and possibilities
from green relations forged with communities fighting coal mining in Australia and
India, between the politics of the Indigenous arms of both resistances and between
the framing of the national environmental problem by ENGOs in Australia and
Greenpeace in India. It suggests how the global North–South divide in environmental activism to keep coal in the ground demonstrated through the case studies
can be bridged, particularly through articulating Indigenous land justice as a crucial human right. It reflects on this book's contribution to political ecology and
suggests areas for future research.

Notes

1 For a 50:50 chance to keep global warming within 2 degrees, 88% of the world's coal reserves, 52% gas reserves and 35% oil reserves need to be considered 'unburnable fuel' and left in the ground (Steffen 2015).
2 This occurred because the Chinese demand for coal that used to account for around half the world's coal consumption fell in 'absolute terms' around this time (Climate Analytics 2019).
3 The Forest Rights Act (2006) concerns the entitlements of forest-dwelling Adivasi and non-Adivasi communities related to land and other resources. These communities had been denied their natural and historic rights over forestlands and resources even in post-independence India, owing to the continuation of British-colonial era forest laws.

References

Adve, N. 2019, 'Impacts of global warming in India: narratives from below', in N. Dubash (ed.), *India in a* Warming World, Oxford University Press, pp. 65–78.

Agarwal, B. 1997, 'Environmental action, gender equity and women's participation', *Development and Change*, vol. 28, pp. 1–43.

Baer, H. 2016, 'The nexus of the coal industry and state in Australia: historical dimensions and contemporary challenges', *Energy Policy*, vol. 99, pp. 194–202.

Balsiger, J. 2007, *Uphill Struggle: The Politics of Sustainable Mountain Development in the Swiss Alps and California's Sierra Nevada*, PhD diss., Department of Environmental Science, Policy and Management, University of California at Berkeley.

Bouma, G. & Atkinson, G. 1995, *Handbook of Social Science Research*, Oxford University Press, Oxford, UK.

Braun, B. 2002, *The Intemperate Rainforest*, University of Minnesota Press, Minneapolis, MN.

Brown, B. & Spiegel, S. J. 2017, 'Resisting coal: hydrocarbon politics and assemblages of protest in the UK and Indonesia', *Geoforum*, vol. 85, pp. 101–111.

Brown, B. & Spiegel, S. J. 2019, 'Coal, climate justice and the cultural politics of energy transition', *Global Environmental Politics*, vol. 19, no. 2, pp. 149–168.

Bryant, R. L. 1998, 'Power, knowledge and political ecology in the third world: a review', *Progress in Physical Geography*, vol. 22, no. 1, pp. 79–94.

Bryant, R. L. & Bailey, S. 1997, *Third World Political Ecology*, Psychology Press.

Bulkeley, H. 2000, 'Discourse coalitions and the Australian climate change policy network', *Environment and Planning C: Government and Policy*, vol. 18, pp. 727–748.

Carney, J. A. 1996, 'Converting the wetlands, engendering the environment: the intersection of gender with agrarian change in Gambia', in R. Peet & M. Watts (eds), Liberation Ecologies: Environment, Development, Social Movements, Routledge, London, pp. 165–187.

Central Electricity Authority. 2019, *Monthly Reports*, CEA.

Climate Analytics. 2019, Evaluating the Significance of Australia's Global Fossil Fuel Carbon Footprint, July 2019, viewed 20 March 2020, <https://climateanalytics.org/media/australia_carbon_footprint_report_july2019.pdf>.

Commonwealth of Australia. 1992, *National Greenhouse Response Strategy.*

CSIRO. 2016, *Australia's Changing Climate*, The Australian Climate Change Science Program – An Australian Government Initiative, Canberra.

Dodson, P. 1997, "Finding common ground', habitat Australia', *Australian Conservation Foundation Magazine*, April 1997, no. 4.

Doyle, T. 2005, *Environmental Movements in Majority and Minority Worlds: A Global Perspective*, Rutgers University Press, New Brunswick, NJ.

Duus, S. 2013, 'Coal contestations: learning from a long broad view', *Rural Society*, vol. 22, no. 2, pp. 96–110.

Eckersley, R. 2009, 'Understanding the interplay between the climate regime and the trade regime', in *Climate and Trade Policies in a Post-2012 World*, UNEP, Geneva.

Edwards, G. A. S. 2019, 'Coal and climate change', *Wiley Interdisciplinary Reviews: Climate Change*, vol. 10, no. 5.

Environment Justice Atlas. 2014, 'Oil extraction forces Ogoni to consume benzene water for survival, Nigeria', *Atlas of Environment Justice*, viewed 20 March 2020, <https://ejatlas.org/conflict/oil-extraction-forces-ogoni-to-consume-benzene-water-for-survival-nigeria>.

Environmental Law Australia. 2016, Carmichael Coal ("Adani") Mine Cases in Queensland Courts, viewed 15 March 2020, <http://envlaw.com.au/carmichael-coal-mine-case/>.

Friedmann, J. & Rangan, H. (eds) 1993, *In Defense of Livelihood: Comparative Studies in Environmental Action*, Kumarian Press, West Hartford, CT.

Ghai, D. and Vivian, J. M. (eds) 1992, *Grassroots Environmental Action: People's Participation in Sustainable Development*, Routledge, London.

Goodman. 2016, 'The "climate dialectic" in energy policy: Germany and India compared', *Energy Policy*, vol. 99, no. C, pp. 184–193.

Government of India. 2008, *National Action Plan in Climate Change*, Prime Minister's Council on Climate Change, New Delhi.

Government of India. 2014, *India's Intended Nationally Determined Contribution: Working Towards Climate Justice*, viewed 20 March 2020, <https://www4.unfccc.int/sites/ndcstaging/PublishedDocuments/India%20First/INDIA%20INDC%20TO%20UNFCCC.pdf>.

Greenpeace Australia. 2012, Boom Goes the Reef: Australia's Coal Export Boom and the Industrialisation of the Great Barrier Reef, March 2015, viewed 20 June 2020, <www.greenpeace.org.au/news/boom-goes-the-reef/>.

Guha, R. 1989, *The Unquiet Woods: Ecological Change and Peasant Resistance in the Himalaya*, Oxford University Press, New Delhi.

Guha, R. & Martinez-Alier, J. 1997, *Varieties of Environmentalism: Essays North and South*, Routledge.

Hammersley, M. 2006, 'Ethnography: problems and prospects', *Ethnography and Education*, vol. 1, no. 1, pp. 3–14.

Harvey, D. 1996, *Justice, Nature and the Geography of Difference*, Wiley-Blackwell.

Haynes, J. 1999, 'Power, politics and environmental movements in the third world', in C. Rootes (ed.), *Environmental Movements: Local, National and Global*, Frank Cass, London, pp. 222–242.

Hazra, S., Ghosh, T., DasGupta, R. & Sen, G. 2002, 'Sea level and associated changes in the Sunderbans', *Science and Culture*, vol. 68, no. 9–12, pp. 309–321.

Healey, N., Stephens, J. C. & Malin, S. A. 2019, 'Embodied energy injustices: unveiling and politicizing the transboundary harms of fossil fuel extractivism and fossil fuel supply chains', *Energy Research and Social Science*, vol. 48, pp. 219–234.

Jewitt, S. 1995, 'Europe's 'Others'? Forestry policy and practices in colonial and post-colonial India', Environment and Planning D: *Society and Space*, vol. 13, pp. 67–90.

Joekes, S. 1995, 'Gender and livelihoods in northern Pakistan', *IDS Bulletin*, vol. 26, no. 1, pp. 66–74.

Klein, N. 2014, *This Changes Everything: Capitalism vs. the Climate*, Simon and Schuster, New York.

Klein, N. 2015, 'On climate, 'leadership must come from below'', interview in Il Manifesto, 12 December 2015, viewed 20 March 2024, <https://ilmanifesto.it/naomi-klein-on-clim ate-leadership-must-come-from-below>.

Kousis, M., Della Porta, D. & Jimenez, M. 2008, 'Southern European environmental movements in comparative perspective', *American Behavioural Scientist*, vol. 51, no. 11, pp. 1627–1647.

Kumarappa, J. C. 1945, *Economy of Permanence,* Sarva Seva Sangh Prakashan, Rajghat, Varanasi.

Lahiri-Dutt, K. 2016, 'The diverse worlds of coal in India: energising the nation, energising livelihoods', *Energy Policy*, vol. 99, pp. 203–213.

Lawhon, M. 2013, 'Situated, network environmentalism: a case for environmental theory from the south', *Geography Compass*, vol. 7, no. 2, 128–138.

Levien, M. 2011, 'Special economic zones and accumulation by dispossession in India', *Journal of Agrarian Change*, vol. 11, no. 4, pp. 454–483.

Martinez-Alier, J. 2023, 'Environmental conflicts and the making of world movements for environmental justice', Editorial in *Economia Politica,* vol. 40, pp. 765–779.

Martinez-Alier, J., Temper, L., Del Bene, D. & Scheidel, A. 2016, 'Is there a global environmental justice movement?' *Journal of Peasant Studies*, vol. 43, no. 3, pp. 731–755.

McCarthy, J. 2001, 'Environmental enclosures and the state of nature in the American West', in N. Peluso & M. Watts (eds), Violent Environments, Cornell University Press, Ithaca, NY, pp. 117–145.

McCarthy, J. 2002, 'First world political ecology: lessons from the wise use movement', *Environment and Planning A: Economy and Space*, vol. 34, no. 7, pp. 1281–1302.

O'Neill, K. 2012. 'The comparative study of environmental movements', in P. F. Steinberg & S. D. VanDeever (eds), *Comparative Environmental Politics: Theory, Practice and Prospects*, MIT Press, pp. 115–142.

Padel, F. 2016, 'Investment induced displacement and the ecological basis of India's economy', in S. Venkateshwar and S. Bandyopadhyay (eds), *Globalisation and the Challenges of Development in Contemporary India*, Springer, Singapore, pp. 147–169.

Pai, D. S., Guhathakurta, P., Kulkarni, A. & Rajeevan, M. N. 2017, 'Variability of meteorological droughts over India', in M. N. Rajeevan & S. Nayak (eds), *Observed Climate Variability and Change over the Indian Region*, Springer Geology, Singapore, pp. 73–87.

Peet, R. & Watts, M. (eds) 1996, *Liberation Ecologies: Environment*, Development, Social Movements, Routledge, London.

Peluso, N. L. 1992, *Rich Forests, Poor People: Resource Control and Resistance in Java*, University of California Press.

Peluso, N. L. & Watts, M. (eds) 2001, *Violent Environments*, Cornell University Press, Ithaca.

Plows, A. 2007, 'You've been framed: why publics mistrust the policy process', *Genomics Network Newsletter*, vol. 6, pp. 22–33.

Plows, A. 2008, 'Social movements and ethnographic methodologies: an analysis using case study examples', *Sociology Compass*, vol. 2.

Prudham, W. S. 2003, 'Taming trees: capital, science and nature in Pacific Slope tree improvement', *Annals of the Association of American Geographers*, vol. 93, pp. 636–656.

Raghunandan, D. 2019, 'India in international climate negotiations', in N. Dubash (ed.), *India in a Warming* World, Oxford University Press, pp. 187–204.

Robertson, M. 2000, 'No net loss: wetland restoration and the incomplete capitalisation of nature', *Antipode*, vol. 32, pp. 463–493.

Robertson, M. 2004, 'The neoliberalization of ecosystem services: wetland mitigation banking and problems in environmental governance', *Geoforum*, vol. 35, pp. 361–373.

Rocheleau, D., Barbara, T. S. & Esther, W. 1996, *Feminist Political Ecology: Global Issues and Local Experiences*, Routledge.

Rosewarne, S. 2016, 'The transnationalisation of the Indian coal economy and the Australian Political economy: the fusion of regimes of accumulation?', *Energy Policy*, vol. 99, pp. 214–233, viewed 20 June 2020, https://doi.org/10.1016/j.enpol.2016.05.022.

Rosewarne, S., Goodman, J. & Pearse, R. 2014, *Climate Action Upsurge: The Ethnography of Climate Movement Politics*, Routledge, Abingdon, Oxon, UK and New York.

Roxy, M. K., Ghosh, S., Pathak, A., Athulya, R., Majumdar, M., Murtugudde, R., Terray, P. & Rajeevan, M. 2017, 'A threefold rise in widespread extreme rain events over central India', *Nature Communications*, vol. 8, no. 708, pp. 1–11.

Roxy, M. K., Kapoor, R., Terray, P., Murtugudde, R., Ashok, K. & Goswami, B. N. 2015, 'Drying of Indian subcontinent by rapid Indian Ocean warming and a weakening land–sea thermal gradient', *Nature Communications*, 6 June, pp. 1–10.

Roy, B. & Schaffartzik, A. 2021, 'Talk renewables, walk coal: the paradox of India's energy transition', *Ecological Economics*, vol. 180, February 2021, no. 106871.

Schroeder, R.A and Neumann, R. P. 1995, 'Manifest Ecological Destinies: Local Rights and Global Environmental Agendas', *Antipode*, vol. 27, no. 1, pp. 321–324.

Somanathan, E., Somanathan, R., Sudarshan, A. & Tewari, M. 2017, *The Impact of Temperature on Productivity and Labor Supply: Evidence from Indian Manufacturing*, Working Paper, EPIC-India.

Steffen, W. 2015, Unburnable Carbon: Why We Need to Leave Fossil Fuels in the Ground, Climate Council of Australia, May 2015, viewed 20 March 2020, <www.climatecouncil.org.au/uploads/a904b54ce67740c4b4ee2753134154b0.pdf>.

Talukdar, R. 2018a, 'Democracy Zindabad! A day in the life of an anti-coal resistance in India's energy capital', *New Matilda*, 10 February, viewed 20 March 2020, <https://newmatilda.com/2018/02/10/democracy-zindabad-day-life-anti-coal-resistance-indias-energy-capital/>.

Talukdar, R. 2018b, 'Sparking a debate on coal: case study on the Indian government's crackdown on Greenpeace', *Cosmopolitan Civil Societies: An Interdisciplinary Journal*, vol. 10, no. 1, pp. 47–62.

Talukdar, R. 2019, 'Long shadow of Adani's Australia Mining Project', Newsclick, 12 September, viewed 20 March 2020, <www.newsclick.in/long-shadow-adanis-australia-mining-project>.

Taylor, B. (ed.) 1995, *Ecological Resistance Movements: The Global Emergence of Radical and Popular Environmentalism*, SUNY Press, Albany.

Varadhan, S. 2019, 'India plans $330 billion renewables push by 2030 without hurting coal', Reuters, 4 July, viewed 20 March 2020, <www.reuters.com/article/us-india-renewables-coal-idUSKCN1TZ18G>.

Vihma, A. 2011, 'India and the global climate governance: between principles and pragmatism', *Journal of Environment and Development*, vol. 20, no. 1, pp. 69–94.

Wainwright, J. 2005, 'The geographies of political ecology: after Edward Said', *Environment and Planning A*, vol. 37, pp. 1033–1043.

Watts, M. 1983, *"Silent Violence": Food, Famine and Peasantry* in Northern Nigeria, University of California Press, Berkeley.

Williams, G. & Mawdsley, E. 2006, 'Postcolonial environmental justice: government and governance in India', *Geoforum*, vol. 37, pp. 660–670.

Wilson, R. 1999, "Placing nature': the politics of collaboration and representation in the struggle for La Sierra in San Luis, Colorado', *Ecumene*, vol. 6, no. 2, pp. 122–128.

2 Environmentalism of the global North and South

Historic divisions and potential for common ground

This chapter sets out what historically constituted the North–South divide in environmentalism, and how this divide is relevant even in today's era of climate activism. These divisions, and the possibilities of finding intersections across these divisions, need to be kept in mind while reading the Indian and Australian chapters. The sites in this research – the Mahan forests in central India and the Galilee basin in central Queensland in Australia – are the meeting grounds for multiple streams of resistances to coal mining. The nature of the relationships and political alliances formed between the various strands of resistances to coal mining, can give us insights into the possibilities for finding common ground across environmentalism's historic divisions that existed even within the same geography.

Environmentalism, a term that broadly stands for a collection of ideologies, politics, and actions towards the environment, has essentially been variously realised by movements emerging from different socio-economic and socio-ecological contexts. The conceptualisation of wilderness-centric environmentalism, practised in the United States and similar cultural geographies such as Australia, has been critiqued from the perspectives of human-centric environmental justice of marginalised communities, such as the environment justice movement in North America, and the environmentalism of the poor in the Global South (Martinez-Alier 1995). By deeming places from where Indigenous presence has been removed by the violent process of settler colonialism as pristine, wilderness-centric environmentalism has not only disregarded but also been complicit in perpetuating historical injustice.

The first section delineates the historic criticisms of wilderness-centric Northern environmentalism, particularly American environmentalism that has cultural bearings on Australian environmentalism, from the perspectives of various subaltern politics of eco-social justice. The second section specifies the politics and narratives of Australian environmentalism between the 1970s and 1990s. The third section specifies the politics and narratives of the environmentalism of the poor in India during a similar time period, in contrast to Northern environmentalism. The first three sections set out the historical contexts and divisions across the various concepts of environmental justice and their politics.

Climate change has made it imperative for Northern environmentalism to transform towards a politics of solidarity with human-centric modes of environmental

DOI: 10.4324/9781003410416-2

justice, based on a realisation that 'what we are fighting for [now] is each other' (Stephenson 2015, p. xv). The fourth section analyses the tensions as well as the possibilities for common ground between Northern and Southern perspectives of environmentalism in the era of climate change through three subsections. First through a discussion of the dominant Southern criticisms of Northern approaches towards 'burden sharing' and historic responsibility for climate change; then by discussing how the quest for climate justice has reconstituted the politics and perspectives of various modes of environmental resistance; and finally suggesting a bridge across environmentalism's historic divide through a common yet diverse understanding of climate justice.

The conclusion argues that a movement towards a shared understanding of climate justice signifies a resolution of the historic divide between environmentalism's Northern and Southern approaches. The Indian and Australian cases in this book offer valuable insights as to how Northern environmentalism now inter-relates with human-justice-centric environmentalisms of marginalised groups, and how a globally representative perspective on climate justice can emerge by investigating the activisms and their respective contexts, in both the North and the South.

2.1 A critique of wilderness-centric Northern environmentalism

Both the popular and radical streams of scientific ecology emerged as practices of environmentalism in the United States and similar cultural geographies. Popular wilderness environmentalism emerged as a post-war cultural phenomenon through an expansion of state and national parks for the aesthetic appreciation of a new consumer class (Hays 1987). Wilderness was consequently attributed a high place in national cultural identity (Nash 1982). The creation of the national parks systems in the United States also led to the removal of Indigenous communities to construct a wilderness that was 'uninhabited as never before in the history of the place' (Cronon 1996, p. 15).

Growing enthusiasm for outdoor recreation and an awareness of environmental degradation through works such as Rachel Carson's *Silent Spring* (1962) increased the popularity of the wilderness (Novotny 2000). Wilderness environmentalism therefore demonstrated settler colonial societies' paradoxical ability to 'devastate the natural world and at the same time mourn its passing' (Ekirch 1963, p. 189). It imagined remaining forests and uncleared landscapes as spaces preserved from industrialisation's contamination: 'As pervasive a problem as DDT was, and is, one could, and can always imagine that somewhere a place existed free of its taint' (McKibben 1989, p. xii).

In contrast, the radical eco-centric movement viewed the world as an interconnected web of relations, including human-nature relations and non-human communities (Eckersley 1992). Its philosophy and politics developed as an unstable and paradoxical amalgam of scientific and romantic traditions in America and similar cultural geographies (Hillier 2010). The paradox of radical ecology can be seen in the dichotomy between its philosophy and its objective. The former was constituted of the interdependence of humans with nature, influenced both by the natural sciences, particularly the writings of Aldo Leipold and James Lovelock, and the humanities, through the writings of Murray Bookchin and Joanna Macy amongst others. Paradoxically, its objective remained to maintain a separation between society and nature (Milton 1999).

Radical environmentalists argued that their activism borrowed elements of nature-romanticism from eastern religious traditions while not regressing into mystical faiths, striving instead to preserve wilderness based on a scientific understanding of ecology (see Devall and Sessions 1985; Naess 1973). Going back to nature was intended to create a profound cognitive change. The radical purpose behind this approach was to break Judaeo-Christianity's disconnect and domination of nature to generate a cultural and social realisation of nature's intrinsic values (Naess 1973).

Historian Roderick Nash's thesis *Wilderness and the American Mind* (1982) identified nature conservation through national parks as America's distinctive cultural contribution to the world, one that 'less developed nations may eventually evolve economically and intellectually towards'. Social scientist Ronald Inglehart's postmaterialist thesis (1977, 1990, 1997) further claimed that environmentalism, defined as an appreciation of wilderness spaces, was a new value born from intergenerational cultural turns in industrialised post-war industrial societies.

Wilderness-centricity and postmaterial values were shared by both radical and popular environmentalisms, and were demonstrable through the declaration of national parks, 'an area where the earth and its community of life are untrammelled by man, where man himself is a visitor who does not remain' (Wilderness Act 1964). America's Yellowstone, established in 1872 was the world's first national park, followed a close second by Australia's The Royal National Park.

Both the popular and scientific expressions of wilderness-centricity were underpinned by deep ecology's vision of wilderness as the only authentic essence of nature because it predated human occupation (Foreman 1998). The assertion of scientific conservation's universal relevance and the political and social implications of its assumed supremacy have been challenged by cross-disciplinary scholarship and activist writings.

The wilderness thesis made it evident that who got to define nature and consequently how nature was constructed was essentially about who had power (White 2004). By excluding urban and industrial concerns, environmentalism deflected awareness from everyday places and their realities (Cronon 1996). Therefore, even as American environmentalism took on big new challenges in the 1970s, it ended up excluding certain kinds of people. Such exclusions assumed racial overtones against the backdrop of historic settler colonialism and structural social inequalities (Purdy 2015).

The scientific conservation model of the North also carried portentous consequences when applied to the global South, by forcing the displacement of nature-dependent communities for the creation of national parks. Dispossessing Indigenous people of their lands to create Tiger Reserves in India is considered a clear example of the deleterious effect of the Northern conservation model in Southern geographies (Guha and Martinez-Alier 1997). On the basis of this effect,

historian Ramachandra Guha (1998) has argued that the practice of scientific conservation can result in a direct transfer of resources from the poor for the benefit of
the rich, causing critical social injustices.

Guha made a further distinction between the philosophy of Arnae Naess, the
founder of deep ecology, which he found to reflect a concern for social inequality
(see Naess 1973), and later writings on deep ecology that informed American environmentalism (see Devall and Sessions 1985). Guha (1989b, 1998) argued that the
latter demonstrated a lack of concern for the ecological burden of the poor. Overall,
critics found that the wilderness thesis informing the politics of American environmentalism did not allow the environmentalism of the poor, which could arise
both from within and beyond Northern geographies, to be expressed (Martinez-
Alier 1995).

Texts such as *Varieties of Environmentalism; Essays North and South* (Guha
and Martinez-Alier 1997) challenged the universal relevance of American environmentalism through a conceptual analysis of empirical cases of socio-nature
relations and politics from diverse geographies. It argued for clearly articulated
environmentalisms in the global South, with entirely different imperatives and
politics from that of the North. The Southern environmentalisms of the poor can
be understood as dominated by both the materialities of the disproportionate ecological burden from industrial activities and the disadvantage experienced by the
poor in accessing natural resources (Martinez-Alier 2002). Southern environmentalism remained preoccupied with 'shallow' ecologies such as pollution control
and agro-ecology, and aimed to secure a just share of the commons – land, water,
fisheries, and forests – that were vital for the subsistence of communities (Gadgil
and Guha 2000).

Scientific ecology's claim to universality has been challenged even within the
same geography on account of the presence of 'different kinds of societies' (Harvey
1996, p. 44) that express different environmentalisms. In particular the paradox
of what Doyle (2005) describes as the majority worlds of the poor within the
minority worlds of the affluent. In the 1980s, while radical eco-centrism aimed to
preserve wilderness through scientific ecology, the emergent environmental justice
movement in North America, which represented the struggles of low-income classes
and vulnerable communities of racial minorities, including Native Americans,
aimed to stop governments and corporations from turning poor neighbourhoods
into hazardous waste sites (Baer and Singer 2020). Being concerned with the direct
impacts of social and economic inequalities, it aimed for a future based on justice,
empowerment and accountability. For affected communities, the environment
occurred where they 'live(d), work(ed) and play(ed)' (Gottlieb 2005, p.34).[1] Instead
of standing apart, nature was understood as enmeshed in the political ecology of
social relations (Escobar 1999; Peet and Watts 1996; Zimmerer 2000).

Critics also questioned the relevance of radical eco-centrism in addressing the
underlying causes of environmental degradation. Carson's expose of the chemical
industry through *Silent Spring* (1962) and the *Limits to Growth* report by the Club
of Rome (Meadows et al. 1972), which identified resource depletion, overpopulation, and pollution as threats to humanity's future, brought industrial development

in its entirety into environmentalism's focus. *Limits to Growth* introduced terms such as the earth's carrying capacity into its lexicon. Environmentalism's response was to develop an eco-centric philosophy. Paradoxically, even though eco-centric environmentalism's narratives 'rail(ed) against the destruction of the world', its politics turned on an 'enlightened individualism' instead of collective social action to transform destructive global capitalism (Hillier 2010).

Guha (1989b) contended that this approach demonstrates that environmentalism's values were shaped by a unique environmental history, and not the universality that the 'New Ecologists' of the 1980s laid claim to. In *The Trouble with Wilderness: Or Getting Back to the Wrong Nature* (1996), historian William Cronon also similarly argues that the notion of wilderness as separate from us was created at a particular moment in history from a culture 'whose relation to the land was already alienated' (p. 17).

Scholarship across the humanities and social sciences has challenged the notion of society and nature as non-overlapping domains of reality (see Braun and Castree 1998; Castree 2005; Davison 2008; Haraway 1991; Latour 1993; Macnaghten and Urry 1998; Plumwood 1993; Soper 1995; Williams 1972). A critical debate around the universal relevance of postmaterial values (Abramson 1997; Brechin and Kempton 1994; Dunlap and Mertig 1995, 1997; Inglehart 1997; Kidd and Lee 1997; Martinez-Alier 1995; Pierce 1997) has finally accepted that Southern varieties of environmentalisms exist, and they are distinct processes arising from different contexts to those of Northern environmentalisms (Brechin 1999).

Environmentalism now accepts within its discourses the concepts of intra-generational and inter-generational justice along with inter-species justice. It is now studied as a phenomenon that occurs across class, racial, and ethnic divides and as a movement whose actions can be motivated by a variety of imperatives from loss of community resources and ancestral lands, the threat of toxic emissions, preserving biodiversity, or protecting future generations from the risk of climate change (O'Neill 2012).

2.2 Australian environmentalism

As in the United States, Australia's modern environment movement was shaped by the post-war economy's focus on resource extraction and industrialisation that involved land clearing on a massive scale. Wilderness concerns dominated over other manifestations and discourses of environmentalism, and a large part of the environment movement practised a politics of conserving nature that the colonial settler society had still left untouched (Doyle 2000).

The 1980s were defined by the wilderness wars, a series of movements centred around stopping logging and dam building at forested sites that involved sustained protests and blockades at these remote locations. In Tasmania's southwest wilderness, seven years of campaigning to stop the damming of the Franklin River culminated in non-violent river blockades and the arrests of around 1300 protestors in 1983. The blockades, and a colour spread of an iconic photograph of a misty bend in the River in *The Age* newspaper, influenced voters ahead of the 1983 federal elections. The newly elected Prime Minister Bob Hawke suspended the Tasmanian government's hydroelectric project using the external affairs power of the Federal government.

The Franklin was a high-profile campaign because of its publicity and the success in applying political pressure (Thompson 1984). It set a precedent for federal interference in development projects and was followed by other successful wilderness struggles in the 1980s and early 1990s such as in Queensland's wet tropics, Tasmania's southern forests, and the Kakadu wetlands in the Northern Territory (Toyne 1994). As a result of these wilderness campaigns, Prime Minister Bob Hawke's federal Labor government established these landscapes as World Heritage Areas, giving them protection under Australia's national parks system (Christoff 2016).

Grassroots mobilisations had started growing within the environment movement from the 1970s out of frustration at the bureaucratisation of big NGOs like Greenpeace and the Australian Conservation Foundation (Cianchi 2015). The success of the Franklin campaign was preceded by an unsuccessful Tasmanian grassroots movement to save Lake Pedder 'with a fringing beach of white quartzite sand', from being flooded for hydroelectricity in 1972 (Hay 1994, p. 5). Soon after the failure to save Lake Pedder, the world's first Green Party, the United Tasmania Group (UTG), was formed in 1972 to put wilderness protection on the political agenda (Rainbow 1992).[2] In 1992, a national Greens confederation was instituted by uniting various state and territory sections that had formed since the 1980s. The Green Party regarded itself as forward-looking, and transcending the traditional left-right divide by rejecting class struggle (Hillier 2010).

Across Western democracies, modern environmentalism had formed as a new social movement in the late 1960s based on a shared philosophy of protest, social action, and radical critique (Gottlieb 2005). However, although the Australian and German Green Parties were formed in a similar time period and shared a radical critique of capitalist ecological destruction, they demonstrated distinctly different political impulses. Doherty and Doyle (2007) differentiated the politics of the German and the Tasmanian Greens as 'post-industrial with a new left-derived analysis of power versus a postmaterialistic environmentalism prevalent in the minority countries of the new world' (p. 707).

While Naess and deep ecologists in the United States and Australia emphasised shifting people's values toward nature, Rudolph Bahro, the founding philosopher of the German Green Party, emphasised changing the patterns of production, consumption and distribution, and acknowledged the historic links between industrialisation, militarisation, and colonisation in creating the global ecological problem (Bahro 1982). Aiming for a radical reversal of the capitalist industrial system, the German Green Party adopted policies on wealth redistribution and anti-militarism (Guha 2000).

Despite diverse outlooks within the Australian Greens due to the diverse political evolutions of the various state units, its early approach was still dominated by a conservative eco-centrism, under the influence of the Tasmanian section. Conservative eco-centrism also found strong support from tertiary educated, left-of-centre, secular and dominantly urban Australians, who constituted the largest membership of Australian environmental nongovernmental organisations (ENGOs) (Pakulski and Tranter 2004). The wilderness approach had a bearing on how society and the media categorised various environmental issues; while wildlife preservation and preventing logging were regarded as 'green' issues that environmentalists fought for, pollution and waste disposal were regarded as brown issues (Pakulski and Crook 1998).

Veteran environmentalist Dr Bob Brown, co-director of UTG and the first leader of the Australian Greens, attempted to unite diverse values within the party through an anti-consumption political narrative:

> The global ecological crisis unleashed by capitalism and the political vacuum created by the Australian Labor Party's embracing of economic rationalism made the rise of a Green Party inevitable...like spontaneous combustion from the rotting haystack of an overblown consumerist society.
>
> Brown and Singer (1996, p. 20)

Australia felt the global influences of the anti-nuclear and peace movements in the late 1970s through mobilisations in cities by grassroots organisations such as Friends of the Earth (Martin 1982). Due to these and internal influences from its left-leaning Western Australian and New South Wales sections, some of the Australian Green's ideologies resonated with its German counterpart. Beyond this, the understanding of environmental processes, political histories and geographical specificities that underpinned efforts at political change, differed between Europe and America (Guha 1989b, 2000) and by extension Australia.

The Greens and the environment movement did not become a working person's movement. However tactical alliances were built with the union movement such as when a relatively socially aware labour movement became active in the anti-nuclear campaign, leading to an alignment between the environment and labour movements (Martin 1982). The Green Bans movement that started in Sydney within the Australian Builders' Labourers' Federation (ABLF) in the 1970s, set a global precedent for progressive unionism on ecological issues by refusing to work on building projects that harmed environmental and cultural heritage. Under trade unionist Jack Mundey's leadership, Green Bans collaborated with resident action groups against destructive local developments and built a grand coalition that cut across the class divide, and traditional antagonism between environmentalists and workers (Burgmann 2008).

The environment movement's fixed ontology and antagonistic legacy made collaborations challenging (Pickerill 2018). The antagonism between farmers and environmentalists owing to fundamental differences in values, beliefs, motivations and worldviews, is well regarded (Brummans et al. 2008). The Land Care movement that started in 1989 was a collaboration between the national ENGO Australian Conservation Foundation and the peak farming body National Farmers Federation. It was regarded as 'one of the best national examples of rural partnerships and group formation' (Pretty et al. 2001, p. 278). Land Care proved an exception to the antagonism in farmer-ENGO relations in Australia. It was made possible by politically acceptable framings of natural resource management for farmers, such as self-reliance and participation (Lockie 2004).

Owing to settler colonialism and the creation of new meaning in landscapes from where Indigenous presence has been removed, American and Australian environmentalism faced comparable challenges towards reconciling eco-centricity with Indigenous land rights and economic justice. The entry of the *Native Title Act 1993* (NTA) into Australia's legal system stimulated inclusiveness towards Indigenous claims to Country in the approaches of environmental groups. Although officially Indigenous groups have the same access to formal structures of democracy as non-Indigenous Australians, a small and highly dispersed population means they have little electoral power and few options for political action, except via the 'language and discourse of white liberal democracy' (Sawer and Zappala 2001, p. 290). In cases of contesting mainstream development, Indigenous groups have either mobilised civil society and public opinion or appealed to the international rights standards of prior and informed consent (Altman 2012).

In the 1990s, public awareness of Indigenous land rights contributed to the success of the Jabiluka Action Group against uranium mining on the lands of the Mirrar people, adjacent to the Kakadu National Park in the Northern Territory (Hintjens 2000). By withholding free prior and informed consent enshrined in the United Nations Declaration of the Rights of Indigenous People (UNDRIP), Australia's native title system had limited Aboriginal people's say on what happens on their land, often compelling them to enter into contractual alliances with environmentalists to further land rights against the State–corporate compact (Vincent 2016). Indigenous–green relations posed the question of how to build an equitable and sustainable system by respecting sovereignty as enshrined in the UNDRIP (Esposito and Neale 2016).

For Indigenous groups, the need to collaborate with environmentalists was driven by their own limitations in resources and legislative capacities. Upon losing their native title claim over the Barmah–Millewa River Redgum forests at the Victoria–New South Wales border in 1998 the Yorta-Yorta people formed an alliance with Friends of the Earth to campaign for a jointly managed National Park (Atkinson 2004).[3] Between the 1970s and 1990s, environmental groups forged proximal but unstable tactical alliances with Indigenous groups in places with strong Indigenous presence and continuing traditional practices (Vincent and Neale 2016). Green–Black alliances sought to navigate their differences through collaborations, informal agreements and negotiations at multiple scales and complexities that the public narratives of prominent ENGOs failed to reflect (Christoff 2016).

As environmentalism's collaborations with Indigenous land management in Australia have occurred within regulatory contexts that are geographically specific, they have produced unique definitions and environmental co-management methods. However, owing to a 'colonial paternal sense of responsibility' and 'unexamined social norm', collaborations between natural resource managers or environmentalists and Indigenous groups often created dualisms that denied commonality and created tensions (Weir 2016, p. 137).

Australia's water management culture forced Indigenous people into either co-option or marginalisation on account of Indigenous knowledge being deemed traditional, local, spiritual, emotional and culturally specific as opposed to Western conservation science, which was deemed modern, universal, technical, rational, and culturally neutral (Plumwood 2002). Attempts at conserving entire catchments in the manner of national parks in Queensland through the controversial *Wild Rivers Act 2005 (QLD)* clashed with priorities of land use for economic development on Aboriginal-owned land, disrupting long-term relations between the Cape York Land Council and ENGOs (Neale 2016).

The bias and marginalisation inherent in such conceptual frames needed to be countered through a 'double movement' or a gesture both of solidarity and the other's difference as an entity to be engaged on their own terms (Plumwood 2002, p. 138). Indigenous–environmentalist alliances needed to overcome the problematic rhetoric of Black versus green, of environment versus economy, and of green colonialism versus Indigenous autonomy (Pickerill 2018). Most of the Indigenous–green interactions between the 1970s and 1990s indicated a fundamental mismatch of visions (Dodson 1997).

From the 1990s, the need to accept climate change has forced Australia's environment movement to reconsider the simple binary of its narrative, of nature as either pristine or transformed. At the same time, the emergence of the native title system and the beginning of the process of returning Indigenous lands since 1993, has given Aboriginal Australians a voice on the issue of mining-related environmental conflicts and compelled environmentalist narratives to reflect a pragmatic approach. Together, these two factors have set a future framework for Australian environmentalism after the era of the wilderness wars (Christoff 2016).

2.3 Indian environmentalism

Although currents such as wildlife conservation and middle-class appreciation of the aesthetics of national parks also characterise Indian environmentalism, most environmentalists subscribe to the environmentalism of the poor in the Indian context (Lele 2012; Baviskar 2002; Mawdsley 2006). Scholars of Indian environmentalism mostly draw from a combination of the Gandhian vision for alternative development based on the idea of self-sufficient villages and critiques of industrialisation and global anti-development discourses (Gandhi 1937; Escobar 2001).[4]

In the context of India's postcolonial development in a highly unequal society, environmental injustice pertains to unequal access to a shrinking pool of natural resources by nature-dependent subsistence communities, whom Dasmann (1988) called the ecosystem people. Agarwal (1986) argues that as the sections of society most affected by disruptive development, the primary concern of peasants, women, and India's Indigenous Adivasi ('native dweller') communities is that they should benefit from environmental resources. Guha (2000) argues that the struggles of marginalised groups are often marked by a powerful indigenous ideology of social justice. Such struggles employ tactics of peasant resistance – strikes, road blockades, protest marches and hunger strikes – collectively termed as 'weapons of the weak' (Spodek 1971; Scott 1985).

Political ecology in India has revolved around the use and control of natural resources by different groups. The New Delhi-based Centre for Science and Environment's (CSE) *State of the Environment* reports (1983, 1985) highlighted that the overexploitation of resources by commercial interests was creating a disproportionate burden on ecological communities. Guha (2000) considers this to be the fundamental claim of the environmentalism of the poor in India. With political parties being largely indifferent to environmental destruction and its social consequences, civil society groups built non-party political formations to organise ecological refugees (Kothari 1984). The environmentalism of the poor, India's dominant environmentalism, has been conceptualised and articulated on the basis of such practices.

Chipko Aandolan, a social movement of peasant communities formed in the Himalayas in the 1970s, is considered the starting point of India's modern environment movement (Guha 1989a). *Chipko* ('to hug') became the definitive word for the movement when in a critical moment in the struggle village women hugged trees to prevent contractors from logging them. The standoff between the State and villagers forced a community-oriented forest management plan for the region. Whether ecological protection was implicit or incidental to Chipko has been much debated. Guha (1989a) has argued that as the imperative for the standoff was their being denied a fair share of forest resources, protection of catchments and forests was a consequence rather than explicit objectives of the struggle. Shiva (1988) has contended, however, that the politics of the Chipko movement reflects the long-held knowledge of village women – 'that forests sustain the earth and all she bears' – making it an explicitly ecological and feminist movement (p. 76).

Chipko was one piece in an entire landscape of resistance in post-independence India: consisting of peasants and Adivasi people opposing displacements from large dams; artisanal fishers resisting trawler fishing, resistance to commercial forestry activities; and downstream peasants and fishers opposing upstream industrial pollution (Agarwal 1984; Gadgil and Guha 1995). From the 1980s, as the Indian government's plans to generate large-scale hydroelectricity to power the industries of the postcolonial economy started taking effect, various organised but mutually disconnected oppositions to large dams erupted across the Indian landscape alongside movements of dam-displaced people for rehabilitation and compensation (Centre for Science and Environment 1985).

The Narmada Bachao Aandolan (NBA), India's largest anti-dam movement was launched in the 1980s. It challenged the drowning of lands and displacement of communities in 193 villages in central India. Called the world's greatest planned environmental disaster, the World Bank-funded Sardar Sarovar hydroelectricity scheme proposed 30 major dams, and 135 medium and 3000 minor dams along the Narmada River (Kothari and Singh1988). Adivasi people were most affected by this grand dam scheme that commenced in 1979. The NBA spread across multiple states. It received support from movements and networks across India as well as international NGOs due to the involvement of the World Bank (Baviskar 1995).

The movement faced repression from governments and was accused of being anti-development. Urban activists leading the movement were accused of wishing to 'keep poor farmers and Adivasis in hunger, illiteracy, and nakedness by denying them the fruits of development' which they themselves enjoyed (Anklesaria 1988).

Urban-based activist groups sought to democratise India's development process to make it sustainable and equitable for rural-based livelihood communities who had no legal rights to challenge the effects of industrialisation on their lands and livelihoods. In 1985, a joint statement from civil society organisations appealing for the participation of livelihood communities in the State's resource management process reflected the impulse of indian environmentalism to democratise development:

> Today, with no participation of the common people in the management of local resources, even the poor have become so marginalised...that they are ready to discount their future and sell away the remaining natural resource for a pittance... Given the changed socio-economic circumstances and greater pressure on natural resources, new community control systems have to be established that are more highly integrated, scientifically sophisticated, equitable and sustainable
>
> Joint statement cited in Guha (2000, p. 67)

The vision of Indian environmentalism made ecological and social harmony contingent on the creation of an economically just society (Gadgil and Guha 1995). Its political approach called for a rethinking of development based on a democratisation of natural resources and the environment (Guha 2000). Its narratives attempted to widen the development debate beyond conventional economics towards inclusivity for ecosystem and livelihood-dependent communities (Kothari and Parajuli 1993).

The political impulse for democratisation in the Southern context was further evident in the transformations of ecological thought in India. In *An Indian Conservation Strategy* (1982), ecologist Madhav Rao Gadgil criticised the paradigm of international scientific conservation – the big continuous wilderness, and a 'hands off nature', 'keystone species' approach to biodiversity protection – and emphasised a decentralised network of small parks as a suitable model for India's peopled natural landscapes. In 1987, Gadgil spearheaded the 'Save the Western Ghats' movement along the ecologically sensitive western rim of the Indian peninsula, with the twin objectives of ecological conservation and equitable access for communities (Guha 2006).

The striving for democracy has thus assumed entirely different purposes in environmentalism's Northern versus Southern contexts. While the wilderness wars in Australia and American radical environmentalism sought democratic representation for the rights of nature, livelihood movements in India appealed for a human-centric justice in the distribution of natural resources by the State. On account of differences in historical contexts, political and economic realities of industrialising and industrialised nations, and effects of environmental destruction on densely populated Southern geographies such as India, the environmentalism of the poor also differed in degrees from Indigenous land rights struggles in Australia and the environmental justice movement in the United States. Collectively, the lived relationships of India's subsistence communities with nature had the twin effect of reconfiguring concepts from both Northern conservation and Northern environmental justice, even though it shared with the latter the principle of social equity as the foundation for ecological justice.

The environmentalism of the poor's ideological formulations have been critiqued on several grounds. New traditionalist discourses of environmentalism of the poor have characterised colonial rule as having introduced alien social, economic and ecological relations that have continued in postcolonial India (see, for example, Gadgil and Guha 1993; Shiva 1988). Such accounts made generalised assertions about the inherent conservation ethic of women and Adivasis (Shiva 1988; Banuri and Marglin 1993; Pereira and Seabrook 1990) and deployed what Brosius (1999) called 'essentialised images'.

Instead of a historical validity of traditional environmental and gender relations, Vandana Shiva's ecofeminist writings indicate an ideological partiality towards a mostly Hindu way of life in the pre-colonial rural India (Rangan 2000), at the same time ignoring the reality of the harshness of everyday life in the village (Mawdsley 2006). A rising Indian middle class is also unlikely to find political action in an idealised rural past (Sridhar 2010). Environmental sociologist Amita Baviskar writes in *In the Belly of the River: Tribal conflicts over development in the Narmada Valley* (1995) that urban leaders of the Narmada Bachao Aandolan were likely to use ecological metaphors such as 'mother earth's children' for Adivasis; evoking imagery that communities themselves might not use while talking about their relationship to the forests and land (p. 213).

Other criticisms of Southern environmentalist writings relate to the local focus of political ecological studies. Scholarly examinations of livelihood movements in the Global South have paid less attention to the political ecology of global issues as seen from a third-world perspective (Adger et al. 2001; Moore et al. 1996). Another related criticism is that with the exception of the State, political ecological studies in the South have not considered the role of non-local actors (Bryant and Bailey 1997). For example, the role of non-governmental organisations (NGOs) remains a much-needed area of study in third-world political ecology (Bryant 1998). Collaborations between grassroots movements and civil society NGOs in India have created an overlap of what used to be previously considered as distinct categories of social action, requiring a conceptual framework that can accommodate multiple claims as well as contradictory politics at sites of environmental conflict (Baviskar 1997; Nambiar 2014).

However, despite the pertinent criticism of political ecology's focus and representations in Indian studies, its role in highlighting the centrality of the State in both creating and resolving environmental injustices makes it a necessary approach for studying environmental conflicts in the Southern context. The environmentalism of the poor in India has been both defined and challenged by the developmental paradigm of the State. Its narratives have asserted that the post-independence developmental State has continued to exclude the environmental knowledge of Adivasis, much like its colonial predecessor (Gadgil and Guha 1993).

The environmentalism of the poor has retained a suspicion of the State, even in the post-independence era, because the Indian government has continued to use colonial-era land and forest laws and perpetuate the maldistribution of ecological resources. An analysis of the contradictions of the postcolonial developmental State is therefore central to understanding how movements frame environmental conflicts in India. Apart from the obvious differences in history, culture, and public discourse, the centrality of the State's role in shaping movements is one of the fundamental differences between the Southern and Northern contexts of environmentalism (Williams and Mawdsley 2006).

2.4 Environmentalism's divisions and common ground in the climate era

The issue of climate change has contributed to making the environment a truly global issue spanning global politics, regulations and movements (Dryzek 2013). Further, the asymmetries of cause and effect in climate change directly reflect global developmental divides, making the question of how to address climate change unalterably a question of justice (Goodman 2009). The issue of global warming has generated new North–South contestations over environmental justice, particularly in relation to responsibility sharing over greenhouse gas (GHG) emissions. Although environmental and climate NGOs from the South have also paid considerable attention to concerns about sharing the burden of emissions and about global economic justice for the South, the epicentre of these North–South contestations has largely been the international climate talks. But for the various environmentalisms discussed in the earlier sections, climate change also created opportunities to find common causes.

Carbon emissions, equity and the North–South divide

Issues of equity and historic responsibility on the issue of climate change have added new dimensions to the North–South divide in environmentalism. One of the most critical concerns from the South has been the North's lack of consideration for social equity and historic injustice in attributing responsibility for carbon emissions. Agarwal and Narain (1991), Meyer-Abich (1993), Mukherjee (1992) and Sachs (1993) have exposed the politics of blame and agenda-setting surrounding the global warming problem and the promotion of first-world controlled environmental management mechanisms as global climate solutions. The Rio Earth Summit in 1992, when the United Nations Framework Convention on Climate Change (UNFCC) that formed the basis for all future climate negotiations was adopted, proved a flash point on the issue of environmental responsibilities across the North–South divide (Rolston 1995).[5]

In the lead up to Rio, a World Resource Institute (WRI 1990) ranking of the carbon emissions of countries was criticised by Southern groups for ignoring the historic responsibility and 'gargantuan consumption' of developed countries, particularly the United States, for causing global warming. The Centre for Science and Environment's (CSE) report *Global Warming in an Unequal World; A Case of*

Environmental Colonialism (Agarwal and Narain 1991) challenged the approach of the WRI, which equated methane emissions from the livestock and paddy fields of subsistence farmers to emissions from gas-guzzling automobiles, on the grounds of justice and morality. Agarwal and Narain (1991) drew a distinction between the subsistence emissions of the poor and the lifestyle emissions of the rich.

In a similar vein to other contentions across environmentalism's global divide, such as between Southern overpopulation (see Ehrlich 1968) and Northern overconsumption (see Galbraith 1958; Guha 1989b), the WRI (1990) report and CSE's response sparked a vigorous debate over the carbon footprints of the rich versus poor, equitable distribution of carbon space between the North and South, and between the ethics of calculating aggregate country level and per capita GHG emissions.[6]

Southern concerns over global equity in climate responsibility were incorporated into the UNFCCC through the Common But Differentiated Responsibility and Respective Capability (CBDR&RC) Principle under article 3.1 in the Convention (Raghunandan 2019). The principle of common but differentiated responsibilities (CBDR) acknowledges the deep inequalities between, and the different priorities of, developed and developing countries, bringing a nuanced approach to international environmental initiatives (Beyerlin and Marauhn 2011). The Preamble to the 1992 UNFCC acknowledged the historical contribution to GHG by developed economies, and the critical need for economic growth and poverty eradication in developing countries (UNFCCC 1992).

However, from the beginning of climate negotiations under the UNFCCC, a global divide has persisted over the nature of commitments from developing countries, and the financial and technical support from developed countries (Dasgupta 2012, 2019). Overall, three approaches towards responsibility for emissions – the national, historic and per capita – have remained as bones of contention between the North and South, and contributed to an impasse in climate negotiations on several occasions (Dubash and Rajamani 2010).

By the mid-2000s, on account of the rapidly growing emissions of emerging major economies, particularly China and India, Northern economies including the United States and Australia refused binding GHG reduction targets without commitments from the South. Separate from the power dynamics of big economies in the North and South, due to the growing demands for climate responsibility from vulnerable small island nations along with the disproportionate impacts of global warming on the global poor, the per-capita approach had begun to look more like a fig leaf than an instrument for genuine equity and global social justice. The UNFCCC framework of equity through the CBDR&RC principle was replaced by the time of the Paris Agreement, which reflected a symmetric treatment of all parties (Kanitkar and Jayaraman 2019).

The ongoing tension on redistributive justice between developed and developing countries – that developed countries should provide climate finance and technical support for developing countries to transform their economies to low carbon and adapt to climate change – remained unresolved after the Paris Climate Summit in 2015. The Paris Agreement did not set a roadmap for how adequate climate funding would be provided to developing countries, or formal targets for Northern countries' financial contributions through the mechanism of the Green Climate Fund (GCF) (Roberts and Weikmans 2015).

Although a 'loss and damage' fund was created at the UNFCCC Conference in 2022 with the aim of addressing 'loss and damage' measures whereby major greenhouse gas-emitting countries take measures to address the needs of lower-emitting nations who nevertheless face the brunt of climate change, critics say strong advocacy is essential to ensure that the fund is effectively operationalised and expanded (Huq and Sultana 2023).

The issue of emission reductions and how emissions are counted has spanned a long period of tensions between the assertions for economic justice from the developing South and the articulations of convenience (and protection of vested interests) from the North. Unfortunately, this climate dispute between the North and the South has masked critical questions that need to be raised about the effects of climate change on vulnerable communities and the large global poor in the South. Such questions have been extensively raised outside the formal processes and negotiating structures of the international climate convention, through global networks for climate justice.

Varieties of climate justice

Outside of formalised international climate negotiations under the UNFCCC, activist networks for climate action and climate justice have articulated different yet overlapping narratives about why the world needs collective and effective action for a safe climate. The emergence of climate change as a global issue has created new meanings for the actions and politics of various modes of environmental resistance – Northern environmentalism, the environment justice movement and indigenous movements in the global North, and the environmentalism of the poor in the South.

The nature of the effects of climate change, which can be felt both globally and locally, has made it imperative that the politics of those who grapple with the big picture find alignment with the activisms for the human justice of marginalised communities (Purdy 2015). Simultaneously, since the interconnections between colonialism and capitalism, and their underlying role in creating and driving climate change are now well regarded, it is imperative to have new conversations about time, and place, and what constitutes meaningful political action (Birch 2016; Bird Rose 2013). It has been argued that 'climate coloniality', while being a material and visceral experience for frontline communities, is also an epistemological site of struggle, implying agency and resistance in various forms that climate justice researchers, activists and policymakers need to critically engage with (Sultana 2022).

The intertwining of human and environmental fates due to climate change has generated an opportunity for wilderness-centric environmentalism to go beyond its past legacy and to address a contradiction that even while it focussed on worldwide problems, wilderness-centricity brought to environmental politics the 'cultural habits of a much more parochial, and sometimes nastier, movement' (Purdy 2015, p. 15). The realities of the present era have raised hopes for the democratisation of Northern environmentalism towards an inclusive and aligned narrative and politics with other environment-related struggles.

For communities that have historically survived environmental injustice, climate change implies yet another set of anthropogenic ecological disruptions they have had no role in creating, and yet, for which they will once again bear a disproportionate burden (Gottlieb 2005). For such communities, the impacts of climate change are not just the global risks of rising temperatures, floods and droughts that have started to occur today, but also the risks of being poisoned, sickened, exploited and abused, which they have historically faced (Quinn-Thibodeau and Wu 2017).

Climate change has therefore compounded the significance of their historic resistance and compelled them to go beyond their local outlook. While continuing to aim for justice, empowerment and accountability as before, the environment justice movement has now gone truly global through the climate justice movement (Dryzek 2013; Gottlieb 2005). Environmental justice and climate justice have been articulated in new social justice upsurges such as the Black Lives Matter (BLM) movement, which, among other pursuits, has pointed to the disproportionate concentration of incinerators and waste facilities in poor, non-white neighbourhoods and called for divesting from fossil fuels (Purdy 2016).

Indigenous perspectives on climate justice enfold a further distinction. For Indigenous people, the present risk of climate change evokes past injustices of colonial dispossession and cultural disruption.[7] Climate justice has therefore become an umbrella issue under which they can articulate many environmental injustices (Whyte 2017). Due to historical injustice, indigenous climate justice is also intricately linked to notions of sovereignty; movements for indigenous environmental justice have therefore remained centred around land rights as before. In Australia, for example, Seed, the indigenous youth coalition that has 'taken the environmental agenda and built it into their worldview', advocates for the deepening and broadening of the past legacy of indigenous environmental justice (Esposito and Neale 2016).

From the perspective of Indigenous rights movements, the democratising of Northern wilderness-centric environmentalism in the climate era needs to also reflect a decolonisation of solidarity by removing the 'paternalism and tension in relations between non-indigenous and indigenous activists' that are fundamentally a product of the colonial condition (Land 2015).

The economic, social and cultural displacement of vulnerable communities would continue to worsen under climate change, exacerbating what Nixon (2011) calls the slow violence of the poor. Based on this understanding, the interpretation of the climate problem by Southern civil society groups remained grounded in the fundamental principles of equity and justice (Lele 2012). In India, the issue of climate change had long remained the domain of a few elite policy and science-based NGOs and foreign policy experts (Dubash 2012). However, the activism of grassroots science and climate justice groups and transnational organisations such as Greenpeace has been instrumental in not only turning the spotlight of climate responsibility inward toward India's highly unequal society and the disproportionate burden of climate change on the poor but also calling for intra-generational equity for the poor (Dubash 2013; Michael and Vakulabharanam 2016; Thaker and Leiserowitz 2014).

Such activisms have criticised the government's developmental frame, which while seeking equity on the global stage, has continued to play on the domestic divisions between the urban rich by hiding their emissions behind the negligible carbon footprint of the rural poor (Adve 2013; Ananthapadmanabhan et al. 2007). The issue of climate change has created opportunities for Indian civil society organisations (CSOs), which are largely urban-based, to extend their actions to help vulnerable communities, which are largely subsistence-based, in adaptation, mitigation and climate-related development measures. But broadly speaking, the Indian CSO network does not use climate change as an overarching narrative frame for a range of eco-social issues, conflicts and actions. Swarnakar (2019) argues that the network lacks a grand narrative on climate justice due to its contradictory political outlook on who needs to take responsibility for climate action; while parts of the network hold the global North accountable for climate injustice other parts hold the Indian State as responsible.

Finally, with regard to the environmentalism of the poor, climate change has not generated mass mobilisations. However, energy projects such as coal mining and thermal power generation have continued to generate discontent amongst peasant and indigenous communities. These have been challenged by the environmentalisms of the poor in response to the loss of their lands and livelihoods, environmental pollution around the project sites and unfair compensation rates by governments. Environmentalisms of the poor against such industrial developments have increased in India's neoliberal era of rapid development (Roy and Schaffartzik 2021).

Climate justice as a common frame?

Two intersecting lines of arguments have been made around why climate justice can hold a dialogic umbrella over various claims to ecological and social justice today. The first argument relates to the advancing climate crisis, due to which other frameworks for eco-social justice are now understood as subsumed by climate imperatives, to produce the climate justice model (Goodman 2009). Naomi Klein sees climate change as the all-encompassing 'human rights struggle of our time' that requires its many movements to be connected (cited in Stephenson 2015, p. 52). The second is on account of the reconfiguration of the significance and politics of various modes of environmental resistances through the notion of climate justice, as discussed in the previous subsection.

The nature and scale of the problem of climate change made it imperative for wilderness-centric Northern environmentalism to transform its fixed ontology and exclusionary approach towards other human–nature relations, expressed by other modes of environmental resistances. In the North, movements challenging environmental injustices on vulnerable communities, as well as movements for

Indigenous justice have taken climate justice as a higher organising frame for present and historic grievances, thus creating a continuous and deeper significance for their politics and actions.

The climate era is understood to hold the possibility of enmeshing various types of environmentalisms to create an inclusive ecological justice and an intersectional politics that is less like (Northern) environmentalism in the previous era and more like human rights, as 'what we are fighting for is each other' (Stephenson 2015, p. xv). It stands acknowledged that to achieve this necessary intersectionality, ecocentric environmentalism will have to decolonise its engagement and establish solidarity with the purpose and vision of movements that are fighting to redress historic injustices. Towards this end, Klein (2016) sees the task of the climate activist as 'overcoming the various disconnections and connecting our various movements'.

However, these possibilities for intersectionality and the notion of climate justice as a common frame have emerged from the global North, and they are indicative of new relational politics and new approaches to engagements between the various concepts and politics of environmental resistance in the North, such as ecocentrism, environmental justice and Indigenous justice. They might not represent the socio-political context of environmentalism in the global South. As the previous subsection indicates, despite the centrality of the environment in their lives, climate justice has not become a mobilising factor for the environmentalism of the Southern poor, such as in the case of India.

Although climate change has engaged urban-based civil society groups in India that are constituted of the educated middle class, it has not generated mass environmental movements of the largely rural poor, signifying a highly uneven public sphere and a fragmented society. This obstructs the global narrative of climate justice from being adopted by ecosystems-dependent communities that are the most susceptible to climate change, thereby creating a paradox. Williams and Mawdsley (2006) argue that despite similar patterns of structural marginalisations (along race and class in the United States, and caste, ethnicity, class and gender in India) that cause environmental ills and spark environment justice movements in Northern and Southern contexts, the socio-political contexts of environment justice movements still differ across the North and South in three major ways that I now discuss. The differences can be experienced even in the case of democracies.

Williams and Mawdsley's (2006) first argument is that even marginalised communities in the North can experience a relatively homogeneous public sphere and thus access comparatively effective mechanisms for justice as opposed to their Southern counterparts. In the Southern postcolonial context, such as in India, the presence of what Rudolph and Rudolph (1987) characterise as a weak–strong State, creates ambitious yet incompletely realised government programmes for public good across various areas, including in participatory environmental management and environmental policy. These, coupled with a lack of enforcement of legislative rights of subsistence communities, including towards consent for mining on their lands, have created a systematic lack of recognition of 'ecosystems people' within governance systems. This inequality of recognition makes the non-discursive struggles of 'subaltern counterpublics' (Fraser 1997, p. 81), which fall outside the formal structures of States and institutions, a crucial factor in the process of deliberative democracy for environmental justice in the South (Williams and Mawdsley 2006). Although marginalised communities in the North also experience challenges in accessing rights and justice, such as ongoing experiences of denial of sovereign rights amongst Indigenous Australians (see Moreton-Robinson 2017), the combined realities of poverty and everyday violence from developmental States experienced by Southern subalterns such as Adivasis, including torture and killing (see Baviskar 2012), can make the scale of their structural and procedural injustices incomparable with Northern subalterns.

Second, people's struggles for environmental justice in India have to contend with a highly unequal public sphere that is dominated by the urban-based middle class that is largely responsible for the visibility of the environmentalism of the poor, as well as for how they are represented. This mediated representation can risk their essentialisation, as reflected in political–ecological texts on livelihood struggles discussed earlier.

How Western environment justice research contextualises Southern actors has a further bearing on the challenge of accurate representation of the environmentalism of the poor. Williams and Mawdsley (2006) argue that what David Harvey characterises as a 'sideways looking admiration for those marginalised peoples who have not yet been fully brought within the global political economy of technologically advanced and bureaucratically rationalized capitalism' (Harvey 1996, p. 389), can risk using Southern environmental actors as symbols of distant 'others' within frames of Northern environmental justice, without reflecting the complexity of their contextual realities. Therefore, applying the common lens of climate justice should be accompanied by a critical engagement with the contextual differences of environmental resistances in the global South. Climate justice as an organising plank for multiple grievances can however help to globalise the significance of local livelihood resistances of the environmentalism of the Southern poor.

This is especially important given the third factor, which is that the environment makes a direct contribution to the lives and livelihoods of a large section of the population in the South, bringing urgency to the environment versus development debate in the Southern context. Even in the era of climate change, the environmentalism of the poor in India has continued to express grievances about the loss of lands and destruction of livelihoods from large-scale industrialisation. Williams and Mawdsley (2006) argue that it is not enough to simply see how ideas of justice coming from the South may tactically align themselves within a 'global' environmental justice movement that is framed by Northern contexts; rather, distinct frames of environmental justice emerging from the South should be treated as such in Western environmental justice research.

Conclusion

Through an investigation of environmental activism generated from coal mining-related conflicts in Australia and India, this book raises questions about the historic divide between environmentalism, not only across the North and South, but also within eco-centric and human-centric forms of environmental justice in the North, and whether they can find common ground today. Climate change has brought intimations of an end to the kind of nature that eco-centric Northern environmentalism has striven to preserve – a nature that was unharmed by human intervention – making it imperative to transform from the simple binary of a human–nature divide in its vision towards a democratic and inclusive politics.

The concepts, politics and visions of environment justice and Indigenous justice movements in the global North, while retaining the need for historic and structural justice, have been globalised through the adoption of the climate justice narrative. Through critical reflections on its own legacy and a dialectical process of attempting to achieve shared ground, such as in the case of Australian environmentalism's efforts to secure Green–Black relations, Northern eco-centric and historic human-centric environmental justice movements have by now negotiated some shared principles – decolonising solidarity; acknowledging the inter-connections between colonialism and capitalism; and addressing the historic dispossession of Indigenous peoples in settler colonial society. These principles form the basis for a common understanding of climate justice.

However, the division across the contextual realities in the global North and South remains significant, and it is a crucial factor in environmentalism's North–South divide even today. As seen in the case of the environmentalism of the poor in India, climate change has not become an issue of mass mobilisations for livelihood communities, despite their high level of dependency on nature, creating a paradoxical situation. The high level of dependency on nature by a significant part of the Indian population adds a critical urgency to the environmental assertions of livelihood movements. Despite this urgency, their contestations largely fall outside the formal discursive field due to the nature and functioning of the postcolonial State and its ineffective implementation of rights and institutional representation for vulnerable communities. Livelihood movements also struggle for representation in the public sphere that is dominated by the middle class.

Frameworks of climate justice emerging from the North need to recognise this Southern paradox. Instead of merely seeking to align Southern ideas of justice within its framework and treating Southern environmental actors as distant 'others', Western environmental justice research needs to not only include the critical distinctions arising from Southern contexts, but also fully contextualise Southern actors. This imperative informs the comparative purpose and approach of this research. Through a critical comparison of two collective environmental resistances to coal mining in Australia and India, it responds to questions about environmentalism's historic North–South divide, how these differences can be resolved with the common focus on climate justice, and yet what critical differences still remain, and finally what possibilities for common ground emerge in the present era.

Notes

1 From Dana Alston's speech at the People of Colour Environmental Leadership Summit held at Capitol Hill between 24 and 27 October 1991. Alston was one the key organisers of the summit.

2 The UTG's manifesto, *New Ethic* (1972), stated eight requirements for ethical and sustainable development to 'do minimum damage to the web of life of which we are a part' and 'maintain Tasmania's form and beauty not just for our enjoyment but the enjoyment of all future generations'.

3 A 1998 Federal court judgement had denied the Yorta-Yorta people's native title claim on the grounds of a perceived loss of tradition (Atkinson 2004).

4 According to the Gandhian vision, *Gram Swarajya* (self-sufficient villages) would be achieved through building cottages from locally sourced materials, providing sufficient village commons for the grazing of cattle, locally available education, and local governance through *Panchayats* (village councils). The model villages would also grow their own food, and make their own hand-spun *khaadi* cotton cloth (Gandhi 1937). This vision was born out of Gandhi's critique of industrial development: 'the blood that is today inflating the arteries of the cities runs once again in the blood vessels of the villages' (Gandhi 1946).

5 The debates were characterised by contentions around the issues of the developmental rights of Southern nations versus the responsibilities of the North, overpopulation in the South versus overconsumption in the North, and between public and private interests (Rolston 1995).

6 While the WRI warned that if just China and India increased emissions to the global average per capita rate, the earth's net GHG levels would rise by 28%, the CSE argued, based on principles of sustainable development, that the remaining 'global atmospheric common' be shared equally on a per capita basis.

7 *'Climate change is yet another rapid assault on our way of life. It cannot be separated from the first waves of changes and assaults at the very core of the human spirit that has come our way'*, Sheila Watt-Cloutier, interviewed by the *Ottawa Citizen* (Robb 2015).

References

Abramson, P. 1997, 'Postmaterialism and environmentalism: a comment on an analysis and a reappraisal', *Social Science Quarterly*, vol. 78, no. 1, pp. 21–23.

Adger, W. N., Benjaminsen, T. A., Brown, K. & Svarstad, H. 2001, 'Advancing a political ecology of global environmental discourses', *Development and Change*, vol. 32, no. 4, pp. 681–715.

Adve, N. 2013, 'Another climate change event', *The Hindu*, June 21, viewed 20 September 2019, <www.thehindu.com/opinion/op-ed/another-climate-change-event/article4834 485.ece>.

Agarwal, A. 1984, 'Politics of environment', in *The State of India's Environment: A Citizens' Report*, pp. 1984–1985.

Agarwal, A. 1986, 'The 5th World conservation lecture: human nature interactions in a third world country', *The Environmentalist*, vol. 6, no. 3, pp. 165–183.

Agarwal, A. & Narain, S. 1991, Global *W*arming in an *U*nequal *W*orld: *A C*ase of Environmental Colonialism, Centre for Science and Environment, viewed 20 March 2020, <www.cseindia.org/global-warming-in-an-unequal-world-3126>.

Altman, J. 2012, 'Indigenous rights, mining corporations, and the Australian state in the politics of resource extraction 2012', in S. Sawyer & E. Gomez (eds), *The Politics of Resource Extraction: Indigenous Peoples, Multinational Corporations, and the State*, Palgrave McMillan, London, pp. 46–74.

Ananthapadmanabhan, G., Srinivas, K. & Gopal, V. 2007, *Hiding Behind the Poor*, Greenpeace India, Bangalore, viewed 20 March 2020, <www.greenpeace.org/india/Glo bal/india/report/2007/11/hiding-behind-the-poor.pdf>.

Anklesaria, S. A. 1988, 'Narmada project: government-opposition alliance against Amte', *Indian Express*, 30 October.

Atkinson, H. 2004, 'Yorta Yorta co-operative land management agreement: impact on the Yorta Yorta Nation', *Indigenous Law Bulletin*, vol. 6, no. 9, pp. 23–25.

Baer, H. & Singer, M. 2020, 'The anti-coal and anti-coal seam gas campaigns as components of the climate movement in Australia: responses to corporate hegemony', Capitalism Nature Socialism, vol. 32, no. 3, pp. 1–19.

Bahro, R. 1982, 'Capitalism's global crises', *New Statesman*, 17–24 December.

Banuri, T. & Marglin, F. A. 1993, 'A systems-of-knowledge analysis of deforestation, participation and management', in *Who Will Save the Forests?: Knowledge, Power and Environmental Destruction*, Zed Books, pp. 1–23.

Baviskar, A. 1995, *In the Belly of the River: Tribal Conflicts over Development in the Narmada Valley*, Oxford University Press, New Delhi.

Baviskar, A. 1997, 'Tribal politics and discourses of environmentalism', *Contributions to Indian Sociology*, vol. 31, no. 2, pp. 195–223.

Baviskar, A. 2002, *The Politics of the City Seminar: A Symposium of the Changing Contours of Indian Environmentalism, Seminar,* vol. 516, pp. 40-42.

Baviskar, A. 2012, 'Extraordinary violence and everyday welfare: the state and development in rural and urban India', in S. Venkatesan & T. Yarrow (eds), *Differentiating Development: Beyond an Anthropology of Critique*, Berghahn Books, New York, pp. 126–144.

Beyerlin, U. & Marauhn, T. 2011, *International Environmental Law*, Hart, Oxford.

Birch, T. 2016, 'Climate change, mining, and traditional indigenous knowledge in Australia', *Multidisciplinary Studies in Social Inclusion*, vol. 4, no. 1.

Bird Rose, D. 2013, 'Slowly – writing into the anthropocene', *TEXT*, vol. 20, pp. 1–14.

Braun, B. & Castree, N. 1998, *Remaking Reality: Nature at the Millennium*, Routledge, London.

Brechin, S. R. 1999, 'Objective problems, subjective values, and global environmentalism: evaluating the postmaterialist argument and challenging a new explanation', *Social Science Quarterly*, pp. 793–809.

Brechin, S. R. & Kempton, W., 1994, 'Global environmentalism: a challenge to the postmaterial thesis?', *Social Science Quarterly*, vol. 75, no. 2, pp. 245–269.

Brosius, P. J. 1999, 'Analysis and interventions: anthropological engagements with environmentalism', *Current Anthropology*, vol. 40, no. 3, pp. 277–309.

Brown, B. & Singer, P. 1996, *The Greens*. Text Publishing, Melbourne.

Brummans, B., Putman, L., Gray, B., Hanke, R., Lewicki, R. J., & Wiethoff, C. 2008, 'Making sense of intractable multiparty conflict: a study of framing in four environmental disputes', *Communication Monographs*, vol. 75, no. 1, pp. 25–51.

Bryant, R. L. 1998, 'Power, knowledge and political ecology in the third world: a review', *Progress in Physical Geography*, vol. 22, no. 1, pp. 79–94.

Bryant, R. L. & Bailey, S. 1997, *Third World Political Ecology*, Psychology Press, London.

Burgmann, V. 2008, ''The Green Bans Movement', worker's power and ecological radicalism in Australia in the 1970s', *Journal for the Study of Radicalism*, vol. 2, no. 1, pp. 63–89.

Carson, R. 1962, *Silent Spring*, Fawcett Crest, New York.

Castree, N. 2005, *Nature*, Key Ideas in Geography Series, Routledge, Abingdon, UK.

Centre for Science and Environment. 1983, *The State of India's Environment*, CSE, New Delhi.

Centre for Science and Environment. 1985, *The State of India's Environment*, CSE, New Delhi.

Christoff, P. 2016, 'Renegotiating nature in the Anthropocene: Australia's environment movement in a time of crisis', Environmental Politics, vol. 25, no. 6, pp. 1034–1057.

Cianchi, J. 2015, *Radical Environmentalism: Nature, Identity and More-than-Human Agency*, Palgrave McMillan, London.

Cronon, W. 1996, 'The trouble with wilderness: or, getting back to the wrong nature', *Environmental History*, vol. 1, no, 1, pp. 7–28.

Dasgupta, C. 2012, 'Negotiating the framework convention on climate change: a memoir', in K. V. Rajan (ed.), *The Ambassadors' Club*, Harper Collins, New Delhi, pp. 61–84.

Dasgupta, C. 2019, 'Present at the creation: the making of the framework convention on climate change', in N. Dubash (ed.), *India in a Warming World*, Oxford University Press, New Delhi, pp. 142–155.

Dasmann, R. 1988, 'Towards a biosphere consciousness', in D. Woster (ed.), *The Ends of the Earth: Perspectives on Modern Environmental History,* Cambridge University Press, England.

Davison, A. 2008, 'The trouble with nature: ambivalence in the lives of urban Australian environmentalists', *Geoforum*, vol. 39, no. 3, pp. 1284–1295.

Devall, B. & Sessions, G. 1985, *Deep Ecology: Living as if Nature Mattered*. Gibbs Smith, Publisher, Utah.

Dodson, P. 1997, 'Finding common ground', *Habitat Australia, Australian Conservation Foundation Magazine*, April 1997, no. 4.

Doherty, B. & Doyle, T. 2007, 'Beyond borders: transnational politics, social movements, an modern environmentalisms', *Environmental Politics*, vol. 15, no. 5, pp. 697–712.

Donella, H. M., Meadows, D. L., Randers, J. & Behrens III, W. W. 1972, *The Limits of Growth: A Report for the Club of Rome's Project on the Predicament of Mankind*, Universe Books, New York.

Doyle, T. 2000, *Green Power: The Environment Movement in Australia*, UNSW Press, Sydney.

Doyle, T. 2005, *Environmental Movements in Majority and Minority Worlds: A Global Perspective*, Rutgers University Press, New Brunswick, NJ.

Doyle, T. & Kellow, A. J. 1995, *Environmental Politics and Policy Making in Australia*, Macmillan Education, Melbourne.

Dryzek, J. S. 2013, *The Politics of the Earth: Environmental Discourses*, Oxford University Press.

Dubash, N. 2012, 'Climate politics in India: three narratives', in N. K. Dubash (ed.), *Handbook of Climate Change in India: Development, Politics and Governance*, Oxford University Press, New Delhi, pp. 197–207.

Dubash, N. 2013, 'The politics of climate change in India: narratives of equity and co-benefits', *WIRES Wiley Interdisciplinary Reviews: Climate Change*, vol. 4, no. 3, pp. 191–201.

Dubash, N. & Rajamani, L. 2010, 'Beyond Copenhagen: next steps', *Climate Policy*, vol. 10, no. 6, pp. 593–599.

Dunlap, R. & Mertig, A. G. 1995, 'Global concern for the environment: is affluence a pre-requisite?', *Journal of Social Issues*, vol. 51, no. 4, pp. 121–137.

Dunlap, R. & Mertig, A. G. 1997, 'Global environmental concern: an anomaly for postmaterialism', *Social Science Quarterly*, vol. 78, no. 1, pp. 21–29.

Eckersley, R. 1992, *Environmentalism and Political Theory: Toward an Ecocentric Approach*, UCL Press, London.

Ehrlich, P. 1968, *The Population Bomb*, Ballantine Books, New York.

Ekirch, A. A. 1963, *Man and Nature in America.* Columbia University Press, New York.

Escobar, A. 1999, 'After nature: steps to an anti-essentialist political ecology', *Current Anthropology*, vol. 40, no. 1, pp. 1–30.

Escobar, A. 2001, 'Culture sits in places: reflections on globalism and subaltern strategies of localization', *Political Geography*, vol. 20, no. 2, pp. 139–174.

Esposito, A. & Neale, T. 2016, 'Never squib the rights issue in favour of conservation', in E. Vincent & T. Neale (eds), *Unstable Relations: Indigenous People and Environmentalism in Contemporary Australia*, UWA Publishing, Crawley.

Foreman, D. 1998, 'Wilderness areas for real', in J.B. Callicott & M.P. Nelson (eds), *The Great New Wilderness Debate*. University of Georgia Press, Athens, GA, p. 395.

Fraser, N. 1997, 'Rethinking the public sphere: a contribution to the critique of actually existing democracy', in N. Fraser (ed.), *Justice Interrupts: Critical Reflections on the 'Postsocialist' Condition*, Routledge, London, pp. 69–98.

Gadgil, M. 1982, 'An Indian Conservation Strategy', Policy Paper, Indian Social Science Institute, New Delhi.

Gadgil, M. & Guha, R. 1993, *This Fissured Land: An Ecological History of India*, University of California Press, Los Angeles, CA.

Gadgil, M., & Guha, R. 1995, *Ecology and Equity: The Use and Abuse of Nature in Contemporary India*, Psychology Press, London.

Gadgil, M., & Guha, R. 2000, 'Ecological conflicts and environmental movements in India', *Development: Challenges for Development*, vol. 6, p. 254.

Galbraith, J. K. 1958, *The Affluent Society*, Houghton Miffin Co., Boston.

Gandhi, M. K. 1937, Harijan, vol. 9–10, no. 66, pp. 169–170.

Gandhi, M. K. 1946, *Gandhian Constitution for Free India*, Kitabistan, Allahabad.

Goodman, J. 2009, 'From global justice to climate justice? Justice ecologism in an era of global warming', *New Political Science*, vol. 31, no. 4, pp. 499–514.

Gottlieb, R. 2005, *Forcing the Spring: The Transformation of the American Environmental Movement*, Island Press, Washington DC.

Guha, R. 1989a, *The Unquiet Woods: Ecological Change and Peasant Resistance in the Himalaya*, Oxford University Press, New Delhi.

Guha, R. 1989b, 'Radical American environmentalism and wilderness preservation', *Environmental Ethics*, vol. 11, no. 1, pp. 71–83.

Guha, R. 1998, 'Deep ecology revisited', in J. B. Callicott and M. P. Nelson (eds), *The Great New Wilderness Debate*. University of Georgia Press, p. 271.

Guha, R. 2000, *Environmentalism: A Global History*, Longman, New York.

Guha, R. 2006, *How Much Should a Person Consume?: Environmentalism in India and the United States*, University of California Press, Los Angeles, CA.

Guha, R. & Martinez-Alier, J., 1997, *Varieties of Environmentalism: Essays North and South*, Earthscan, Routledge, London.

Haraway, D. 1991, *Simians, Cyborgs and Women: The Reinvention of Nature*. Free Association, London.

Harvey, D. 1996, *Justice, Nature and the Geography of Difference*, Wiley-Blackwell.

Hay, P. R. 1994, 'The politics of Tasmania's World Heritage Area: contesting the democratic subject', *Environmental Politics*, vol. 3, pp. 1–21.

Hays, S. 1987, *Beauty, Health, and Permanence: Environmental Politics in the United States, 1955–1985*. Cambridge University Press.

Hillier, B. 2010, 'A marxist critique of the Australian greens', Marxist Left Review, no. 1, Spring 2010.

Hintjens H. 2000, 'Environmental direct action in Australia: the case of Jabiluka Mine', *Community Development Journal*, vol. 35, no. 4, pp. 377–390.

Huq, S. & Sultana, F. 2023, 'In 2023 we've seen climate destruction in real time, yet rich countries are poised to do little at CoP28', *Guardian*, 1 November, viewed 10 January 2024, <www.theguardian.com/commentisfree/2023/nov/01/climate-destruction-rich-countries-cop28>.

Inglehart, R. 1977, *The Silent Revolution: Changing Values and Political Styles among Western Publics*, Princeton University Press, Princeton, NJ.

Inglehart, R. 1990, 'Values, ideology and cognitive mobilization in new social movements', in R. J. Dalton & M. Kuechler (eds), *Challenging the Political Order: New Social and Political Movements in Western Democracies*, Oxford University Press, pp. 43–66.

Inglehart, R. 1997, *Modernization and Post-Modernization: Cultural, Economic and Political Change in 43 Societies*. Princeton University Press, Princeton, NJ.

Kothari, A. & Singh, S. 1988, *The Narmada Valley Project: A Critique,* Kalpavriksh, New Delhi.

Kanitkar, T. & Jayaraman, T. 2019. 'Equity in long-term mitigation', in N. Dubash (ed.), *India in a Warming World*, Oxford University Press, pp. 92–113.

Kidd, Q. & Lee, Aie-Rie. 1997, 'Postmaterial values and the environment: a critique and reappraisal', *Social Science Quarterly,* vol. 78, no. 1, pp. 1–15.

Klein, N. 2016, 'Let them drown: the violence of othering in a warming world', *London Review of Books*, 2 June, vol. 38, no. 11.

Kothari, R. 1984, 'Party and the state in our times: the rise of non-party political formations', Alternatives, Global, *Local, Political*, vol. 9, no. 4.

Kothari, S. & Parajuli, P. 1993, *No Nature without Social Justice: A Plea for Cultural and Ecological Pluralism in India [1993]*, Lokayan Centre, New Delhi.

Land, C. 2015, *Dilemmas and Directions for Supporters of Indigenous Struggles*, Zed Books, London.

Latour, B. 1993, *We Have Never Been Modern*, Harvard University Press.

Lele, S. 2012, 'Climate change and the Indian environmental movement', in N. Dubash (ed.), *Handbook of Climate Change and India: Development, Politics and Governance*, Oxford University Press, Delhi.

Lockie, S. 2004, 'Social nature: the environmental challenge to mainstream social theory', in R. White (ed.), *Controversies in Environmental Sociology*, Cambridge University Press, pp. 26–42.

Macnaghten, P. & Urry, J. 1998, *Contested Natures*, Sage Publications, California.

Martin, B. 1982, 'Australian anti-uranium movement', *Alternatives*, vol. 10. no. 4, pp. 26–35.

Martinez-Alier, J. 1995, 'The environment as a luxury good or 'too poor to be green'?', *Ecological Economics*, vol. 13, pp. 1–10.

Martinez-Alier, J. 2002, *The Environmentalism of the Poor: A Study of Ecological Conflicts and Valuation*, Edward Elgar Publishing.

Mawdsley, E. 2006, 'Hindu nationalism, neo-traditionalism and environmental discourses in India', *Geoforum*, vol. 37, no. 3, pp. 380–390.

McKibben, B. 1989, *The End of Nature*, Random House Incorporated.

Meyer-Abich, K. M. 1993, 'Winners and losers in climate change', in S. Wolfgang (eds), *Global Ecology: A New Arena of Political Conflict*, Fernwood Publishing, Halifax, pp. 68–87.

Michael, K. & Vakulabharanam, V., 2016, Class and climate change in post-reform India, *Climate and Development*, vol. 8, no. 3.

Milton, K. 1999, 'Nature is already sacred', *Environmental Values*, pp. 437–449.

Moore, D. S., Peet, R. & Watts, M. 1996, 'Marxism, culture and political ecology: environmental struggles in Zimbabwe's eastern highlands', in *Liberation Ecologies: Environment, Development, Social Movements*, Routledge, London, pp. 125–147.

Moreton-Robinson, A. 2017, 'Citizenship, exclusion and the denial of indigenous sovereign rights', *Australian Broadcasting Corporation Religion and Ethics*, 30 May, viewed 10 June 2021, < www.abc.net.au/religion/aileen-moreton-robinson-citizenship-indigenous-sovereignty/13431896>.

Mukherjee, N. 1992, 'Greenhouse gas emissions and the allocation of responsibility', *Environment and Urbanization*, vol. 4, no. 1, pp. 89–98.

Naess, A. 1973, 'The shallow and the deep, long-range ecology movement: a summary', *Inquiry*, vol. 16, 1973, no. 1–4, pp. 95–100.

Nambiar, P. 2014, *Media Construction of Environment and Sustainability in India*, Sage Publications, India.

Nash, R. 1982, *Wilderness and the American Mind*, Yale University Press, New Haven, CT.

Neale, T. 2016, 'Re-reading the wild rivers act controversy', in E. Vincent & T. Neale (eds), *Unstable Relations: Indigenous People and Environmentalism in Contemporary Australia*, UWA Publishing, Crawley.

Nixon, R. 2011, *Slow Violence and the Environmentalism of the Poor*, Harvard University Press, Cambridge, MA.

Novotny, P. 2000, *Where We Live, Work, and Play: The Environmental Justice Movement and the Struggle for a New Environmentalism*. Greenwood Publishing Group.

O'Neill, K. 2012. 'The comparative study of environmental movements', in P. F. Steinberg & S. D. VanDeever (eds.), *Comparative Environmental Politics: Theory, Practice and Prospects*, MIT Press, pp. 115–142.

Pakulski, J. & Crook, S. 1998, *Ebbing of the Green Tide? Environmentalism, Public Opinion and the Media in Australia*, School of Sociology and Social Work, University of Tasmania, Hobart.

Pakulski, J. & Tranter, B. 2004, 'Environmentalism and social differentiation: a paper in memory of Steve Crook', *Journal of Sociology*, vol. 40, no. 3, pp. 221–235.

Peet, R. & Watts, M. 1996, *Liberation Ecologies: Environment, Development, Social Movements*, Routledge, London

Pereira, W. & Seabrook, J. 1990, *Asking the Earth: Farms, Forestry and Survival in India*, E arthscan, London.

Pickerill, J. 2018, 'Black and green: the future of Indigenous environmentalist relations in Australia', *Environmental Politics*, vol. 27, no. 6, pp. 1122–1145.

Pierce, J. C. 1997, 'The hidden layer of political culture: a comment on "Postmaterial values and the environment": a critique and reprisal', *Social Science Quarterly*, vol. 78, pp. 30–35.

Plumwood, V. 1993, *Feminism and the Mastery of Nature*, Routledge.

Plumwood, V. 2002, 'Decolonising relationships with nature', PAN: Philosophy, Activism, Nature, no. 2.

Pretty, J., Brett, C., Gee, D., Hine, R., Mason, C., Morison, J., Rayment, M., Van Der Bijl, G. & Dobbs, T. 2001, 'Policy challenges and priorities for internalizing the externalities of modern agriculture', *Journal of Environmental Planning and Management*, vol. 44, no. 2, pp. 263–283.

Purdy, J. 2015, 'Environmentalism's racist history', *The New Yorker*, 13 August, viewed 20 September 2018, <www.newyorker.com/news/news-desk/environmentalisms-racist-history>.

Purdy, J. B. 2016. 'Environmentalism was once a social justice movement: it can be again', *The Atlantic*, 8 December, viewed 20 March 2020, <www.theatlantic.com/science/arch ive/2016/12/how-the-environmental-movement-can-recover-its-soul/509831/>.

Quinn-Thibodeau, T. & Wu, B. 2017, 'NGOs and the climate justice movement in the age of trumpism', *Development*, vol. 59, pp. 251–256.

Raghunandan, D. 2019, 'India in international climate negotiations', in N. Dubash (ed.), *India in a Warming World*, Oxford University Press, pp. 187–204.

Rainbow, S. L. 1992, 'Why did New Zealand and Tasmania Spawn the world's first green parties?', *Environmental Politics*, vol. 1, no. 3.

Rangan, H. 2000, *Of Myths and Movements: Rewriting Chipko into Himalayan History*, Verso.

Rangarajan, M. & Shahabuddin, G. 2006, 'Displacement and relocation from protected areas: towards a biological and historical synthesis', *Conservation and Society*, vol. 4, no. 3, pp. 359–378.

Robb, P. 2015, 'Q and A: 'Sheila Watt-Cloutier seeks some cold comfort'', *Ottawa Citizen*, March 27, viewed 20 March 2020, <https://ottawacitizen.com/entertainment/books/q-and-a-sheila-watt-cloutier-seeks-some-cold-comfort>

Roberts, T. & Weikmans, R. 2015, 'The international climate finance accounting muddle: is there hope on the horizon?' *Climate and Development*, vol. 11, no. 2, pp. 97–111.

Rolston, H. 1995, 'Environmental protection and an equitable international order: ethics after the Earth Summit', *Business Ethics Quarterly*, vol. 5, pp. 735–752.

Roy, B. & Schaffartzik, A. 2021, 'Talk renewables, walk coal: the paradox of India's energy transition', *Ecological Economics*, vol. 180, February 2021, no. 106871.

Rudolph, L. I. & Rudolph, S. H. 1987, *Pursuit of Lakshmi: The Political Economy of the Indian State*, Chicago Indian Press.

Sachs, W. 1993, *Global Ecology: A New Arena of Political Conflict*, Zed Books Ltd.

Sawer, M. & Zappala, G. 2001, *Speaking for the People; Representation in Australian Politics*, Melbourne University Press, Melbourne, VIC.

Scott, J. C., 1985, *Weapons of the Week: Everyday Forms of Peasant Resistance*, Yale University Press.

Shiva, V. 1988, *Staying Alive: Women, Ecology, and Survival in India*, Kali for Women, New Delhi.

Soper, K. 1995, *What Is Nature? Culture, Politics, and the Non-Human*, Wiley Blackwell.

Spodek, H. 1971, 'On the origins of Gandhi's political methodology: the heritage of Kathiawad and Gujarat', *Journal of Asian Studies*, vol. 30, no. 2, pp. 361–372.

Sridhar, V. K. 2010, 'Political ecology and social movements with reference to Kudremukh environment movement, social change', *Social Change*, vol. 40, no. 3, pp. 371–385.

Stephenson, W. 2015, *What We're Fighting for Now Is Each Other: Dispatches from the Front Lines of Climate Justice*, Beacon Press, Boston, MA.

Sultana, F. 2022, 'The unbearable heaviness of climate coloniality', *Political Geography*, vol. 99, November.

Swarnakar, P. 2019, 'Climate change, civil society, and social movement in India', in N. K. Dubash(ed.) *India in a Warming World: Integrating Climate Change and Development*, Oxford University Press.

Thaker, J. & Leiserowitz, A. 2014, 'Shifting discourses of climate change in India', *Climate Change*, vol. 123, pp. 107–109.

Thompson, P. 1984, *Bob Brown and the Franklin River*, George Allen and Unwin, Sydney.

Toyne, P. 1994, 'The reluctant nation: environmental law and politics in Australia 1994', *Australian Environmental Law News*, vol. 37, no. 3.

United Nations Framework Convention on Climate Change 1992 (UNFCCC). 1992, 'United Nations Framework Convention on Climate Change, 1771', *United Nations Treaty Series*, vol. 107, p. 165.

Vincent, E. 2016, 'Kangaroo tails for dinner? Environmental culturalists encounter Aboriginal greenies', in E. Vincent & T. Neale (eds), *Unstable Relations: Indigenous People and Environmentalism in Contemporary Australia*, UWA Publishing, Crawley.

Vincent, E. & Neale, T. 2016, *Unstable Relations: Indigenous People and Environmentalism in Contemporary Australia*, UWA Publishing, Crawley

Weir, J. K. 2016, 'Hope and Farce, indigenous people's water reforms during the millenium drought', in E. Vincent & T. Neale (eds), *Unstable Relations: Indigenous People and Environmentalism in Contemporary Australia*, UWA Publishing, Crawley.

White, R. 2004, 'From wilderness to hybrid landscapes: the cultural turn inenvironmental history', *The Historian*, vol. 66, pp. 557–564.

Whyte, K. 2017, 'Way beyond the lifeboat: an indigenous allegory of climate justice', in D. Munshi, K. Bhavnani, J. Foran & P. Kurian (eds), *Reimagining Global Climate Justice*, University of California Press.

Wilderness Act 1964, https://wilderness.net/learn-about-wilderness/key-laws/wilderness-act/

Williams, R. 1972, 'Ideas of nature', in J. Benthal (ed.), *Ecology, the Shaping of Enquiry*, Longman, London, pp. 146–164.

Williams, G. & Mawdsley, E. 2006, 'Postcolonial environmental justice: government and governance in India', *Geoforum,* vol. 37, pp. 660–670.

World Resource Institute. 1990, *World Resources 1990–1991: Global Climate Change Report*, WRI, UNEP & UNDP.

Zimmerer K. S. 2000, 'The reworking of conservation geographies: nonequilibrium landscapes and nature-society hybrids', *Annals of the Association of American Geographers*, vol. 90, no. 2, pp. 356–369.

3 Environmentalism of the poor in neoliberal India

Little Posterity ran on – We're here at the bazaar!
What would you like to buy, the shopkeeper asked.
Brother, little rain, a handful of wet earth,
A bottle of river, and that mountain preserved
There, hanging on that wall, a piece of nature as well.
And why is rain so dear, pray tell?
The shopkeeper said – This wetness is not of here!
It comes from another sphere.
Times are slack, have ordered just one sack.
Fumbling for money in the corner of my sari,
I untied the knot only to see
In place of a few folded rupees
The crumpled folds of my entire being.

Kerketta (2016)

The Greenpeace-Mahan anti-coal movement emerged out of specific political and economic contexts around coal mining in central India, which is discussed in Chapter 4. At the same time, Mahan was also one of a multitude of people's resistances to industrial projects in India under neoliberalism; it faced similar risks of land evictions and livelihoods losses, used similar grassroots tactics, and sought similar legal recourse for justice. This chapter sets the bigger context in India – including the political, economic, social and ecological – from which the Mahan movement arose. Owing to the centrality of the State in shaping the environmentalism of the poor as outlined in Chapter 2, a discussion on the contradictions of India's postcolonial developmental State is central to understanding the nature of India's environmental conflicts and mass environmental movements.

This chapter answers the first of the four main questions about the respective case studies in the book: how have the discourses and politics of environmentalism transformed from the previous era? Essentially, this chapter shows the Indian State's role in shaping today's environmentalism of the poor.

First, by highlighting the significance and the limitations of the Constitution and democracy in enabling the rights of vulnerable communities. Then, through

DOI: 10.4324/9781003410416-3

a historical discussion of how the postcolonial State has changed from the previous socialist era (1947 to early 1990s) to the present neoliberal era (early 1990s onwards). Followed by a discussion on how today's environmental struggles arise from contradictions of the Indian State under neoliberalism. The chapter then draws on some prominent people's environmental struggles in the neoliberal era. The Analysis answers the main question of this chapter. The chapter ends with an argument about the obvious gap between today's articulations of environmental justice in India and dominant climate justice narrative in the global North.

3.1 Constitutional democracy

In their book *An Uncertain Glory: India and Its Contradictions* (2013), development economists Jean Dreze and Amartya Sen grapple with the nature of inequality in Indian society and reflect on the possibility of change through the practice of democracy. Historic and pervasive rich–poor divisions continue to play out along the lines of caste, ethnicity, class and gender in India. The authors describe Indian society as a 'unique cocktail of lethal divisions and disparities' that mutually reinforce one another to create 'an extremely oppressive social system where those at the bottom of these multiple layers of disadvantage live in conditions of extreme disempowerment' (Dreze and Sen 2013, p. 213).

According to United Nations (UN) estimates India is now the world's most populous country. Although urbanisation is increasing as people move away from rural areas to make a living in cities, roughly two-thirds of India's population still live in non-urban areas. This statistic clearly indicates that the greater majority of Indians' lives and livelihoods are associated with the land, and forests, where the case might be. India is the world's largest constitutional democracy and the majority of its eligible voting populations live in non-urban settings. This creates one of the most significant differences in what democracy means in the Indian context as opposed to standard political science postulates around material prerequisites for the functioning of democracy and free and fair elections in a society. As Debasish Roy Chowdhury and John Keane (2021) write, 'Struggling against poverty, they (millions of poor and illiterate people) decided instead that they must become materially fit *through* democracy' (p. 5).

The Indian Constitution was made effective on the 26th of January 1950; it legally enshrined political equality in an economically and socially unequal Indian society. Reflecting on the scale of the challenge of attempting equality across such unevenness, Dr B. R. Ambedkar, the chief architect of the Indian Constitution and leader for civil rights for 'untouchable' Dalits, forewarned in his final address to the Constituent Assembly that developed the national document:

On the 26th of January 1950, we are going to enter into a life of contradictions. In politics we will have equality and in social and economic life we will have inequality. In politics we will be recognising the principle of one man one vote and one vote one value. In our social and economic life, we shall, by reason of our social and economic structure, continue to deny the principle of one man one value.

Ambedkar (1949, para 24)

Constitutional provisions for marginalised peoples

Although it holds all citizens as equal, the Indian Constitution acknowledges historical marginalisation of India's Indigenous people by demarcating them as Scheduled Tribes (ST). And 'untouchable' Dalits, meaning oppressed people, those traditionally ascribed the lowest places in the Hindu caste hierarchy, by demarcating them as Scheduled Castes (ST). It entitles such groups to positive discriminations through reservations for government jobs and in education.[1]

India's Indigenous peoples comprise 8.6% of the country' population according to the 2011 census and are predominantly rural. Nine-tenths of this population live in rural areas across 30 states and union territories. Article 244 of the Constitution enshrines special safeguards for Indigenous land rights through geographically demarcated tribal majority Scheduled Areas where separate legal and administrative frameworks apply. The Fifth Schedule of the Constitution maps out tribal majority areas in 100 districts across 10 states, six of which are in the central Indian region.[2] Tribal populations in central India identify as Adivasis, which means original inhabitants in the Hindi language. It was first used when political movements for separate Indigenous statehood in central India began under British Indian rule in the 1930s. The Sixth Schedule of the Constitution applies to tribal majority areas across four states in the northeast. Scheduled Areas hold some of India's thickest forests and largest mineral reserves including coal.

Limits of Constitutional democracy for Adivasi self-determination

In her paper 'We will teach India democracy: indigenous voices in constitution making' (2023) sociologist Nandini Sundar demonstrates a continuous engagement of Adivasi organisations with lawmakers in the Constituent Assembly in the lead up to the Constitution, despite the dominant impression that Adivasi politics is in opposition to the State and the law. As Sundar points out, the passing of the Constitution proved a definitive point in the relation between Adivasis and the Indian State after independence from the British. The paper reminds of the limitations of the Fifth Schedule of the Indian Constitution. Despite appeals from Adivasi groups to the Constituent Assembly, and owing to other political considerations, the Fifth Schedule did not enable autonomous, village-level decision-making, for self-governance in Adivasi-majority areas in central India. Laws brought in four to five decades after the Constitution, the Panchayat (Extension to Scheduled Areas) Act (PESA) in 1996, and the Forest Rights Act (FRA) in 2006, that I discuss in detail in the third section, attempted to undertake the task of self-governance and resource autonomy for Indigenous peoples in central India. Through provisions for Adivasis to have decision making powers over developmental projects on their lands.

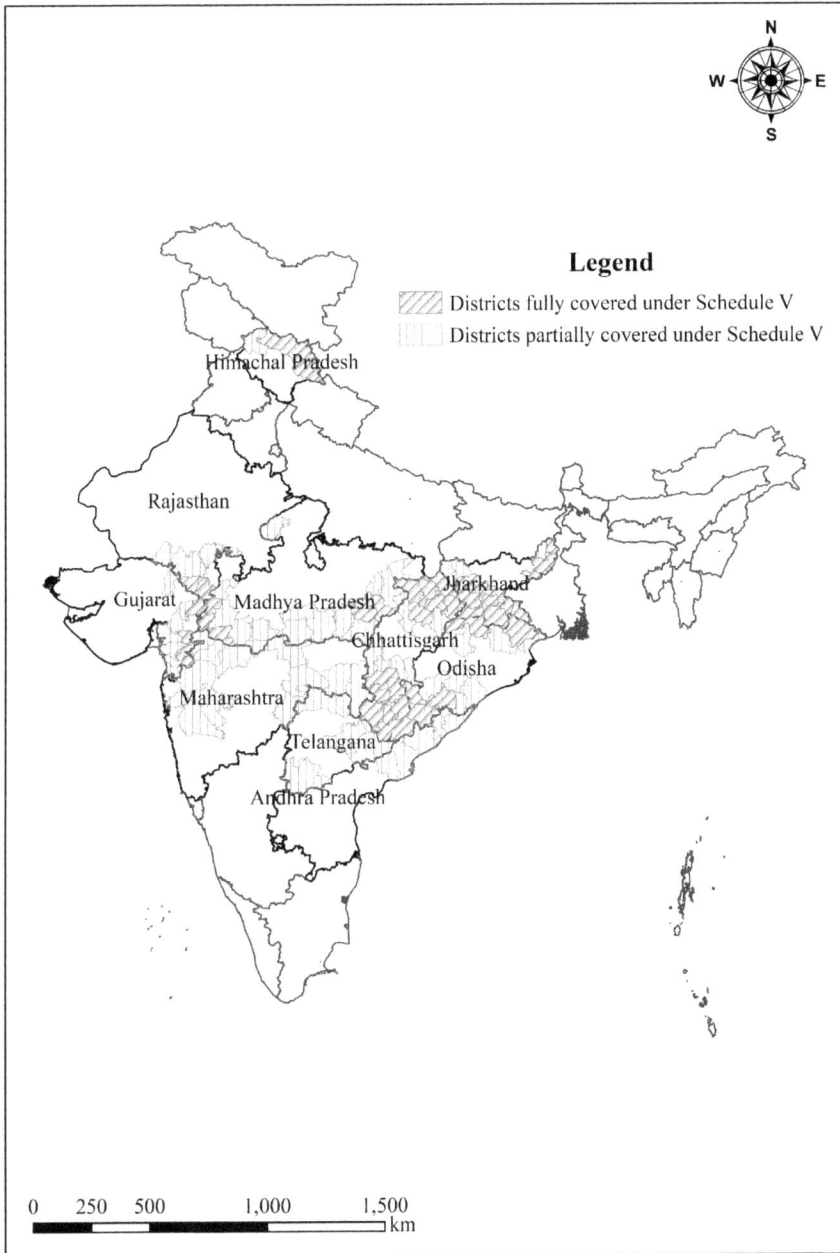

Figure 3.1 Fifth Schedule Areas in India.

This point about the Constitution's limitations in legal empowerment for Indigenous self-determination, despite its intention of safe-guarding Adivasi lands, is necessary to bear in mind for discussions on dispossession and discontent with industrial development in central India in this and the following Indian chapters. At a later state, Adivasi movements in central India have asserted a grassroots democracy, using the subsequently passed PESA and the FRA as legal campaign tools, to resist mining and other extractive projects on their lands and in their forests, as I discuss in this and the following two chapters.

3.2 How the postcolonial State shaped Indian environmentalism

The Constitution's intention of redressing structural marginalisation was however incorporated within a vision of India as a rational-scientific modern State whose so-called backward citizens would be given special assistance to become modern. Mainstream industrialisation was deemed the only possible solution to bring Adivasis into modernity. Economic growth through industrialisation was viewed as the driving force to eradicate mass poverty in a predominantly agricultural society considered as backward. For India's first Prime Minister Jawaharlal Nehru, the economic imperative for poverty alleviation co-mingled with the express ambition to grow as a sovereign nation:

> We are trying to catch up, as far as we can, with the Industrial Revolution that occurred long ago in Western countries'.
>
> Nehru (1954, 93)

In the future that this model of development envisaged for a primarily rural and economically diverse society, the majority of Indians were seen as living in cities and participating in the mainstream economy. Not pursing traditional farming practices that were deemed unproductive and other largely subsistence-based livelihoods that fell outside the formal economy. This approach differed from the Gandhian vision of self-sufficient villages as the social and economic web of life mentioned in Chapter 2. In an essay (and an addendum responding to questions about the essay) 'Whither India' (1933) Nehru had written:

> I believe in industrialisation and the big machine and I should like to see factories spring up all over India. I want to increase the wealth of India and the standards of living of the Indian people and it seems to me that this can only be done by the application of science to industry resulting in large-scale industrialisation.
>
> Nehru ([1933] 1969, p. 36)

Jawaharlal Nehru opted for a Socialist, State-directed model of industrial development, with the intended purpose of allowing both production as well as distribution of wealth and purchasing power; to avoid the concentration of wealth in the hands of a few as under capitalism. To stay free from the vested interests of international private capital, allowing both economic and political freedom in

India (Nehru 1933, p. 15). Centralised industrialisation was initiated through the Five-Year Plans of the National Planning Commission. The Second Five-Year Plan (1956–1961) elevated the State to the 'commanding heights of the economy', to control sectors deemed critical to build India's industrial economic base – mineral resources, energy, heavy industries, infrastructure, transportation, public utilities, telecommunications, defence and trade. State-owned and run public sector companies were developed to grow India's major economic sectors under subsequent Five-Year Plans. India set up a mixed economy containing liberal–democratic institutions and combining precapitalist and capitalist modes of production in agriculture, and State and private sectors in industry (Nayar 1989).

The postcolonial Indian State invoked the moral imperative of national interest and poverty alleviation to disrupt people's livelihoods and remove them from their lands for industrial projects. In what is considered the most emblematic example of this phenomenon, Jawaharlal Nehru, talking to potential evictees at the site of India's largest dam project in 1948, infamously said, '…if you must suffer, suffer for the greater common good' (Roy 1999, para 1). At international forums, India drew the link between industrial modernity and environmental protection, projecting economic development and affluence as preconditions for environmental protection. At the United Nations Conference on Human Environment (UNCHE) in Stockholm in 1972, Prime Minister Indira Gandhi reflected on India's environment versus development question:

> On the one hand the rich look askance at our continuing poverty – on the other, they warn us against their own methods. We do not wish to impoverish the environment any further and yet we cannot for a moment forget the grim poverty of large numbers of people. Are not poverty and need the greatest polluters? The environment cannot be improved in conditions of poverty. Nor can poverty be eradicated without the use of science and technology.
>
> Gandhi (1972, para 9)

Such mainstream views that large-scale industrialisation and economic affluence are essential for wellbeing and environmental protection are coming under scrutiny today as the world grapples with multiple ecological crises in the Anthropocene. Mahatma Gandhi had been concerned about large-scale industrialisation in India for both social and environmental reasons:

> Western civilisation is urban. Small countries like England and Italy may afford to urbanise their systems. A big country like America, with a very sparse population, perhaps cannot do otherwise. But one would think that a big country, with a teeming population with an ancient rural tradition which has hitherto answered its purpose, need not, must not, copy the Western model.
>
> Gandhi (1929)

From the point of view of Indigenous self-determination, particularly redressal of the historic wrongs during British colonisation that dispossessed Adivasis of their forests, the post-independence Indian State assumed a paternalistic approach. Under the guise of postcolonial sovereign development, India's vision for economic development perpetuated a colonial pattern of extraction that we will see in the following part of this section.

Displacement and discontent from postcolonial development

Adivasi and subsistence communities were the most affected by the loss of lands and livelihoods from India's one-size-fits-all approach to economic growth through large-scale industrialisation. Even though they make up only 8.6% of the Indian population, Adivasis constitute 40% of all development-induced displacements since independence for dams, mines, other industrial projects, as well as for national parks, tiger reserves and wildlife sanctuaries (Kohli et al. 2018). Sharma and Singh (2009) explain that having lived on their land for generations and often without formal property titles made Adivasis particularly vulnerable to displacement without proper rehabilitation by the State. Overall, an estimated 20–40 million Indians became development refugees in the first 50 years since Indian independence (Dreze et al. 1997).

Kohli et al. (2018) offer a breakdown of the categories of industrial projects that resulted in the displacements: 16.4 million of this larger figure were ousted by dams, 2.55 million by mines, 1.25 million by industrial developments and 0.6 million by wildlife sanctuaries and national parks. Large dams played the biggest role in internal displacements in the 1960s and 1970s, with an estimated 11.5 million largely Adivasi Indians displaced without proper rehabilitation (Fernandes and Ganguly-Thukral 1989). Although not on the scale of industrial projects, large-scale wildlife conservation through tiger reserves since 1973 and expansions to India's Protected Areas Network through the Indian Wildlife Act (1972) also displaced Adivasis (Rangarajan and Shahabuddin 2006; Shahabuddin and Bhamidipati 2014). An estimated 100,000 Indians have been displaced by the creation of protected areas between 1970 and 2008 (Lascorgeix and Kothari 2009).

Forests cover 22% of the Indian landscape. Home to large Indigenous populations, India's forests have served as sites of conflict since colonial times. The colonial-era Indian Forest Act was passed in 1878 and amended in 1927. It restricted forest-dependent communities from accessing their forest commons and forest produce under the guise of scientific forestry.[3] Denial of their customary rights through a lack of access to forests presented a direct threat to livelihood and food sources and risked causing hunger and famine. Historian Ramachandra writes that these moves caused 'a deep feeling of injustice and resentment' towards the colonial government (Guha 2000, p. 55). The Forest Act and the Forest Department that started in 1864 owe their origin to the expansion of the colonial railway that extensively deforested peninsular India for fuel and timber.

The British colonial government also brought in the *Land Acquisition Act 1894* (LAA) that continued unchanged for nearly a century till 1984.[4] The LAA vested arbitrary powers in the State for land acquisitions, using a justification of common good or public purpose as the objective behind such acquisitions. The doctrine of eminent domain enshrined in the LAA established that 'land may be taken because the State holds a superior layer of property rights' (Reynolds 2010, p. 2).[5] Several scholars argue in a similar vein to anthropologist-activist Felix Padel (2016) that in several respects, independent India continued aggregating resources and land for industrialisation in the same vein as the erstwhile colonial State. Ramesh and Khan (2015) show that the postcolonial State compounded its own contradiction towards India's Adivasis by asserting eminent domain over their lands and forests as integral to sovereignty.

The 1970s and 1980s brought a shift in India's outlook towards the environment and forests and the State recognised the environmental crisis. The Constitution was amended following the 1972 UNCHE in Stockholm to include articles 48A and 51A that made the State and citizens responsible for environmental protection. Article 48A states that 'the state shall endeavour to protect and improve the environment and to safeguard the forests and wild life of the country' (Constitution of India Part IV). Chakravarty (2006) reminds us that this amendment gave the State a constitutional mandate for environmental protection, making the Indian Constitution one of the very few in the world to enshrine protection and improvement of the environment.

Prominent laws brought in during this period include the *Forest Conservation Act 1980* (FCA), *Water (Prevention and Control of Pollution) Act 1974* and the *Air Pollution Act 1981*. The Bhopal Gas Disaster of 1984 acted as a trigger for various environmental policies starting with the issues of corporate accountability and toxic waste management (Reich and Bowonder 1992). The *Environment Protection Act 1986* (EPA) was enacted to implement the decisions of the UN conference relating to the protection and improvement of the human environment.

However, environmental reforms ushered in from the 1970s did not alter the disenfranchisement of communities through industrialisation. In some cases, they exacerbated land dispossessions such as through the creation of sanctuaries as mentioned earlier in this section. Decision-making on forests was further centralised through the constitutional amendment and passing of the FCA in 1980. Prakash Kashwan writes that 'this had the effect of removing the contested question of forestland rights from the arena of state-level electoral politics' (2017, p. 78). The era of the State at the commanding heights did not see the emergence of significant land and rights-based reforms for Indigenous peoples and communities in India.

Civil society movements did question the lack of distributive justice in India's development process. Environmentalisms of the poor sometimes succeeded in asserting another vision of development that prioritised sustainability and livelihoods, as seen through the Chipko movement in Chapter 2. But on the whole, ideas such as dams as the 'temples of modern India', as proclaimed by Prime Minister Nehru while starting the construction of the massive Bhakra-Nangal dam in Punjab 1954 (Khilnani 1998, p. 61), dominated India's political–economic discourse under State capitalism.

The dominance of such ideas sustained a selective understanding of the notion of the greater common good: owing to the non-privatised nature of development characterised by public sector corporations and direct control of resources by the

State, economic growth was largely believed to be serving the national interest. Disproportionate impacts across India's class, ethnicity, caste and gender divides were accepted as necessary sacrifices for the 'greater common good'. This account by Adivasi leader Keshavbhau Vasave in the mass movement *Narmada Bachao Aandolan* that began in the mid-1980s to protest large-scale dams on the river Narmada in Madhya Pradesh, Maharashtra and Gujarat, tells a poignant story. Of how, often, as the State machinery went about the task of sacrificing communities at the altar of development, it did so by leaving no room for justice:

> But they never followed that verdict. *(Laughs)*. They submerged us first. *(Laughs)*. So, even the directive of the Supreme Court was not observed by the Gujarat government. And though the verdict went against us, and though it was in their favour, they did not allow it. The Court had decided that the rehabilitation of the people falling within those 5 meters had to be completed, and only then could the dam height be increased by 5 meters. What the government did was exactly the opposite. They destroyed every one of us.
>
> Oral History Account/Interview with
> Keshavbhau Vasave in Oza (2022, p. 95)

With the selective notion of public good dismantled through economic liberalisation, in the next section we will see how the State's role towards communities changed as it brokered privatised economic growth, and communities claimed justice through newly enshrined rights-based laws. And how this put democracy to the test in India.

3.3 How the neoliberal State shapes environmentalism of the poor

Changes to India's economy after the 1990s brought the selective definition of public interest into question by introducing private players into various economic sectors. Although in the making from the mid-1980s, the definitive transition to liberalisation is attributed to sweeping reforms in 1991 aimed at relieving India's international debt and under persuasion from international financial institutions. Liberalisation ushered in transformations to the relationship of the State and economy through a combination of outward looking reforms to align with the global market and private capital, and internal reforms to allow the entry of Foreign Direct Investments (FDIs) into core sectors like education, telecommunications, healthcare, and energy. The entry of private companies into sectors previously operated by State-run corporations was facilitated through setting up public–private partnerships (PPPs). The Indian State decentralised by devolving powers to state governments and local bodies. Gupta and Sivaramakrishnan (2011) amongst others have argued that changes during the first decade of liberalisation altered the character of the Indian State more than any changes instituted since independence.

Although the term neoliberalism has been variously described, across texts, they all share certain common perspectives. David Harvey approaches neoliberalism as a 'theory of political, economic practices proposing that human well-being can be advanced by the maximisation of entrepreneurial freedoms within an institutional framework characterised by private property rights, individual liberty, unencumbered markets, and free trade' (Harvey 2007, p. 22). The process entails the commodification even of previously uncommodified 'natural benefits' and the non-human natural world, including land, water, ecosystem services, as well as environmental pollution, creating ethical and moral issues around the commodification of everything (Harvey 2007).

Broadly speaking, the State in the neoliberal context is understood as playing an active role in enabling and defending the market, private property rights, and commodification (Peck 2001). Outside of this role, the State is understood to withdraw or reconfigure its functions through rescaling of governance and fiscal and administrative cuts (McCarthy and Prudham 2004). There has been a substantial discussion on the nature and function of the Indian State since the 1991 reforms, particularly around what is distinctly neoliberal about the present State.

While reflecting on the question about how Indian environmentalism has changed historically, my respondents from Indian civil society networks placed a lot of emphasis in delineating changes to the Indian political economy since liberalisation, which shaped the broader context from which environmental resistances arose since the 1990s. Their responses also reflected on the possibilities for environmental justice that arose from crucial land and forest reforms passed in the second decade after economic liberalisation. The following two subsections are themed as per these reflections. In the first subsection I discuss a range of interconnected themes that characterise the neoliberal Indian State, its actions under neoliberalism and their significance for peasant and Adivasi communities whose lands, forests and livelihoods came under the purview of neoliberal development. In the second subsection I discuss one of the most critical possibilities for land justice and forest rights for communities that emerged during the neoliberal era, and, drawing on themes from the first subsection, analyse the actions of the State, which was caught between favouring private capital and responding to the needs of communities.

Indian State under neoliberalism

A major shift for the Indian State was in relation to moving away from being a welfare state under the pressures of global finance. While India's exports increased, critical schemes such as the Public Distribution System (PDS) that serves as India's food security system, and sectors such as agriculture, were deprioritised through the reduction in fertile land for food cultivation (Shrivastava and Kothari 2012).[6] But despite reducing welfare, the Indian State still could not follow the standard neoliberal narrative of slashing public infrastructure. Gupta and Sivaramakrishnan (2011) explain that this is a peculiar outcome in India's populous democracy where poor and rural groups far outweigh the urban electorate and governments are subjected to popular pressures. The State's high visibility in the lives of millions of Indians in the decades following 1991 can be seen through state-level midday meal schemes and subsidized rice for below poverty line (BPL) families amongst other programmes (Munster 2012).

Scholars and activists point out many 'contradictions' in the India State under neoliberalism, owing to the reality of a large rural and poor population, and how, instead of a withdrawal from their lives, neoliberalism reconfigured and problematized the relationship of the Indian State with the Indian poor, rural and Indigenous communities. I thematically discuss the major contradictory characteristics and actions of the India State under neoliberalism.

Uneven development

Economically, the most visible difference was through a significant increase in India's Gross Development Product (GDP). From a low average 3.5% economic growth rate till the 1980s, India became the second fastest growing economy after China from the mid-1990s, and then overtook China as the fastest growing major economy in 2015–2016 with a growth rate of 7.6% even as China's economic activity slowed down. But Harriss and Corbridge (2010) contend that a high growth rate has not led to a corresponding reduction in poverty, which requires a systematic focus on vital sectors and meeting developmental indicators. Governments also did not keep the needs of India's rural economy at the centre; instead, they redistributed revenues obtained from high growth sectors aligned to the global market amongst indigent sections of the Indian population. Critics point to the case of Gujarat, hailed as the poster child of India's neoliberal growth. The state of Gujarat clocked the country's highest GDP by concentrating large investments in resource extraction, chiefly by private corporations. But what has come to be known as the Gujarat model of growth failed to improve social and developmental indicators in the state (Jafferlot 2016).[7]

It has been argued that neoliberal development contradicted India's vision of inclusive growth by integrating urban India into the global economy while leaving rural India largely behind, widening the socio-economic divide.[8] Shrivastava and Kothari (2012) show that mining in particular has not led to a corresponding economic benefit or improvement in welfare for locals in the mining-dependent eastern states of Jharkhand, Odisha and Chhattisgarh. These states continue to have lower per capita incomes and higher food insecurity than other states. This situation is worsened by declining employment in mining, although the value of mineral production increased fourfold between 1991 and 2004 employment in mining dropped by 30% (Shrivastava and Kothari 2012).

The 'trickle down theory' of development economics – where the benefits of economic growth that accrued to urban-dwelling upper economic classes were expected to 'trickle down' to the poor and the rural once they too became part of the economic development story – as the approach to reducing economic inequality now stands disproven (Basu and Mallick 2008).[9]

Crony capitalism

The Indian State's relations with influential political classes have been reconfigured under neoliberalism. Political historian Partha Chatterjee (2008) contends that the precarious balance between the three dominant classes – industrialist capitalists, rich farmers and the salaried white collars – over bargaining for State power proposed by Bardhan (1984) has been replaced by an exclusive favouring of industrial capitalists by governments. Further, the State–business nexus in India is now characterised by a 'narrow alliance of business and political elites' creating the phenomenon of 'crony capitalism' under which undue favours are extended to corporations at the neglect of the public interest (Kohli 2007; Guhathakurta 2015).

Critics interpret the success of the Adani Enterprises, now the world's largest private coal developer and India's largest private port operator, as a clear example of crony capitalism. Privileges made available to Adani Enterprises by the Gujarat government under the Chief Ministership of Bharatiya Janata Party's (BJP) Narendra Modi for a private coal port, thermal power plant, and Special Economic Zone at Mundra included land offered at throwaway prices (Nayar et al. 2014). The Indian government reportedly tweaked rules for special economic zones that specially favoured the Adani Group; it enabled the Adani Power Limited based in the Mundra SEZ to reap a financial bonanza (Guhathakurta et al. 2017).[10]

Developed under the *Special Economic Zones Act 2005*, Mundra in coastal Gujarat was one of many SEZs dotting the Indian landscape after 1991 where full play of the market was enabled through a suspension of national laws.[11] Previously a region undisturbed by industrialisation, Mundra was transformed by severe environmental degradation, and disruption to farming and traditional fishing livelihoods (Narain 2013).[12] Neoliberal industrialisation, exemplified through special economic zones, transformed landscapes on a geographic scale, so massive were they. They wrought critical social, environmental and economic impacts for local communities whose livelihoods were disrupted. Opposition to SEZs formed the first wave of resistances to neoliberalism in India; activists criticised governments for failing to deliver in the public interest and eroding the (thinly held) notion of 'greater common good' in India's earlier Socialist economic model (Sharma and Singh 2009).

Affecting unprecedented environmental change

Concentrated industrialisation and resource extraction led to unprecedented environmental changes in what historian Ramachandra Guha calls India's age of ecological arrogance (Guha 2014). The State dismantled its elaborate bureaucratic red tape known as the 'Licence Raj' to facilitate ease of business. This drastically reduced environmental approval and social impact assessment (SIA) time periods, particularly for mining projects. Weakening environmental safeguards became a prominent trend in the second decade of neoliberal growth in the mid-2000s and corresponded with the time period when the national GDP grew to a record 7–8% of annual growth. But the extent of weakening of environmental protections from 2014 under the Narendra Modi government is considered as unprecedented (Nayar 2016).[13]

To put the Environment Ministry's approval timescales from 1982 to 2016 into perspective: between 1982 and 1999, it took the ministry an average of 5 years to approve mining projects but between 2000 and 2004, this window fell to 3 years (Rajshekhar 2012). What followed since can only be described as a landslide: in the two terms of the Congress-led United Progressive Alliance the clearance window first fell to 17 months (2004–2009) and then to 11 months (2009–2014). In 2016, the Modi Government publicised that it would reduce environmental clearance periods to just 180 days (Press Trust of India (PTI) 2016b), claiming it will unlock Rs 10 trillion (A$200 billion) worth of investments through the clearance of 2000 projects in 2 years (PTI 2016a).[14] But while India attempted to send a clear message to global investors about ease of doing business, its environmental protection record appeared 'dismal', with a Yale University Environmental Performance Index ranking India 168 out of 180 countries (Srinivas 2020; epi.yale.edu/epi-country-report/IND).

In *Green Wars*, journalist Bahar Dutt puts into perspective the scale of environmental destruction from proposed energy projects by 2014:

> In North India, the Upper Gangetic Basin has been earmarked for over 300 small and big hydropower dams. Once these dams are constructed, almost 70% of the Ganga and its tributaries will flow through tunnels, submerging large swathes of rich Himalayan forests…Combined with other projects like shipyards, ports and coastal mining, this (fifteen coal fired power projects) implies there would be big infrastructure projects every 20–25 km along the (western) Konkan coast.
>
> …An estimated 182 large dams, power plants and chemical treatment plants will be set up in the biodiversity hotspot of the Western Ghats, just recognised by the UN as a World Heritage Site, home to over 5,000 species of plants and 100 species of mammals, with many new species yet to be discovered. Now take a look at this model of development on India's forests. Since 1980, over 1.5 lakh (150,000) hectares of forestland have been diverted for the cause of India's development, 50% of that figure in the last ten years…(Environment) ministry that should have played a protectionist role now plays the role of a distributor. The ministry has itself admitted that almost 95% of the projects that come to it are cleared.
>
> Dutt (2014, p. xiii)

In their book *Churning the Earth: The Making of Global India* economist Aseem Shrivastava and ecologist Ashish Kothari point out that weakening environmental safeguards enabled mining of coal and other minerals to rise by 75% between 1993–1994 and 2008–2009 (Shrivastava and Kothari 2012). A growing proportion of land was acquired and forests were cleared to pave the way for extensive mining. The mining industry in general and coal mining in particular were granted the highest share of environmental and forest clearances (Shrivastava and Kothari 2012).

Land broker for private resource extraction

The neoliberal State began acting as a 'land broker' to acquire a growing proportion of land for private industrialisation. Scholars and activists consider this as a
clear example of the neoliberal State favouring private corporations. Arguments
by Oskarsson (2015) that the reforms specifically reoriented the State's behaviour in favour of private resource extraction, and Levien (2011) that through this
reorientation the State began facilitating land acquisition for private mining, further qualify the neoliberal State's role in driving environmental conflicts and community discontent.

The government in the mineral rich and Adivasi majority state of Jharkhand
formulated the Land Bank Policy, under which it pooled common lands in villages
to attract private companies often without free, prior and informed consent from
communities (Pingali 2022). Lawyer and Trade Unionist Sudha Bharadwaj (2018a)
writes that government rhetoric surrounding the transfer of communities' lands for
private mining tends to characterise affected peasants and Adivasis as stumbling
blocks not stakeholders.

The neoliberal State's actions have therefore been compared to colonial expansion during the late 19th and early 20th century that Karl Marx characterised as
primitive accumulation. David Harvey's concept of accumulation by dispossession builds on the phenomenon of primitive accumulation by depicting its ongoing
nature as well as its extension through privatisation and commodification (see
Harvey 2003). Reoriented towards private extraction, the neoliberalising process
in India is characterised by 'accumulation by dispossession' in which the State
plays a key role by acting as a land broker, exposing its corporate bias and acting
against the public interest.

Land grabs

A report on the unfinished task of land reforms in India published in 2009 by the
Indian Rural Development Ministry found that landlessness had increased from
40% in 1991 to 52% in 2005 in India's rural areas.[15] The report also described the
dispossession of Adivasis as 'the biggest grab of tribal lands after Columbus' in
which the State is complicit (State Agrarian Relations Committee 2009). The term
land grab has been specifically used to describe land acquisition after the 1990s.

Under neoliberalism, the appetite for land speculation amongst investors
increased. Land acquisitions often acquired a brutal character with the State even
using police machinery to tackle farmers' efforts to protect their lands. State
governments recategorised the condition and value of land to be able to transfer
them over to private corporations, for mining, other industrial projects, as well
as so called environmental programs. As an example, The Centre for Science and
Environment (CSE 2016) *Report Card* on the Narendra Modi Indian government's
environmental performance warned that government proposals to transfer so-
called 'degraded forest lands' to the private sector for afforestation threatened

the livelihoods of over 20 million farmers involved in farm forestry, and violated provisions of the FRA that I discuss in the next subsection (CSE 2016). Sudha Bharadwaj describes that state governments acquiring an increasing proportion of land for private mining in mineral-rich central India, including in designated Scheduled areas, caused a ground-clearing of Adivasis (Bharadwaj 2018a).

Managing disaffection through land reforms

Scholars and activists point to the paradox of ameliorative legislations and programmes introduced by the neoliberal State to manage peoples' disaffection from land dispossession, loss of livelihoods and environmental destruction from industrial projects. A Press Release by the Ministry of Home Affairs on 27 April 2007 titled 'Coordination Centre meeting on Naxalism held' emphasised that:

> On the development front, the states were advised to review their Resettlement and Rehabilitation policies on a priority basis. The need to put special focus on the implementation of Backward Regions Grant Fund (BRGF), *Panchayat (Extension to Schedule Areas) Act 1996* (PESA), the National Rural Employment Guarantee Program (NREGP), and the *Scheduled Tribes and other Traditional Forest Dwellers (Recognition of Forest Rights) Act 2006* (FRA) was emphasized.
> Government of India (GOI) (2008, p. 23)

The term *Naxal* comes from the village *Naxalbari* in the state of West Bengal where an armed peasant uprising occurred in 1967. Formed by radical-left communist leaders oriented towards the Maoist ideology, the Naxalite movement spread to rural and Adivasi-dominated parts of central India by the 1970s. Discontent from illegal alienation of Adivasi lands protected under the Indian Constitution formed the basis for the movement taking hold in many designated Scheduled Tribes regions. Armed rebellion against the State has been one of the approaches of the movement.

The movement undertook ecologically sustainable alternative development activities in rural and tribal regions that often faced the brunt of industrial development but, however, did not benefit from economic growth. Adivasi communities often found themselves caught in the cross fire between the State and the Naxalite movement, but preferred to side with the movement as they found it gave them a better chance to access resources and fulfil their customary rights, writes anthropologist Alpa Shah in an ethnographic account from the state of Jharkhand in *In the Shadows of the State* (2010).

The Naxalite movement has been the strongest in areas that are also rich in coal and other minerals, where governments have been eager to attract private corporations for mining in the neoliberal era. Prime Minister Manmohan Singh described the Naxalite movement as 'the single largest internal security threat' to India (Roy 2009). In 2009, the Congress-led Indian government initiated 'Operation Green Hunt' against 'Maoist rebels' headquartered in the forests of central India, sending paramilitary troops and suspending democracy in certain parts (Sethi 2010).

In the paper 'The Rule of Law and Citizenship in Central India: Postcolonial dilemmas' Political Sociologist Nandini Sundar (2011) explains that the neoliberal State's attempts at mitigating the social impacts of land acquisition in a fashion repeats the pattern of the colonial State in India, which was compelled to mitigate the effects of repressive laws through the passage of countervailing protective legislations.[16] I discuss the countervailing legislations of LARR and FRA in the next subsection. At the same time that the Indian government operationalised 'Green Hunt' in parts of central India, it paved the way for land reforms through taking to the task of implementing the provisions of the newly passed FRA 2006 and preparing to overhaul a colonial land acquisition mechanism that I discuss in the next subsection.

In another prominent example, Partha Chatterjee (2008) points to the paradox of the *Mahatma Gandhi National Rural Employment Guarantee Act 2005* (MGNREGA), the world's largest social security scheme, created after liberalisation, as evidence that the neoliberal development is robbing the majority of Indians of their land-based livelihoods.

In *The Great Transformation* (1944), Karl Polanyi conceptualised double movement as the dialectical process of marketisation and push for social protection against marketization in reference to the historical European case in the 19th and early 20th centuries. Harriss (2011) suggests that instead of exceptions to the process of neoliberalisation, significant State interventions in India can in fact be understood as what can be called a 'new' Polanyian double movement.

Uneven democracy

Sudha Bharadwaj writes that state governments acquired an increasing proportion of land for private mining in mineral-rich central India including in designated Scheduled areas, often by violating provisions for prior consent and individual and community rights over forestlands enshrined in new rights-based laws (Bharadwaj 2018a). Therefore, with the introduction of rights-based legislations from the second decade onward since India's economic liberalisation, Adivasis in central India have increasingly found themselves at the crossroads of the market and the democracy, albeit with a highly 'uneven' experience of democracy and the ability to assert their legal-democratic rights.

As communities who face the brunt of today's developments have asserted their rights, conflicts have multiplied. According to the Global Atlas of Environmental Justice, a global database of environmental conflicts, India has the world's highest number of environmental conflicts (ejatlas.org/country/india). A 2020 report by Land Conflict Watch, an independent network of journalists and researchers, presents highly representative data on 703 ongoing cases of various types of land conflicts across the country, corroborating that 'tribes and those living in resource-rich but violence-afflicted areas are disproportionately impacted by land and resource conflicts' (Worsdell and Shrivastava 2020, p. 3). Compared to other parts of the country, it found higher concentrations and intensities of conflicts pertaining to mining and the violation and non-implementation of the FRA in Adivasi majority Fifth Schedule Areas compared to other parts of the country (Jamwal 2020).

Finally, these interconnected patterns of the State's behaviour towards communities versus corporations reflects a huge imbalance of power in a democracy that Dr B. R. Ambedkar warned about, as discussed in the first section of this chapter. Partha Chatterjee underlines that the ideals of 'individual freedoms and equal rights irrespective of distinctions of religion, race, language or culture' that underpin a framework of democratic accountability in Western liberal democracies are rarely met in the global South (Chatterjee 2004, p. 4).

Democracy has therefore become a constant push and pull between private and community interests, with communities being at the receiving end of the onslaught of disruptive development, a withholding of their legal rights, and strong-arm tactics of the State against their assertion of dissent and protest.

This phenomenon also explains what Harvey (2005) calls the contrast between the ideology and the practice of neoliberalism. Even though neoliberalism's stated goal is the freedom and wellbeing of all citizens through the withdrawal of State intervention in the market, in most cases neoliberal policies are accompanied by the restoration or formation of class inequality. The State then intervenes with positive[17] or negative consequences for communities, thereby either deepening or threatening democracy. The State can resort to fear or persuasion that is often expressed through nationalistic sentiments (Munster and Strumpell 2014). This has been noted in many cases of the Indian State's narrative legitimising its crackdown on civil society's dissent against industrial and particularly energy projects.

Wide-scale curbing and scrutinising of the activities of India-based civil society organisations since 2014 is a case in point. In 2014, the newly elected Modi government cancelled the registration of as many as 9000 non-governmental organisations (NGOs) that received foreign-funding on the pretext that they did not comply with the Indian tax codes (Kaushal 2015). Critics viewed the move as a first step in fast-tracking development (Ranjan 2014). The Intelligence Bureau, India's domestic security and intelligence agency, deemed many of these groups, civil society networks, and community mobilisations a risk to India's national economic security (Rowlatt 2015). The underlying reason behind this is that these groups resisted large industrial projects. Many scholars and activists are of the view that the contradictions of neoliberal development have effectively created a crisis of democracy in India. Next, the second subsection under Section 3 of this chapter discusses the nature of this 'crisis of democracy' through the contest between the possibilities and promises of progressive legislations – the LARR and FRA – and the failure of their implementation to favour corporate interests.

Land, forests and laws: between empowerment and dispossession

The colonial era LAA continued as the main instrument for land acquisitions till long after Independence. In addition, the *Coal Bearing Areas (Acquisition and*

Development) Act 1957 (CBA) applied specifically to coal-mining-related acquisition (Bedi 2013). Democratic reforms such as the *Forest Rights Act 2006*, or the FRA, and the *Right to Fair Compensation and Transparency in Land Acquisition, Rehabilitation and Resettlement Act 2013* (RFCTLARR), or simply LARR, were passed by the Indian parliament during consecutive terms of the Congress-led United Progressive Alliance Indian government, to redress historic injustices toward Adivasi groups and their access to forestlands, and to overhaul land acquisition's colonial legacy of dispossession.

These laws were also brought with the intent to balance a growing discontent over peoples' lands and livelihoods being disrupted by massive land acquisition drives of governments (Sundar 2011). With provisions of consent, consultation and compensations, these legal instruments allowed communities to have a say on mining and other industrial developments on their lands unlike before. They allowed for a democratic process in determining industrial projects. But given the unevenness of democracy in the Indian context and increasing corporate bias of neoliberal Indian governments, both their enactment and ongoing enforcement face multiple challenges, as the following subheadings show.

Land Acquisition, Rehabilitation and Resettlement Act 2013

The *Land Acquisition, Rehabilitation and Resettlement Act 2013* (LARR) gave communities a say in how the State should deal with their lands by making it mandatory to seek approvals from affected communities through the clauses of consent and social impact assessment[18] as well as to resettle and rehabilitate land title-holders and those who lose their livelihoods (Ramesh and Khan 2015). It set the compensation formula at four times the value of rural and twice that of urban land, and also contained provisions related to return of unused lands and food security (Kohli et al. 2018). But although the LARR imposed checks and balances on the power of the State to acquire people's lands through the principle of eminent domain, it continued to exhibit the colonial mindset of the previous legislation by equating private projects designed and implemented to generate corporate profits with that of the public interest (Ramesh and Khan 2015).

While passing the LARR legislation, the Indian parliament acknowledged that the State's land acquisition drive had created injustices for communities (Sundar 2011). But the passing of the act generated outrage among companies pursuing large-scale industrial projects (Ranjan 2017). The BJP supported the LARR while in opposition, but once in government in 2014 under Prime Minister Narendra Modi, it introduced the LARR Ordinance in parliament with the purpose of providing land cheaply and quickly to investors (Kohli and Gupta 2017).[19] Along with amendments envisaged in the original law, the LARR Ordinance created a special category of projects that would be exempt from requiring consent and a social impact assessment, review by an expert group, and bars on acquisition of agricultural land.[20,21,22] As most land acquisitions fell within these categories, it effectively nullified the safeguards of the 2013 law and defeated the LARR's original aim to empower communities vulnerable to land displacement (Ramesh and Khan 2015).[23]

Although the Central Government failed to pass its proposed dilutions, several state governments amended their own Land Acquisition Acts to substantially replicate the Ordinance, and used constitutional provisions to dilute the progressive clauses of LARR.[24] Dilutions included exempting projects from the mandatory social impact assessment and consent of landowners, and reducing the compensation amounts for land acquisition (Kohli and Gupta 2017).

The dilution of the progressive LARR further aggravated the circumstances for forest-dependent communities attempting to claim their rights through the FRA and participate in decision-making on mining on their lands through PESA that I discuss under the next subheading, given that such legal provisions were already being violated by states in mineral-rich parts of India (Ranjan 2017).

In a definitive turning back of the clock on democratic land reform, in 2021, the Narendra Modi-led Indian government acquired land for coal mines in the mineral rich state of Chhattisgarh in central India based on the earlier CBAA that turns on the principle of eminent domain and is designed for a 'heavy state'; instead of the new democratic LARR that requires community consent through autonomous village-level decision-making on land use for mining (Kodiveri and Rathore 2021).

Significance of and risks to the Forest Rights Act 2006

The Scheduled Tribes and Other Traditional Forest Dwellers (Recognition of Forest Rights) Act 2006 (FRA) was enacted by the Indian Parliament in recognition of decades of mobilisations by forest-dwelling communities. It was also driven by the need to alleviate grassroots unrest from land dispossession from the first decade of neoliberalism, especially to contain Naxalism in forested central India as I previously discussed. It aimed to redress injustices towards forest-dependent communities since the inception of the colonial Forest Act in 1878, which was instrumental in displacing thousands of Adivasis and absorbing their forest commons into state-owned Reserved Forests.

A sustained movement for forest rights converged under the umbrella of the 'Campaign for Survival and Dignity' and made forest rights an issue for the 2004 Indian general elections on account of the Adivasi vote. The FRA was passed into law under the Congress led United Progressive Alliance government. The movement leading up to the FRA generated alternative discourses of legitimacy and the inclusion of marginalised voices in representative democracy (Kumar and Kerr 2012). Essentially, it is a hard-won democratic right for hundreds of millions of forest-dependent people in India (Kothari 2016).

RECOGNITION AND LIMITATIONS

The preamble to the FRA acknowledged that the rights of Adivasis and other traditional forest-dwelling people, and their centuries old lived relationships with their forests, had not been 'adequately recognised in the consolidation of state forests during the colonial period as well as in independent India, resulting in historical injustice' (Dubey and Saxena 2023). The preamble recognises that such communities are integral to the survival and sustainability of forest ecosystems. This acknowledgement is supported by global evidence that community-governed forests are often associated with higher bio-diversity, because, as opposed to the State or corporations, communities govern forests for multiple functions – the social, political, cultural, economic, environmental and spiritual (IPBES 2019, p. 33). For these reasons community governance of forests is understood as an effective safeguard from climate impacts globally (Thompson et al. 2009).

The FRA recognises both community-based and individual rights over forestlands (Government of India (GOI) 2006). It provided statutory backing for community driven forest governance in India for the first time, giving communities a transparent and participatory opportunity for managing biodiversity, water catchments and ecologically sensitive local resources (Kumar and Kerr 2012). Provisions for community forest rights in the FRA initiated a critical movement towards re-emphasising and re-establishing forest commons of village communities (Tyagi and Das 2018). Critics however point out that the FRA attempts to limit and individualise ownership over forestlands (Bharadwaj 2018a).

In terms of gender-responsiveness, the FRA repeats the problem of other Indian legal institutions by failing to take women's social positions and ongoing domestic responsibilities into account: although the FRA provisions the participation of women in local-level decision-making of forest and resource governance through participating in village councils the same as men; in reality women's participation in decision-making is challenged by existing inequalities within social structures and ongoing requirements of domestic care and responsibilities that exert economic and emotional pressures (Adhikari 2022).

The FRA was predated by the *Panchayat (Extension to Schedule Areas) Act 1996* (PESA), that extended certain powers of governance to *Panchayats* or village councils, and recognised *Gram Sabhas* or village assemblies as decision-making bodies in Scheduled Areas. PESA mandated that *Gram Sabhas* be consulted before land could be acquired for industrial projects. *Gram Sabhas* are the only officially recognised spaces where Adivasi and forest-dwelling people can participate in the State's decision-making process on mining on their own lands (Chowdhury 2016). Overall, the PESA and FRA transformed the discourse around forest ownership in favour of communities and made village councils spaces where people could make decisions about mining on their lands, through stipulations such as that forest clearance for mining should not be granted before the process of recognising communities' rights has been completed.

It is worth noting that although the FRA reflected normative changes in international Indigenous rights through presupposing free, prior and informed consent of communities, the FPIC clause was only subsequently appended through a 2009 circular issued by the Ministry of Environment (Sigamany 2020). The circular directed state governments to show documented proof that communities' forest rights had been settled and that *Gram Sabhas* had given their free, prior and informed consent, before industrial projects could be approved in forests. It specified a minimum quorum of 50% attendance in village council meetings for decisions to be made (MoEF 2009).

By giving decision-making powers to forest-based communities, the FRA offered an alternative paradigm to centralised forest management and forest-based climate action. An independent 2015 report titled *Potential for Recognition of Community Forest Resource Rights Under India's Forest Rights Act* (Rights and Resources Initiative et al. 2015) highlighted how the proper implementation of the FRA could transform lives: it has the potential to restore rights of forest dwellers over at least 400 million hectares or 100 million acres of forestlands in 170,000 villages, or one-fourth of all villages in India; recognition of rights under the FRA can benefit an estimated 150 million Indians including 90 million Indigenous people.

However, land reforms passed to strengthen communities' decision-making powers regarding mining on their lands have either been violated or poorly implemented, in many cases to expand mining.

VIOLATIONS

Journalist Chitrangada Chowdhury (2016) writes that the provisions of the FRA often act as a double-edged sword for communities in central India: state governments who are under obligation to comply with the process for community consent while also being eager to earn mining revenues, are often likely to forge *Gram Sabha* resolutions (for mining) on Adivasi lands. Trade Unionist and Lawyer Sudha Bharadwaj (2018b) writes that in Chhattisgarh, village councils were either never informed, or the people were terrorised into consent. Even worse, sometimes a 'No Objection' certificate indicating the community's consent for mining was illegally obtained.[25] Journalist Nitin Sethi (2016) reports that the state government of Chhattisgarh even removed already granted community forest rights to facilitate private coal mining by the Adani Group, setting a dangerous precedent for Scheduled Areas around India.[26]

Researcher Kanchi Kohli and Chhattisgarh-based people's movement leader Alok Shukla (2016) summarise that violations of the Forest Right Act's provisions for community consent for mining are a recurring feature in central India. And based on closely observed cases journalists Chitrangada Chowdhury and Aniket Aga write in their paper 'Manufacturing Consent' (2020, p. 77) that the bureaucratic process of securing consent renders the consent provision of the FRA rife for sabotage by the state-capital nexus. A Centre for Science and Environment (2016) report highlights that risks of compromise of the provisions of the FRA and the rights of communities also arise from the shortening of approval timelines for environmental clearances, exempting linear infrastructure projects from requiring *Gram Sabha* consensus, and relaxing the requirement of public consent for coal-mining projects under the FRA.

SLOW AND FLAWED IMPLEMENTATION

Slow and flawed implementations of the FRA, with rejection rates of all forest-rights claims filed being as high as 46% nationwide and 50% in Chhattisgarh, which prides in having the best record of FRA implementations among states, and rejection of individual forest rights (IFRs) claims being 61% in Madhya Pradesh which has the highest rejection rates among states, have been another driver behind withholding forest rights of forest-dwelling communities (Ministry of Tribal Affairs 2018; Bharadwaj 2018a; Vats 2023).

Analysis of the implementation of the FRA in Madhya Pradesh, the state with the largest tribal population in the country, has shown that only 10% of the potential area for CFRs has been recognised. The state government has not granted rights to *gram sabhas* to conserve and manage forests as stipulated under the FRA (Lele et al. 2020). Analysis of the FRA's implementation across India in the first 10 years has found exceptionally low rates of recognition particularly for community forest rights, at a mere 3% of the potential for community forest rights (excluding India's north-eastern region) (CFR-LA 2016). The analysis shows that overall only 14.5% of the minimum potential forest areas for forest rights have been recognised, indicating that the opportunity for forest rights in India has largely been lost.[27]

Bureaucracies in central Indian states are loath to give up control over forests, owing both to a persistent colonial mindset and for mining revenues, which act as a disincentive for implementing the FRA (Chowdhury and Aga 2020). Another trend in flawed implementations of the FRA can be observed through state governments granting significantly less forestland to families than sought through IFRs claims, as seen in Chhattisgarh. An analysis of Chhattisgarh's land tenure grant records by the independent data-journalism agency IndiaSpend found that sometimes claimants received ownership certificates for 'barely enough land to build a house on', threatening livelihoods and survival (Trivedi 2023).

A policy paper titled *Securing Climate Justice for India's Forest-Dependent Communities* (Agarwal and Dash 2022) highlights that apart from political, economic and social marginalisation, India's forest-dependent communities are amongst the most vulnerable to climate change, owing to higher exposures to extreme climate events and dependence on natural resources. The paper highlights the need to secure forest rights for climate justice for forest-dependent communities.

LEGAL UNDERMINING

Central India's old natural forests that are the traditional lands and sources of livelihoods of Adivasis and forest-dwelling communities, act as powerful carbon sinks to sequester carbon (Deshpande 2021). And yet, in addition to patterns of violation and slow and flawed implementation of the FRA, the governance of natural forests in central India by Indigenous people is undermined through recent legislative changes. To boost coal mining, India loosened forest clearance processes in 2020 that compromised the autonomy of *Gram Sabhas* (CFR-LA 2020). Changes to India's Forest Conservation Rules in 2022 that allow forest clearance for industrial projects (including coal mines) on Adivasi lands even before securing consent from *Gram Sabhas* in contradiction to the FRA's provisions (Joshi and Sethi 2022).

Compensatory afforestation, the mechanism through which state governments implement afforestation programmes to compensate for destruction of forests from mining and other large industrial projects, has also been found to cause land conflicts in central India. State governments receive compensatory payments made by mining and other industrial project proponents for cutting down forests through the 'compensatory afforestation fund'. A report published in 2020 by Land Conflict Watch found that land conflicts from afforestation are most likely to occur when governments afforest community and individual lands that have been secured under forest rights without consultation or consent. It is worth noting that while the *Compensatory Afforestation Fund Act 2016* mentions that forest dwellers should be consulted before beginning plantations on their lands it does not specify how they should be consulted or what the consequences of not respecting their decisions are. Such legal loopholes in legislations enable state governments to violate the consent provision in the FRA.

Last but not the least, through amendments to the Forest Conservation Act 1980 in 2023 with the stated intent of streamlining the definition of forests and the process for issuing forest clearance for mining and other industrial projects under the FCA, the Indian government is potentially exposing large forested areas for destruction especially in India's Northeast (Nandi 2023).[28]

Chowdhury and Aga (2020, p. 80) conclude that the sabotage of the consent process points to a bigger conflict between self-determination for communities and eminent domain exercised by the State–corporate complex that needs to be addressed.

Diluting democratic provision and laws

The first two decades of post-reform growth also witnessed the creation of other tools for participative democracy such as the *Right to Information (RTI) Act 2005* and the National Green Tribunal (NGT) enacted through the *National Green Tribunal Act 2010.*[29] However, such legal progresses as was made in the first two decades of post-reform growth started unravelling during the third.[30] Although the Congress led UPA government started diluting provisions for public consultation for mining, especially coal, the BJP took unparalleled steps towards diluting democratic provisions in legislations, weakening environmental guidelines and regulations, and reducing environmental clearance times.

Both LARR and FRA were passed following significant social movements by grassroots movements, activists, academics and others (Arora-Jonsson 2013, p. 61). Nandini Sundar (2011) calls these social movements legal mobilisations. Sundar qualifies that legal mobilisations play an important part in today's social mobilisations by campaigning for new laws that can fulfil democratic aspirations in the postcolonial era, in addition to often seeking to protect existing legislation from new ones that risk diluting or abrogating hard won rights (Sundar 2011). Section 3.4 of this chapter discusses prominent legal mobilisations in neoliberal India.

Ironically, reforms to India's colonial era practice of land acquisitions were shaped by social movements, but subsequently weakened to favour private investments and further disadvantage communities (Kohli et al. 2018). In addition, as an Amnesty report highlights and this subsection shows, a mosaic of laws with vastly different mandates, for example between the CBAA and the LARR, and contradictory provisions in conservation legislations, has created a legal conundrum leaving critical gaps in safeguarding Adivasi land rights in the face of mineral extraction, especially coal mining (Amnesty India 2016). Finally, the politics surrounding the passage and subsequent weakening of the FRA and LARR speak to a pattern of oscillation by the Indian State between private interests and people's mobilisations through implementing or diluting democratic legal provisions.

To conclude Section 3.3 of this chapter, my interviewees' responses, which were primarily around people's mobilisations against land, forest and environmental loss and degradation, indicated the need to explicate the role of the State and the state of democracy in neoliberal India, particularly central India that is an ethnographic focuses of this book. Essentially to thickly delineate the context from which today's environmentalisms of the poor arise. The two subsections together delineate this context by discussing the significance of the State's actions under neoliberalism towards peasant and Adivasi communities whose lands, forests and livelihoods came under the purview of neoliberal development. They focus on the mechanisms and processes of environmental destruction under neoliberalism and the swinging of the pendulum of State-Adivasi relations through assertion and violation of democratic rights through the LARR, PESA and most importantly the FRA. The State's actions are contested by today's environmentalisms in many cases by deploying an overarching narrative framework around asserting democratic rights, as we will now see in Section 3.4.

3.4 Environmentalism of the poor in neoliberal India

Many activists and scholars of social movements see the various resistances by communities to protect their lands, forests, livelihoods in today's India, not as disparate and unconnected, but as a collective challenge to neoliberal development. Nielsen and Oskarsson (2016) suggest that such resistances raise a common question: 'who benefits and who loses out from the current ways in which resources are governed?' (p. 4). Levien (2007) draws on the example of the National Association of People's Movements (NAPM), founded in 1992, as an organisational attempt at collectivising and adding coherence to a variety of people's struggles against neoliberal globalisation.

Responding to my question around significant environmental struggles in the two decades of neoliberalism in India, my respondents from Indian civil society networks frequently mentioned four exemplary movements that I have outlined in this section. These movements have resisted dispossession using new legal tools. They have used diverse arguments and narratives that are context specific, demonstrating heterogeneity of social and economic representations in today's movement spaces. They have also faced repression from the State owing to the high political and economic stake involved in the industrial projects being opposed. The four movement cases below demonstrate their use of new legal tools and tactics, their claims and counterclaims to discourses of State and Corporations, and their politics of protest.

In response to my question around what has changed in people's environmental struggles in neoliberal India compared to the previous era, respondents simultaneously reflected on two themes: First how legal tools like the FRA are allowing communities to assert their rights and reject industrial projects like they could not before. The second on how the State, acting in the interest of Corporations, is countering today's legally empowered environmental resistances.

Through the interviews and by looking at political narratives of protest movements across various sources, I realised that collective sense making and collective narrativising from an 'irruption' of environmental protests across the Indian landscape under neoliberalism has been an iterative process. In this process movements empowered with new legal tools challenge destructive industrialisation on their terms. States, acting as a land broker for Corporations, can (often) act in violation of its own laws and against the interests of communities, deeming movements as a threat to the national interest. Against the State's obvious corporate bias, movements assert their act of dissent as a fundamental democratic right.

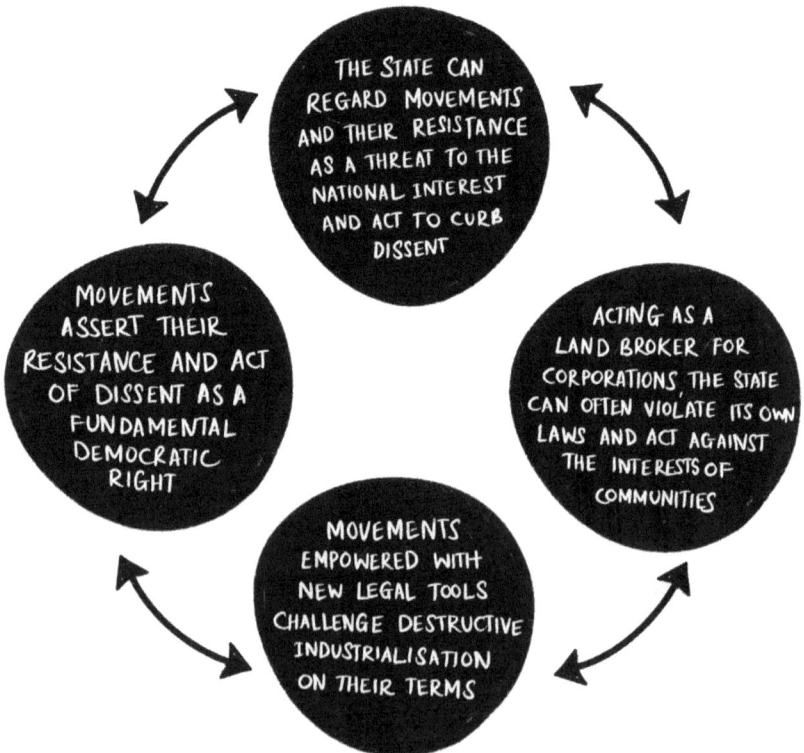

Figure 3.2 Iterative process of movement making in neoliberal India.

The first subsection, through the two movement cases of farmers' resistance to a steel project, and Dongriya Kondh Indigenous peoples' resistance to bauxite mining, both in Odisha, demonstrates the role of legal tools in today's movements through legal activism. The second subsection, through the two movement cases of the resistance to a nuclear power project in Tamil Nadu, and an Indigenous resistance to uphold Constitutional rights in Jharkhand, indicates discourses through which the State delegitimises people's protests. The final subsection indicates how civil society actors make political sense from people's protests and the State's crackdowns. And why democracy is a crucial link in the collective argument made for and by environmentalism of the poor in today's India.

Legal mobilisations: right to livelihood and democracy

The emergence of legal tools and platforms in the neoliberal era marks one of the most significant changes from earlier movements. A senior environmental lawyer told me during an interview in New Delhi that:

> The existence of legal platforms and activists being more aware make a big difference now. When the Narmada Bachao Aandolan went to court they did not have the same extent of legal opportunities. But now we have the NGT, a specialised court where such claims are heard…Now success for communities can be possible through legal tools and resources, like in Kashang in Himachal Pradesh where there was a movement against the Hydropower Project but it was the Forest Rights Act that finally helped to assert their rights.
>
> Interview (06/05/2018)

The FRA is considered a 'weapon for democracy on the ground' as communities previously unable to have a say in industrial projects that affect their lands and livelihoods and environments are now using it to assert their decision-making powers in these matters, exercising a form of direct democracy on the ground (Kothari 2016). Sociologist Nandini Sundar (2011) offers further perspective on how today's environmentalisms of the poor are engaging with the new laws. Broadly speaking, today's movements either demand the fair and proper implementation of democratic provisions such as consent in the face of lack of implementation and legal violations by the State. Or they agitate to safeguard democratic legal provisions in the face of constant dilutions by the State.

Gopalkrishnan's quote in Sundar (2007) further explains that peoples' environmental struggles in neoliberal India, unfolding with the aid of legal tools LARR and FRA, are redefining what fundamental right to property signifies in the law:

> The law is premised on the notion that *the right to livelihood*, even where it seems to be a property right, as in the forests, is not in fact a right in the liberal, legal sense at all. It is rather, a shorthand phrase, for the collective struggle

over resources. It is a right whose exercise in a capitalist and unequal society is impossible without collective organising.

(p. 52)

Struggles unfolding today with the aid of legal tools are subverting the notion of property as monetised and market-aligned, and effectively radicalising the idea of property. The first movement case study discussed here is an outstanding example of how the right to livelihood and property was radicalised. The second is considered one of the most powerful and earliest assertions of Indigenous self-determination with regards to decision on mining on Indigenous lands.

Livelihoods: farmers against land acquisition for POSCO

In 2005, the Korean Pohang Steel Company Ltd (POSCO) signed a memorandum of understanding with the state government of Odisha in eastern India to establish a steel plant in the coastal Jagatsinghpur district. The project was billed as India's largest foreign direct investment. Farming communities put up a sustained resistance against land acquisition for the project.[31] Questions of endangered livelihoods, threatened local economy, and restricted access to communal swamplands, forest produce and water, emerged as the core concerns for the movement (Krishnan and Naga 2017). And *Dhan-paan-meen* or rice-betelnut-fish became the movement's slogan, referring to the dominant subsistence livelihoods of the communities that were resisting land acquisitions.

The POSCO project was a critical investment for both state and central governments. The senior environmental lawyer points out that 'the Prime Minister gave assurances that the project will go ahead even before the forest clearance process had been completed' (interview 06/05/2018). However, over a 12-year period, the company and state government's attempts to acquire land faced a concerted opposition from farming communities. These communities had been accessing these forestlands for betelnut cultivation for close to 100 years (Das 2017). The company finally withdrew in 2016 after having failed to obtain the land necessary for the project (Sethi 2016). One of the possible reasons for the movement's success is that affected people were able to intervene in the early stages of land acquisition, the senior environmental lawyer tells me (interview 06/05/2018).

In *Resisting Dispossession: The Odisha Story* (2020), Ranjana Padhi and Nigamananda Sadangi connect the POSCO resistance to over tens and thousands of other struggles across the globe against loss of lands and livelihoods, stating that 'resisting dispossession has emerged as the biggest struggle of our times across the globe' (p. 296). In the chapter 'The Betel Smiles' on the anti-POSCO movement they write:

If the eastern coast of Odisha is not yet dotted by a mega steel plant, if the coffers of global capital are a wee bit impoverished, and most significantly, if people continue to remain in the land they belong to, it is solely due to people's sustained resistance.

(p. 290)

Indigenous cultures: Dongriya Kondhs against Vedanta's bauxite mine

The Dongriya Kondh, categorised as a Particularly Vulnerable Tribal Group (PVTG) living in Odisha's Fifth Schedule designated Rayagada and Kalahandi districts in Odisha, protected their sacred mountain by unanimously rejecting Vedanta Corporation's bauxite mine in 2013.[32] As this case shows, it was a first of its kind of way in which communities used the provisions of the FRA to reject mining.

At the core of the Kondhs' resistance was the sacred ecology of their forested mountains, Niyamgiri, meaning mountains of law, worshipped as the abode of their ancestor Niyam Raja (king of law). For the community, the loss of Niyamgiri equalled the loss of their identity. A Dongriya Kondh village priestess is quoted as having said while making a deposition that 'your temples are made of bricks and cement, ours are these hills, forests, leaves and streams…if you dig these, we will die with our gods' (quoted in Bera 2015). They demanded that the entire mountain range be protected under a single title in the name of Niyam Raja (see Tatpati et al. 2016).

In 1997, the state government of Odisha entered into a memorandum of understanding with Sterlite, the Indian subsidiary of the London Stock Exchange listed Vedanta, for bauxite mining in Niyamgiri and establishing an aluminium refinery in adjoining Lanjigarh. Constructions and land-acquisitions began in 2002–2003.[33] The movement against mining Niyamgiri grew into a trans-local and diverse environmental justice movement, drawing in participants at different levels: from the local Dongriya Kondh Adivasis under the Niyamgiri Surakshya Samiti (NSS) (Niyamgiri protection group), to state and national allies like the People's Union of Civil Liberties (PUCL), to international human rights NGOs Amnesty, Action Aid and Survival International, that targeted Vedanta's London headquarters (see Kumar 2014).

A long legal campaign highlighting violations of the FRA and risks to the region's intact biodiversity met success in 2013 when the Supreme Court questioned the impacts of bauxite mining in Niyamgiri.[34] The judgement laid the grounds for India's first environmental referendum by directing the state government to hold *Gram Sabhas* to let Adivasi Kondh communities decide whether mining will harm their religious and cultural rights. All 12 villages selected for the referendum unanimously voted against bauxite mining in their council meetings.

As Padhi and Sadangi (2020) quote women from the farming communities resisting POSCO, saying 'How can steel production ever benefit us?' (p. 280). To conclude this subsection, in the context of struggles to protect lands, livelihoods, customary laws and culture against high-stakes industrial projects, the conventional notion of property enshrined in the law transforms into human rights. And the assertion of such legal human rights becomes a critical assertion of democracy, and for a development that is democratic.

Discourse of State-crackdown: environmentalism as anti-nationalism

Central and state governments have often responded to resistance movements against high-stakes industrial projects by invoking the national interest clause and casting disaffected civilians as anti-nationals. Through section 124A of the Indian Penal Code (IPC) India still retains a colonial Sedition Law passed in 1870 to control the activities of groups and individuals fighting for freedom from British colonisation. Governments now use this draconian law to crackdown on resistances to energy projects (Biswas 2016). Another common tactic of the State to repress dissent is through accusing movements as being agents of foreign interests. This is ironic in the context of contested industrial projects funded through foreign direct investments.[35]

The movement against a large nuclear power project in coastal Tamil Nadu, and the Pathalgari (stone slab) movement of Adivasi communities in mining-affected parts of Jharkhand for their constitutional rights, stand as outstanding cases of how the State cracked down on dissent in the neoliberal era. That they were labelled as anti-national gives a sharp perspective into the pitted political–economic context surrounding today's environmentalisms in India. I discuss them under the following subsections.

Anti-nuclear movement

The People's Movement Against Nuclear Energy (PMANE) against the Koodankulam nuclear power plant on the coast of the southern state of Tamil Nadu is often described as one of the strongest demonstrations of non-violent peoples' power. The bulk of the PMANE was made of local fishing and farming communities who were concerned over the impacts from the nuclear plant on their lands, waters, and livelihoods. Yet governments alleged foreign interference to justify a severe crackdown, as movement leader Udaykumar has pointed out (quoted in Subramaniam 2013).

The Indo-Soviet Koodankulam Nuclear Power Project was signed in 1988 but stalled soon after owing to the Soviet Union's collapse. It was revived in 1997 when Russia supplied two nuclear reactors. Nuclear power began to be seen as a 'clean fuel' to potentially meet India's growing energy needs with the advent of climate change (Abdul Kalam 2011). The project proposal met with local opposition owing to the memory of Chernobyl where similar reactors had been used. The movement against the nuclear plant gathered momentum after 2001, when four more nuclear reactors were proposed with the aim of supplying power to all the four southern Indian states (Radyuhin 2002). PMANE was able to connect with national and international organisations (Udaykumar 2004).

The Fukushima nuclear disaster following the 2011 tsunami caused a fresh eruption of protests from coastal communities who had been severely impacted by the 2004 Boxing Day tsunami less than a decade earlier. A PMANE member said:

People saw the scale of the harm that it can cause…in August 2011 people in the village of Idinthakarai went on a day long fast, from then on they decided to fast indefinitely till their demands were met. It was coordinated at a mass-scale:

over hundred people went on an indefinite hunger strike. Every day 30,000 to 40,000 people protested.

Interview (06/05/2018)

But in 2012, a state High Court order gave the go ahead for the expansion of Koodankulam project. This event sparked the movement's most critical phase in which people laid siege on the nuclear plant as a last resort. The PMANE member described the government's clampdown that followed:

The state government sent 10,000 cops, who guarded every access road to the plant, but the fisher folk took the sea route to the site. People protested in the sea, slept in the cemetery, and buried themselves in the sand to camouflage. All food, including children's milk, came by boats because the police had blocked road accesses. This went on for three months starting on September 11. They planted 349 cases on 250,000 people; 9000 were charged with sedition, and 12,000 for 'waging war on the country'...We lost the battle in Koodankulam, but we won the war on nuclear in India...the debate on nuclear became a wildfire...no other nuclear plant has started since then...In the end, ordinary fisher folk challenged India's mighty Department of Atomic Energy for the first time in 80 years... where is the foreign hand in this?

Interview (06/05/2018)

In *Kudankulam: The story of an Indo-Russian Nuclear Power Plant* (2020), Raminder Kaur writes that finally the people's movement reminds us of two things. The first is how democratic development should (or should not be) conducted: at Koodankulam people behaved peacefully while the State acted 'lawlessly'. Second, that democracy is experienced very unevenly across various demographics of Indians (see p. 80, 81). Both these observations hold water for all four movements in this chapter and demonstrate how communities have to fight to assert their fundamental democratic rights to livelihoods and lands in the neoliberal era.

Adivasi constitutional rights movement

Increasing land alienation from mining, and diluted and weak implementation of laws that protect the special status of Adivasis triggered a mobilisation for constitutional rights in the forested parts of the eastern state of Jharkhand, home to significant Adivasi population. Even by conservative estimates, Adivasi's consist over 40% of the 1.5 million Indians who had been alienated from their lands between 1950 and 1995 in Jharkhand (Anwar 2019b).[36]

Jharkhand was carved out in 2000 as a result of a long history of Adivasi struggle over 300 years to protect identity, autonomy, lands and natural resources. But the new State still disenfranchised Adivasi rights and livelihoods. The government attempted to weaken the power of *Gram Sabhas* in the *Chotanagpur Tenancy Act 1908* (CNTA) and *Santhal Parganas Tenancy Act 1949* (SPTA); these acts were passed subsequent to long Adivasi struggles in eastern India during British colonisation. The acts restricted mortgages and prohibited the transfer or purchase of Adivasi lands by non-Adivasis.

Acting as a land broker, the Jharkhand government formulated the Land Bank policy to pool lands owned by *Gram Sabhas* to attract private companies. The phenomenon of land pooling often occurred without the free and prior informed consent of communities. An estimated 2,100,000 acres of common lands that enclosed people's forestlands and sacred groves had added to the land bank without free and prior informed consent of communities to attract private companies (Anwar 2019a). Apart from human displacements, large-scale coal mining without Adivasi consent had caused extensive deforestation, reduced ground water levels, and increased conflict between humans and wildlife, particularly with herds of elephants whose migratory routes have been broken by mining (Bharadwaj 2018a).

The Constitution gives Adivasis primary rights over their natural resources of *Jal, Jangal, Jameen* (water, forests, land). Through the Pathalgadi movement, communities set up stone plaques declaring the special status, rights, and autonomy of Adivasis under the Fifth Schedule of the Constitution at the entrance to their villages. The slabs also quoted sections of the PESA and FRA that recognise *Gram Sabha*s as the basis for self-rule in Adivasi-dominated areas. The movement asserted Adivasi sovereignty by demanding that companies and the state administration seek the *Gram Sabha's* permission to enter the villages (Bharadwaj 2018b). The state responded by arresting the movement's leaders, vandalising villages, and imposing sedition charges on 30,000 locals or approximately 10% of the population of the region. Permanent military camps were set up in the Khutni district that had grown to become one of the epicentres of Pathalgadi and where 86 villages had joined the movement (Anwar 2019b).

Khutni's political history is worth noting. It is the birthplace of Adivasi hero Birsa Munda who led the Munda Adivasis in the Ulgulan land revolt against the British in 1899. In the neoliberal era, a forum for the protection of Adivasis was formed here in 2007 in the wake of a proposed private steel factory. Residents from 42 villages gathered under this front to protect their lands (Sunil and Dungdung 2013). The ongoing struggle to protect their lands and customary rights, and the extent to which the State goes against communities to protect private economic interests today, tells a chequered story of Adivasi–State relations and a history of State-exploitation of Mundas and other Adivasi peoples in Jharkhand, like what anthropologist Alpa Shah shows through the book *In the Shadow of the State: Indigenous Politics, Environmentalism, and Insurgency in Jharkhand, India* (2010).

Right to dissent: how the neoliberal State shapes environmentalism

The State's attack on civil society, motivated by corporate bias and made possible by strong-arm tactics to facilitate projects, is considered a major driver of environmental conflicts. A member of the environmental action NGO Kalpavriksh tells me during a phone interview that:

What we are seeing today is an unprecedented scale of government crackdowns, even using terrorism as a label, on groups and movements…democratic spaces are now under attack, the current climate has left very little space for civil society organisations. The geographic spread of resistance movements appears to have increased because governments and corporations continue behaving without regard for environmental and human rights, and in this obviously biased way. Otherwise, so many movements and civil society groups may not even have been necessary!

Interview (05/05/2017)

State hostility towards movements challenging the developmental agenda is understood as acting as a catalyst to merge ideological differences within Indian environmentalism. The Kalpavriksh member adds that:

There is a long history of various ideologies working in their individual spheres; there was a territoriality so to speak with ideological differences (real or perceived) between orthodox Marxists, Gandhians, Ambedkarites, and others playing out. Of late there has been a greater realisation for the need for convergence from the side of NGOs as well as community and grassroots movements, caused by the external threat of State-repression. Movements have to generate allies on the ground and cannot be rigid purists in their ideological positions.

Interview (05/05/2017)

The greater spread of resistances around the country today is also attributed to increased awareness amongst communities as opposed to the 1970s and 1980s. A senior environment law and policy researcher points out during an interview in New Delhi that many projects that are now being contested are expansions of older projects rather than Greenfield projects. Communities are therefore better informed in some cases. Livelihood movements nowadays have additional 'hooks' and 'stakes' to pursue their claims through environmental regulations; the Forest Rights Act in particular 'has been an organising plank, giving people a voice' (interview 05/05/2017).

The space of Indian environmentalism has effectively become more democratised through the emergence of a variety of new contestations. But simultaneously, the State's corporate bias and its repression of civil society movements has begun to create what many practitioners in India's civil society networks are calling a 'crisis of democracy'. In this conflicted context, community protests for the environment in neoliberal India are being understood as assertion of the democratic right to dissent. The senior environment law and policy researcher concludes that:

In the current political and economic context, with democratic processes challenged, when communities choose protection of environment and livelihoods over development, their actions can also be understood as exercising their democratic right to dissent.

Interview (05/05/2017)

To conclude this section, the issue of a highly uneven democracy in a highly unequal Indian society is a crucial factor in determining the extent to which communities fighting to protect their lands and livelihoods, and Adivasis fighting to protect their *Jal, Jangal, Jameen*, using new democratic tools and constitutional protections for their lands, are able to assert their rights. But additionally, increasing intolerance by the Indian State towards resistances against high-stakes energy, mining and other industrial projects is shrinking the space of democratic dissent in Indian society. Despite private gains by corporations from today's energy, mining and other large-scale industrial activities, the neoliberal State still invokes the moral imperative of poverty alleviation and the national interest to disrupt people's livelihoods and dispossess them of their lands. But movements challenge this narrative and the State's bias towards corporations, and assert their land, forest, and democratic rights as their critical right to dissent in a democracy.

Analysis: environmentalism's journey from democratising development to dissent as democracy

A healthy functioning democracy especially at the grassroots, upholding Constitutional provisions, and being able to exercise progressive democratic rights, hold the possibility to improve the conditions of Adivasis and marginalised communities; and for them to have a say in industrial development on their lands and in their forests. But the uneven nature of democracy in a highly unequal Indian society has created new challenges towards democratising development in the neoliberal era; in the neoliberal era the rights of communities enshrined in new democratic legislations pose a direct risk to private interests in high-stakes industrial projects and the State acting as a 'land broker' for private extractive development.

The postcolonial Indian State has invoked the moral imperative of national interest and poverty alleviation to disrupt people's livelihoods and remove them from their lands for industrial projects, both in the pre- and post-neoliberal era. The pre-neoliberal era in the 1970s and 1980s were marked by a mosaic of resistances for fair distribution of resources that exposed the myth behind the one-size-fits-all model of development as seen in Chapter 2. Industrialisation that resulted in the displacement and loss of livelihood of vulnerable communities was justified through a selective notion of the public interest that was nevertheless largely accepted by civil society owing to India's State-driven model of economic growth.

The scale of reconstruction of space through neoliberal development since the mid-1990s remains unprecedented in the 70 years of India's postcolonial development. The State's collusion with business interests has generated new forms of accumulation by dispossession of vulnerable groups in the neoliberal era (Nielsen 2010; Bannerjee-Guha 2013). They were directly affected through loss of land, livelihoods and environmental degradation, and indirectly through lack of employment in the new economy owing to the jobless nature of contemporary growth. Consequently, privatised growth intensified what Levien (2011) describes as the 'various small wars against land acquisition' across the Indian landscape (p. 66). Progressive legislations that aim for historic reparations and to democratise development by allowing vulnerable communities to secure forest rights and have a say on development on their lands have been passed in the neoliberal era. But their effectiveness is threatened by rampant violations, slow and flawed implementations, and dilutions.

Contemporary environmentalism of the poor movements emerged and took shape against this conflictual background. The threat of loss of land and livelihoods via land grabs and the violation and dilution of laws plays out as a continuous pattern in central India. Associated destitution continues despite India's high economic growth, making the Scheduled Areas in central India pockets of chronic poverty (Shah and Guru 2004). Deepening State bias towards private corporations and the cumulative effects of environmental (and social) destruction due to concentrated resource extraction pose a significant challenge to meeting both human rights for historically marginalised groups and environmental protections enshrined in the Constitution.

To answer the main question this chapter poses – how have the discourses and politics of environmentalism transformed from the previous era? – today's movements both differ from and resemble previous-era movements. The latter were characterised by a resource justice approach and a political impulse towards democratising development. The justice that environmentalism of the poor sought earlier through equitable distribution of natural resources is now enshrined as a series of rights in progressive legislations such as LARR and FRA that afford a legal basis to collective livelihood struggles. In some cases, the increased visibility achieved by today's movements through translocal and even transnational networks, make them more empowered than yesterday's movements.

Although India has passed a reasonably comprehensive suite of environmental legislations since the 1980s, legal reforms for democratising resource ownership and environmental protection have been marred by conceptual weakness due to an 'unwillingness to question the holy cow' of unlimited economic growth by governments (Kothari 2004, p. 4724). But delegitimisation of the notion of the greater common good in the sphere of civil society movements and the emergence of a new language of rights in the neoliberal era has had the double effect of rendering the struggles of ecologically dependent livelihood movements as both critical to avoid ecological crises and as critical assertions of democracy. Today's struggles to protect livelihoods from development, as well as to prevent the dilution of legislations and ensure their proper implementation, have become critical declarations of democracy.

The present mass movements protesting the disregard of special protections guaranteed under the Indian Constitution are labelled anti-national and violently crushed. The 2018 Human Rights Watch Report (www.hrw.org/world-report/2018) indicated that the government used draconian sedition and criminal defamation laws to curb the freedom of expression of government critics across a range of human rights issues, and not solely on occasions of environmental and Adivasi land rights conflicts. The *Freedom in the World 2020* report by the Washington-based Freedom House ranked India among 'least free' democracies (Freedom House 2020). As the facilitator of projects that dislocate Adivasi and peasants from their lands for private gain, and as the suppressor of human rights of movements, the neoliberal State plays a double role in shaping the environmentalism of the poor. State-repressions have catalysed the formation of movement narratives that signify the actions of the environmentalism of the poor as dissent that is critical for democracy.

The extent to which movements have transformed in India's neoliberal era reflexively represents the extent to which the State has altered from its earlier Socialist iteration while still retaining its postcolonial developmental agenda.

The tension between the legal-democratic assertions of movements and the private interests that the State favours makes the political context of today's environmentalism a highly charged one. The socio-ecological impacts of two decades of concentrated economic growth can therefore be regarded as a crisis of democracy. Against this context of environmental and resource politics in the world's most populous democracy, assertions of livelihood, land, and forest rights, against India's neoliberal mission find a common bond under the democratic right to dissent roof.

To conclude this chapter, the bigger absence of climate change as a generative concern for livelihood movements indicates a fundamental North–South difference in today's environmentalisms. This difference points to a continued need to understand the distinctive social, political and economic contexts of movements in the Global South. There is a need to understand the social–political context produced during India's neoliberal industrialisation, how the relation of the postcolonial State with communities and Adivasis has been transformed in this period, and how it shapes the narratives of the environmentalism of the poor today.

Land alienation remains the strongest contention of the environmentalism of the poor in India. The difference in today's movements is that their narratives are shaped within a new legal–democratic context of rights to land, livelihood, culture and tradition. Their narratives are also determined by the double movement – towards favouring private corporations on the one hand and repressing community dissent over industrial projects on the other – of the neoliberal State. As in the first four decades of postcolonial economic growth, the environmentalism of the poor in neoliberal India has reconfigured concepts from both northern conservation and northern environmental justice. This is owing to its different historical context, differing political and economic realities from industrialised nations, and on account of the continuing lived experience and livelihood dependency of India's subsistence communities on land. I further discuss the difference in the elements of global North versus global South environmentalisms in Chapter 9. This chapter which sets the context from which today's environmentalisms arise and the challenges they face in securing justice lays the necessary background against which we can better understand the politics of the Mahan coal mine and the community's struggle for forest rights in Chapters 4 and 5.

Notes

1 Dalit is a self-adopted term by the community. The exact origin of the term remains unknown, although records show that it was used by anti-caste social-reformer Jyotiba Phlue (1827–1890). Dr. B. R. Ambedkar (1891–1956) used the term extensively in his writings.

2 Fifth Schedule areas span across the states of Andhra Pradesh, Jharkhand, Chhattisgarh, Himachal Pradesh, Madhya Pradesh, Gujarat, Maharashtra, Odisha, Rajasthan and Telangana.

3 The Indian Forest Act 1878 demarcated the majority of India's forests as fully government controlled reserved forests. The other two categories of control and ownership set down by the Act included protected forests that were partly government controlled, and village forests that were controlled by villages adjoining forests.

4 It was amended and finally replaced by the *Right to Fair Compensation and Transparency in Land Acquisition, Rehabilitation and Resettlement Act 2013* (RFCTLARR) or LARR.

5 The term eminent domain refers to the overarching power vested in the State to compulsorily acquire private property or land for public use without consent.

6 The PDS was reduced as per terms of agreement with the International Monitory Fund for structural adjustment loans.

7 The significant social costs of the Gujarat model included the neglect of crucial sectors like agriculture (Rajagopal 2010). Neglect of healthcare was found to accentuate poverty despite Gujarat's relatively rich status compared to most other states (Mahadevia 2000).

8 The vision for inclusive growth has most recently been publicised through Narendra Modi's catchy Hindi slogan *sab ke saath sab ka vikas* which means 'development with and for everyone'.

9 The expression is known to have been first used by Jawaharlal Nehru to describe a 'vertical flow of wealth from the rich to the poor' (Nehru [1933] 1969, p. 24).

10 This chapter does not detail special favours made available to the Adani Group by the Indian government since the beginning of the Prime Ministership of BJP's Narendra Modi in 2014. Such a discussion falls outside the scope of this research that focuses on the Essar Corporation in India (see Chapter 4). However, a mention of Adani is unavoidable in the context of 'crony capitalism'. The Mundra Port and SEZ warrants a discussion as one of the most prominent and earliest examples of special favours by the State towards large industrial projects that are maximised for private profit and while ignoring community concerns and socio–ecological harms.

11 The Indian government approved 439 SEZs between 1991 and 2008 (Sharma 2009).

12 At Mundra, extensive industrialisation was coupled with widespread environmental breaches by the Adani Group. A federal inquiry found that Illegal coastal pollution and large-scale mangrove destruction by Adani had caused a drastic decline in fish catch. Illegal port developments obstructed access to traditional fishing grounds (Narain 2013).

13 The Centre for Science and Environment's (CSE) *Report Card* on the Modi government's environmental governance summed up the streamlining effort as involving the setting up of a single window for environmental, forest and wildlife clearances, standardising terms of reference for various sectors, devolving project clearance to state, district and regional authorities, diluting public consultation in the environmental clearance process by requiring fewer projects to undertake it and relaxing the need for community consent before destroying forests. It further undermined community concerns, especially around contentious mining projects, compared to the previous Congress-led UPA government (see CSE 2016).

14 The Government however failed to substantiate its own claim of one million new jobs being created from this dramatic easing of clearance windows (Dutt 2016).

15 The Report of the Committee on State Agrarian Relations and the Unfinished Task in Land Reforms brought out by the Ministry of Rural Development is a comprehensive policy intervention on the unfinished task of land reform in India (State Agrarian Relations Committee 2009).

16 The British Government enacted the Chotanagpur Tenancy Act (CNTA) 1908 that recognised customary forest and land rights in eastern India in a region which is today the state of Jharkhand. The Santhal Pargana Tenancy Act (SPTA) was enacted in 1949 following India's independence. These laws were passed in response to a long history of Adivasi-peasant uprisings demanding 'all land belongs to tribals'; notably the Santhal Rebellion (1855–1857), the Birsa Munda-led rebellion (1895–1900) and the Kol rebellion (1931).

17 Chatterjee (2008) argues that the State's interventions are guided primarily by a fear of the dispossessed turning into dangerous classes and affecting political stability.

18 According to the law, private projects required 80% of the affected community's consent while public–private partnerships (PPP) required 70%, to be able to proceed with land acquisition.

19 The Modi Government claimed that the Act stalled investments up to Rs 20 lakh crore (A\$400 billion) especially in rail, steel, mining and roads sectors (Jitendra 2015).

20 The law exempt 13 laws pertaining to acquisition for priority projects in sectors such as railways, national highways, atomic energy and electricity from fulfilling the LARR's stipulated conditions on a temporary basis. With the requirement that these laws be amended within a year to bring their compensation, resettlement and rehabilitation clauses at par with the new law (Ramesh and Khan 2015).

21 Items in the special category included defence, rural infrastructure, affordable housing, industrial corridors and infrastructure projects that included PPPs with government ownership of land (Kaushal and Tewari 2015).

22 The Ordinance introduced nine amendments to the original law. Other dilutions carried out by the BJP government's Ordinance included the retrospective clause, the definition of compensation paid, accountability for defaulting bureaucrats, and provision for return of unutilised land. It also extended the special powers of the government and expanded the definition of private entity (Kaushal and Tewari 2015).

23 An analysis of all land related disputes brought before the Supreme Court of India between 1950 and 2016 showed that the exempt categories account for half the land conflicts (Wahi et al. 2018).

24 It was passed in the Lok Sabha (Lower House) owing to the BJP's majority, amidst protests by the opposition. It was defeated in the Rajya Sabha (Upper House) where the ruling BJP did not have a majority (Jitendra 2015). Outside parliament the Ordinance triggered mass protests by farmers and civil society groups.

25 Sudha Bharadwaj (2018b) writes about the complicity of the judicial system in the violation of the FRA and PESA that: the state government of Chhattisgarh violated Adivasi rights with impunity despite PESA stipulating prior informed consent of village councils for mining projects. But when cases were brought to court, the judiciary tended to 'wink' at the transfer of forestland to corporations, thereby legitimising land grab in Scheduled Areas.

26 The state government of Chhattisgarh granted community forest rights to the village of Ghatbarra in the northern part of the state in 2013 but revoked them in 2015, arguing that the locals were misusing their rights to disrupt mining activities. Adani Mining acting as

the mine operator for a Rajasthan state-owned power utility RVUNL is undertaking coal mining in the forestlands adjoining Ghatbarra.

27 A 2021 circular by the Indian government to state governments gives the latter the responsibility to review and facilitate the proper of implementation of forest rights, stating that 'despite a considerable lapse of time since it (FRA) came into force, the process of recognition of forest rights is yet to be completed' (Nandi 2021).

28 The Forest Conservation Amendment Act 2023 allows forests to be diverted for projects deemed to be of strategic national importance and concerning national security within 100 kilometres of India's borders or lines of control without requiring forest clearance. This potentially risks forestlands in India's Northeast, a region comprising of seven north-eastern Indian states. The 2023 amendments narrow the interpretation of forests under the FCA, potentially exposing large, forested areas that are not legally notified as forests under any Act, but where communities lead forest-ecology-dependent livelihoods.

29 The NGT is an independent body with special adjudicative powers for environmental protection through which communities and their legal representatives can seek information on developmental projects, as well as take issues to special interest courts for redressing. It was created to implement the provision of Article 21 of the Constitution that assures Indian citizens a healthy environment as a part of their fundamental rights.

30 For example, the NGT has been undermined through the ceasing of four of its regional benches from operating due to non-appointment of members, making it difficult for challenges to be filed, heard and examined (Chowdhury and Srivastava 2018).

31 For a timeline of the resistance, see *Timeline of the POSCO Project Prepared by the Campaign for Survival and Dignity* at: https://forestrigthsact.com/corporate-projects/the-posco-project/timeline-of-events-relating-to-forest-rights-in-posco-area/.

32 The Dongriya Kondh are categorised as a Particularly Vulnerable Tribal Group (PVTG) by the Indian Government. Their customary occupation has been reported as a complex system of agro-forestry where patches of land are cleared in rotation, collecting minor forest produce for sustenance and medicine, and rearing livestock for meat and ritual sacrifices (Saxena et al. 2010).

33 For a full timeline of the movement see Amnesty International's report, p. 11: www.amnesty.org/download/Documents/36000/asa200012010en.pdf.

34 The same court had earlier in 2008 dismissed the Environment Ministry's concerns and declared that development was necessary (see Kumar 2014).

35 Raising the spectre of the foreign hand has been prevalent since the national emergency in 1976 when Prime Minister Indira Gandhi passed the *Foreign Contribution (Regulation) Act 1976* (FCRA) to curb international interference in Indian politics from other political parties and non-political institutions.

36 The figure indicates land displacements for the Adivasi majority region of the eastern state of Bihar that went on to form the separate state of Jharkhand in 2000.

References

Abdul Kalam, A. P. J. 2011, 'Kudankulam nuclear plant safe', *BBC News*, 7 November, viewed 20 September 2019, <www.bbc.com/news/world-asia-15616019>.

Adhikari, D. 2022, *Indigenous Women's Struggle for Forest Rights in India*, Australian National University, 25 July, viewed 20 November 2022, <https://crawford.anu.edu.au/news-events/news/20455/indigenous-womens-struggle-forest-rights-india>.

Agarwal, S. & Dash, T. 2022, *Policy Brief: Securing Climate Justice for India's Forest-Dependent Communities*, viewed 20 November 2022, <www.fra.org.in/document/Revi sed%20Policy%20brief%20-%20Securing%20climate%20justice%20for%20forest-dependent%20communities.pdf>.

Ambedkar, B. R. 1949, *Dr. B. R. Ambedkar's speech in the Constituent Assembly on 25th November 1949*, viewed 20 September 2019, <www.roundtableindia.co.in/babasaheb-dr-b-r-ambedkar-s-speech-in-the-constituent-assembly-on-25th-november-1949/>.

Amnesty International India. 2016, *"When Land Is Lost, Do We Eat Coal?": Coal Mining and Violations of Adivasi Rights in India*, Bangalore, viewed 20 June 2021, <www.amne sty.org/download/Documents/ASA2043922016ENGLISH.PDF>.

Anwar, T. 2019a, "Jharkhand's Land Bank', injustices to Adivasis continue', *Newsclick*, 18 June, viewed 20 September 2020, <www.newsclick.in/Jharkhand-Land-Bank-Adiva sis-Tribes-Revenue-Land-Reforms>.

Anwar, T. 2019b, 'Jharkhand: every 10th Adivasi in Khunti dist charged with sedition for resorting to Pathalgadi', *Newsclick*, 22 July, viewed 20 September 2020, <www.newscl ick.in/jharkhand-every-10-adivasi-khunti-dist-charged-sedition-resorting-pathalgarhi>.

Arora-Jonsson, S. 2013, *Gender, Development, and Environmental Governance: Theorising Connections*, Routledge, London/New York.

Bannerjee-Guha, S. 2013, 'Accumulation and dispossession: contradictions of growth and development in contemporary India', *Journal of South Asian Studies*, vol. 36, no. 2, pp. 165–179.

Bardhan, P. K. 1984, *The Political Economy of Development in India*, Basil Blackwell, Oxford.

Basu, S. & Mallick, S. 2008, 'When does growth trickle down to the poor? The Indian case', *Cambridge Journal of Economics*, vol. 32, pp. 461–477.

Bedi, H. P. 2013, 'Special Economic Zones: National Land Challenges, Localised Protest', *Contemporary South Asia*, vol. 21, no. 1, pp. 38–51.

Bera, S. 2015, 'If you dig these, our gods will die', *Down to Earth*, 11 June, viewed 20 September 2019, <www.downtoearth.org.in/coverage/niyamgiri-answers-41914>.

Bharadwaj, S. 2018a, *The Legal Face of the Corporate Land Grab in Chhattisgarh*, Janhit People's Legal Resource Centre.

Bharadwaj, S. 2018b, 'The New Land Acquisition Act, does it help the tribal people?', *Journal of Resource, Energy and Development*, vol. 15, no. 1–2, pp. 53–61.

Biswas, S. 2016, 'Why India needs to get rid of its sedition laws', *BBC News*, 26 August, viewed 20 September 2019, <www.bbc.com/news/world-asia-india-37182206>.

Centre for Science and Environment. 2016, *Report Card: Environmental Governance Under NDA Government*, 22 June, viewed 20 September 2019, <www.downtoearth.org.in/cover age/governance/report-card-environmental-governance-under-nda-government-54359>.

Chakravarty, B. K. 2006, 'Environmentalism: Indian constitution and judiciary', *Journal of the Indian Law Institute*, vol. 48. no. 1, pp. 99–105.

Chatterjee, P. 2004, *The Politics of the Governed; Reflections on Popular Politics in Most of the World*, Columbia University Press.

Chatterjee, P. 2008, 'Democracy and economic transformation in India', *Economic and Political Weekly*, vol. 43, no. 16, pp. 53–62.

Chowdhury, C. 2016, 'Making a hollow in the Forest Rights Act', *The Hindu*, 7 April, viewed 20 September 2019, <www.thehindu.com/opinion/columns/Making-a-hollow-in-the-Forest-Rights-Act/article14226592.ece>.

Chowdhury, C. & Aga, A. 2020, 'Manufacturing consent: mining, bureaucratic sabotage and the Forest Rights Act in India', *Capitalism Nature Socialism*, vol. 31, no. 2, pp. 70–90.

Chowdhury, N. & Srivastava, N. 2018, 'The National Green Tribunal in India: examining the question of jurisdiction', *Asia Pacific Journal of Indian Law*, vol. 21, no. 2, pp. 190–216.

Community Forest Rights-Learning and Advocacy (CFR-LA). 2016, *Promise and Performance, Ten Years of the Forest Rights Act in India*, Citizen's Report on the Promise and Performance of Scheduled Tribes and Other Traditional Forest Dwellers (Recognition of Forest Rights) Act 2006.

Community Forest Rights-Learning and Advocacy (CFR-LA). 2020, Community Forest Rights and the Pandemic: Gram Sabhas Lead the Way, October, viewed 20 September 2022, <https://rightsandresources.org/wp-content/uploads/2020/10/CFR-and-the-Pand emic_GS-Lead-the-Way-Vol.2_Oct.2020.pdf>.

Constitution of India Part IV Article 48A. *Protection and Improvement of Environment and Safeguarding of Forests and Wildlife*, viewed 20 September 2019, <www.constitutionofin dia.net/articles/article-48a-protection-and-improvement-of-environment-and-safeguard ing-of-forests-and-wild-life/>.

Corbridge, S. & Harriss, J. 2003, *Reinventing India: Liberalisation, Hindu Nationalism, and Popular Democracy*, Oxford University Press, New Delhi.

Das, P. 2017, 'Where the public resistance is building over POSCO land in Odisha', *The Hindu*, 20 May, viewed 20 September 2019, <www.thehindu.com/news/national/other-states/in-odisha-fresh-row-brewing-over-posco-land/article18516063.ece>.

Deshpande, T. 2021, 'Why is India silent on forests despite its ambitious climate targets', *IndiaSpend*, 11 November, viewed 20 November 2022, <www.indiaspend.com/clim ate-change/cop26-why-india-is-silent-on-forests-despite-its-ambitious-climate-targets-786554>.

Dreze, J., Samson, M. & Singh, S. 1997, *Displacement and Resettlement; Ecological and Political Issues on the Narmada Valley Conflict*, Oxford University Press, New Delhi and Oxford.

Dreze, J. & Sen, A. 2013, *An Uncertain Glory: India and Its Contradictions,* Princeton University Press, Princeton, NJ.

Dubey, S. & Saxena, A. 2023, 'Adivasi and forest feminism: the path to Just Transitions through the Forest Rights Act', in S. Arora-Jonsson, K. Michael & M. K. Shrivastava (eds), *Just Transitions; Gender and Power in India's Climate Politics*, Routledge, London.

Dutt, B. 2014, *Green Wars, Dispatches from a Vanishing World*, Harper Collins.

Dutt, B. 2016, 'Green ministry stands exposed', *Livemint*, 15 January.

Fernandes, W. 2008, 'Sixty years of development induced displacement in India', *India Social Development Report*, pp. 89–102.

Fernandes, W. & Ganguly-Thukral, E. 1989, *Development, Displacement, and Rehabilitation: Issues for a National Debate*, Indian Social Institute.

Freedom House. 2020, *Freedom in the World 2020: A Leaderless Struggle for Democracy*, viewed 20 December 2023, <https://freedomhouse.org/sites/default/files/2020-02/FIW_ 2020_REPORT_BOOKLET_Final.pdf>.

Gadgil, M. & Guha, R. 2000, 'Ecological conflicts and environmental movements in India', *Development: Challenges for Development*, vol. 6, p. 254.

Gandhi, I. 1972, Man and the Environment, Indira Gandhi's Speech at the Stockholm Conference in 1972, 14 June, viewed 20 September 2019, <http://lasulawsenvironmental. blogspot.com/2012/07/indira-gandhis-speech-at-stockholm.html>.

Gandhi, M. K. 1929, 'The curse of industrialism', *Young India*, 25 July, viewed 20 June 2023, <http://bapu.wordpress.com/>.

Ganguly-Thukral, E. (ed.) 1992, *Big Dams, Displaced People: Rivers of Sorrow, Rivers of Change*, Sage Publications.

Government of India (GOI). 2006, *The Scheduled Tribes and Other Traditional Forest Dwellers (Recognition of Forest Rights) Act.*

Government of India (GOI). 2008, *Ministry of Home Affairs; Annual Report 2007-08; Departments of Internal Security, States, Home, Jammu and Kashmir Affairs and Border Management*, viewed 20 December 2017, <www.mha.gov.in/sites/default/files/AnnualR eport_07_08.pdf>.

Guha, R. 2000, *Environmentalism, A Global History*, Longman, New York.

Guha, R. 2014, 'The Indian path to unsustainability', *The Telegraph*, 9 August, viewed 20 September 2019, <http://ramachandraguha.in/archives/the-indian-path-to-unsustainabil ity-the-telegraph.html>.

Guhathakurta, P. 2015, *Gas Wars, Crony Capitalism, and the Ambanis*, New Delhi.

Guhathakurta, P., Palepu, A. R., Jain, S. & Dasgupta, A. 2017, 'Modi government's Rs 500-crore bonanza to the Adani Group', The Wire, 19 June, viewed 20 January 2024, <https:// thewire.in/business/modi-government-adani-group>.

Gupta, A. & Sivaramakrishnan, K. (eds.) 2011, *The State in India After Liberalisation: Interdisciplinary Perspectives*, Routledge, New York.

Harriss, J. 2011, 'How far have India's economic reforms been guided by compassion and justice? Social policy in the neoliberal era', in S. Ruparelia, S. Reddy, J. Harriss & S. Corbridge (eds.), *Understanding India's New Political Economy: A Great Transformation?*, Routledge, London & New York, pp. 127–140.

Harriss, J. & Corbridge, S. 2010, 'The continuing reinvention of India', in C. Sengupta & S. Corbridge (eds), *Democracy, Development and Decentralisation in India*, Routledge, London, pp. 38–39.

Harvey, D. 2003, *The New Imperialism*, Oxford University Press, Oxford.

Harvey, D. 2005, *A Brief History of Neoliberalism*, Oxford University Press, Oxford.

Harvey, D. 2007, 'Neoliberalism as creative destruction', *Annals of the American Academy of Political and Social Science,* vol. 610, no. 1.

Intergovernmental Science-Policy Platform on Biodiversity and Ecosystem Services (IPBES), 2019, *Summary for Policymakers of the IPBES Global Assessment Report on Biodiversity and Ecosystem Services*, United Nations, viewed 20 November 2023, <https://ipbes.net/sites/default/files/inline/files/ipbes_global_assessment_report_summa ry_for_policymakers.pdf>.

Jafferlot, C. 2016, 'What 'Gujarat Model'? – Growth without development – and socio-political polarisation', *Journal of South Asian Studies*, vol. 38, no. 4, pp. 820–838.

Jamwal, N. 2020, 'Tribal districts report more land conflicts compared to other regions in the country', Gaon Connection, 10 March, viewed 20 December 2023, <https://en.gaoncon nection.com/tribal-districts-in-india-report-more-land-conflicts-compared-to-other-regi ons-in-the-country/>.

Jitendra. 2015, 'Government introduces land amendment bill amidst protests, walkout', Down to Earth, 17 August, viewed 20 September 2019, <www.downtoearth.org.in/news/ government-introduces-land-amendment-bill-amid-protests-walkout-48723>.

Joshi, M. & Sethi, N. 2022, 'Government to approve cutting down of forests without con-sent from tribals and forest dwellers', *News Laundry*, 7 July, viewed 20 November 2022, <www.newslaundry.com/2022/07/07/government-to-approve-cutting-down-of-forests-without-consent-from-tribals-and-forest-dwellers>.

Kashwan, P. 2017, *Democracy in the Woods: Environmental Conservation and Social Justice in India, Tanzania, and Mexico*, Oxford University Press.

Kaur, R. 2020, *Kudankulam: The Story of an Indo-Russian Nuclear Power Plant*, Oxford University Press.

Kaushal, A. 2015, 'Government cancels registration of 9,000 NGOs', *Business Standard*, 29 April, viewed 14 August 2019, <www.business-standard.com/article/current-affairs/gov ernment-cancels-registration-of-9-000-ngos-115042800367_1.html>.

Kaushal, P. & Tewari, R. 2015, *Land Bill: Modi Government Gives in, Agrees to Bring Back UPA's Key Provisions*, 4 August, viewed 20 December 2021, <https://indianexpress.com/ article/india/india-others/bjp-takes-u-turn-on-land-bill-agrees-to-bring-back-upas-key-provisions/>.

Kerketta, J. 2016, 'The river, the mountain and the bazaar', in *Angor*, Adivaani, Kolkata.

Khilnani, S. 1998, *The Idea of India*, Farrar Straus Giroux, New York.

Kodiveri, A. & Rathore, V. 2021, 'One Act, Three Proposed Amendments, and an Overhaul of India's Coal Extraction', *The Bastion: Development in depth*, 26 July, viewed 20 September 2022, < https://thebastion.co.in/politics-and/one-act-three-proposed-ame ndments-and-an-overhaul-of-indias-coal-extraction/>

Kohli, A. 2007, 'State, business and economic growth in India', *Studies in Comparative International Development*, vol. 42, no. 1–2, pp. 87–114.

Kohli, K. & Gupta, D. 2017, *Mapping Dilutions in a Central Law*, Centre for Policy Research, 25 September, viewed 20 September 2019, <www.cprindia.org/research/pap ers/mapping-dilutions-central-law-0>.

Kohli, K., Kapoor, M., Menon, M. & Vishwanathan, V. 2018. *Midcourse Manoeuvres: Community Strategy and Remedies for Natural Resource Conflicts in India*, CPR-Namati Environmental Justice Program, New Delhi.

Kohli, K. & Shukla, A. 2016. 'Burying the law to make way for a coal mine', *The Wire*, 19 February, viewed 20 November 2022, <https://thewire.in/politics/bypassing-the-law-to-make-way-for-coal-mine-in-chhattisgarh>.

Kothari, A. 2004, 'Draft national environmental policy: a critique', *Economic and Political Weekly*, vol. 39, no. 43, pp. 4723–4727.

Kothari, A. 2016, 'Decisions of the people, by the people, for the people', *The Hindu*, 18 May, viewed 20 September 2019, <www.thehindu.com/opinion/op-ed/Decisions-of-the-people-by-the-people-for-the-people/article14324692.ece>.

Krishnan, R. & Naga, R. 2017, "Ecological warriors', versus 'indigenous performers': understanding state responses to resistance movements in Jagatsinghpur and Niyamgiri in Odisha', *Journal of South Asian Studies*, vol. 40, no. 4, pp. 878–894.

Kumar, B. 2005, 'Postcolonial state: an overview', *Indian Journal of Political Science*, vol. 66, no. 4, pp. 935–954.

Kumar, K. 2014, 'The sacred mountain: confronting global capital at Niyamgiri', *Geoforum*, vol. 54, July, pp. 196–206.

Kumar, K. & Kerr, J. M. 2012, 'Democratic assertions: the making of India's recognition of Forest Rights Act', *Development and Change*, vol. 43, no. 3, 751–771.

Lascorgeix, A. & Kothari, A. 2009, 'Displacement and relocation of protected areas: a synthesis and analysis of case studies', *Economic and Political Weekly*, vol. 44, no. 49, pp. 37–47.

Lele, S., Khare, A. & Mokashi, S. 2020, *Estimating and Mapping CFR Potential for Madhya Pradesh, Chhattisgarh, Jharkhand and Maharashtra*, Centre for Environment and Development, Ashoka Trust for Research In Ecology and the Environment, viewed 20 December 2020, <www.atree.org/sites/default/files/reports/CFR_Potential_Mapping_Report_compressed.pdf>.

Levien, M. 2007, 'India's double movement: Polanyi and the national alliance of people's movements', *Berkeley Journal of Sociology*, vol. 51, Globalisation and Social Change, 2007, pp. 119–149.

Levien, M. 2011, 'Special economic zones and accumulation by dispossession in India', *Journal of Agrarian Change*, vol. 11, no. 4, pp. 454–483.

Mahadevia, D. 2000, 'Health for all in Gujarat, is it achievable?', *Economic and Political Weekly*, vol. 35, no. 35, pp. 3193–3197.

McCarthy, J. & Prudham, S. 2004, 'Neoliberal nature and the nature of neoliberalism', *Geoforum*, vol. 35, no. 3, pp. 275–283.

Ministry of Environment and Forests. 2009, 'Diversion of forest land for non-forest purposes under the Forest (Conservation) Act, 1980 – ensuring compliance of the Scheduled Tribes and Other Traditional Forest Dwellers (Recognition of Forest Rights) Act 2006', 30 July, viewed 20 June 2019, <https://forestsclearance.nic.in/writereaddata/public_display/sche mes/337765444$30-7-2009.pdf>.

Ministry of Tribal Affairs. 2018, <https://tribal.nic.in/FRA/data/MPRNov2018.pdf>.

Moolakkattu, J. S. 2014, 'Nonviolent resistance to Nuclear Power Plants in South India', *Peace Review, A Journal of Social Justice,* vol. 26, no. 3, pp. 420–426.

Munster, D. 2012, 'Farmers suicides and the state in India: conceptual and ethnographic notes from Wayanad, Kerala', *Contributions to Indian Sociology*, vol. 46, no. 1–2, pp. 181–208.

Munster, D. & Strumpell, C. 2014, 'The anthropology of neoliberal India: an introduction', *Contributions to Indian Sociology*, vol. 48, no. 1, pp. 1–16.

Nandi, J. 2021, 'Review implementation of forest rights: union ministries to state governments', *Hindustan Times,* 7 July, viewed 20 December 2021, < www.hindustanti mes.com/environment/review-implementation-of-forest-rights-union-ministries-to-state-govts-101625562374847.html>.

Nandi, J. 2023, *Why Naga Peoples Front has opposed the Forest Conservation (Amendment) Act 2023*, 9 August, viewed 20 December 2023, <www.hindustantimes.com/india-news/indias-new-forest-conservation-law-faces-opposition-from-naga-people-s-front-deemed-antitribal-101691551815887.html>.

Narain, S. 2013, *Report of the Committee for Inspection of M/s Adani Port and SEZ LTD*, Mundra, Gujarat, Ministry of Environment and Forests, New Delhi, April 2013, viewed 20 September 2019, <http://cdn.cseindia.org/userfiles/adani_final_report.pdf>.

Nayar, B. R. 1989, *India's Mixed Economy: The Role of Ideology and Interest in Its Development*, Popular Prakashan, Bombay.

Nayar, L. 2016, 'Unlock Mantri: Javadekar's Key to Success', *Outlook India,* 5 July, viewed 20 September 2019, <www.outlookindia.com/website/story/unlock-mantri-javadekars-key-to-success/297036>.

Nayar, L., Mukherjee, A. & Arora, S. 2014, 'All along the waterfront', *Outlook India*, 10 March, viewed 20 September 2019, <https://magazine.outlookindia.com/story/all-along-the-waterfront/289708>.

Nayyar, D. 1998, 'Economic development and political democracy: interaction of economics and politics in independent India', *Economic and Political Weekly*, vol. 33, no. 49, pp. 3121–3131.

Nehru, J. [1933] 1969, "Whither India?" India's freedom', Unwin Books, Allen & Unwin Press, London, Reprinted in Nehru, J. 1962, viewed 20 June 2023, <https://archive.org/details/dli.ministry.24624/page/35/mode/2up>.

Nehru, J. 1954, *Jawaharlal Nehru's Speeches*, 2, Publications Division, New Delhi.

Nielsen, K. B. 2010, 'Contesting India's development? Industrialisation, land acquisition, and protest in West Bengal', *Forum for Development Studies*, vol. 37. no. 2, pp. 145–170.

Nielsen, B. & Oskarsson, P. 2016, *Industrialising Rural India: Land, Policy and Resistance*, Routledge, London.

Oskarsson, P. 2015, 'Governing India's bauxite mineral expansion: caught between facilitating investment and mediating social concerns', *The Extractive Industries and Society*, vol. 2/3, no. 2, pp. 426–433.

Oza, N. 2022, *The Struggle for Narmada: An Oral History of the Narmada Bachao Aandolan, by Adivasi Leaders Keshavbhai and Kevalsingh Vasave*, Orient BlackSwan.

Padel, F. 2016, 'Investment induced displacement and the ecological basis of India's economy', in S. Venkateshwar and S. Bandyopadhyay (eds), *Globalisation and the Challenges of Development in Contemporary India*, Springer, Singapore, pp. 147–169.

Padhi, R. & Sadangi, N. 2020, *Resisting Dispossession: The Odisha Story*, Palgrave MacMillan.

Peck, J. 2001, 'Neoliberalizing states: thin policies/hard outcomes', *Progress in Human Geography*, vol. 23, no. 3, pp. 445–455.

Peck, J. & Tickell, A. 2002, 'Neoliberalizing space', *Antipode*, vol. 34, no. 3, pp. 380–404.

Pingali, G. 2022, *Indigenous Question, Land Appropriation, and Development: Understanding the Conflict in Jharkhand, India*, Routledge, London.

Polanyi, K. 2001 [1944], *The Great Transformation: The Political and Economic Origins of Our Time*, Beacon Press, Boston.

Press Trust of India. 2016a, 'Projects worth Rs 10 lakh crore cleared: environment Minister Prakash Javadekar', *Indian Express*, 17 May.

Press Trust of India. 2016b, 'Centre for completing environmental clearances in 100 days', *Indian Express*, 29 June.

Radyuhin, V. 2002, 'Two more reactors for Koodankulam', *The Hindu*, 14 February.

Rajagopal, A. 2010, 'Special political zone, urban planning, spatial segregation and the infrastructure of violence in Ahmedabad', *South Asian History and Culture*, vol. 1, no. 4, pp. 529–556.

Rajshekhar, M. 2012, *The Act That Disagreed with the Preamble: The Drafting of the 'Scheduled Tribes and Other Traditional Forest Dwellers (Recognition of Forest Rights) Act 2006*, viewed 20 September 2019, <https://rightsandresources.org/wp-content/uploads/CommunityForest_July-20.pdf>.

Ramesh, J. & Khan, M. A. 2015, *Legislating for Justice: The Making of the 2013 Land Acquisition Law*, Oxford University Press.

Rangarajan, M. & Shahabuddin, G. 2006, 'Displacement and relocation from protected areas: towards a biological and historical synthesis', *Conservation and Society*, vol. 4, no. 3, pp. 359–378.

Ranjan, A. 2014, 'Foreign-aided NGOs are actively stalling development, IB tells PMO in a report', *Indian Express*, 7 June, viewed 20 August, 2019, <http://indianexpress.com/article/india/india-others/foreign-aided-ngos-are-actively-stalling-development-ib-tells-pmo-in-a-report/>.

Ranjan, R. 2017, 'Unraveling the Narratives of Adivasi Disposession: A Case Study of Land Acquisition in Nagri Village, Jharkhand', *Local/Global Encouters*, vol. 60, pp. 227–234.

Rao, A. & Kadam, K. 2016, '25 years of liberalisation: a glimpse of India's growth in 14 charts', *Firstpost*, 7 July, viewed 20 June 2023, <www.firstpost.com/business/25-years-of-liberalisation-a-glimpse-of-indias-growth-in-14-charts-2877654.html>.

Reddy, V. R. & Reddy, B. S. 2007, 'Land alienation and local communities: case studies in Hyderabad–Secunderabad', *Economic and Political Weekly*, vol. 42, no. 31, pp. 3233–3240.

Reich, M. R. & Bowonder, B. 1992, 'Environmental policy in India: strategies for better implementation', *Policy Studies Journal*, vol. 20, no. 4, pp. 643–661.

Reynolds, S. 2010, *Before Eminent Domain; Towards a History of Expropriation of Land for Common Good,* The University of North Carolina Press.

Rights and Resources Initiative (RRI), Vasundhara & Natural Resources Management Consultants. 2015, *Potential for Recognition of Community Forest Resource Rights under India's Forest Rights Act.* https://rightsandresources.org/wp-content/uploads/Communit yForest_RR_A4Final_web1.pdf.

Rowlatt, J. 2015, 'Why India's government is targeting Greenpeace', *BBC News,* 16 May.

Roy, A. 1999, 'The greater common good', *Outlook India,* 24 May, viewed 20 September 2019, <https://magazine.outlookindia.com/story/the-greater-common-good/207509>.

Roy, A. 2009, 'The Heart of India is under attack', *The Guardian,* 30 October, viewed 20 December 2023, <www.theguardian.com/commentisfree/2009/oct/30/mining-india-maoi sts-green-hunt>.

Roy Chowdhury, D. & Keane, J. 2021, *To Kill a Democracy: India's Passage to Despotism,* Oxford University Press.

Saxena, N. C., Parasuraman, S., Kant, P. & Baviskar, A. 2010, Report of the Four Member Committee for Investigation into the Proposal Submitted by the Orissa Mining Company for Bauxite Mining in Niyamgiri, 16 August, viewed 20 January 2019, <https://cdn.csein dia.org/userfiles/Report_Vedanta.pdf>.

Sethi, A. 2010, 'Green Hunt: the anatomy of an operation', *The Hindu,* 6 February, viewed 20 January 2024, <www.thehindu.com/opinion/op-ed/Green-Hunt-the-anatomy-of-an-operation/article16812797.ece>.

Sethi, N. 2016, 'Chhattisgarh government cancels tribal rights over forest lands', Business Standard, 18 February, viewed 20 September 2019, <www.business-standard.com/article/current-affairs/chhattisgarh-govt-cancels-tribal-rights-over-forest-lands-116021601327_1.html>.

Sethi, N. 2019, 'Modi government's move to amend forest rights takes a giant leap backwards', *Business Standard,* 1 April, viewed 20 March 2020, <www.business-standard. com/article/economy-policy/modi-govt-s-move-to-amend-forest-act-takes-a-giant-leap-backwards-119040101292_1.html>.

Sethi, N. & Shrivastava, K. S. 2019, 'Modi government plans more draconian version of colonial-era Indian Forest Act', *The Wire,* 21 March, viewed 20 March 2020, <https:// thewire.in/rights/modi-government-plans-more-draconian-version-of-colonial-era-ind ian-forest-act>.

Shah, A. 2010. *In the Shadows of the State: Indigenous Politics, Environmentalism, and Insurgency in Jharkhand, India,* Duke University Press.

Shah, A. & Guru, B. 2004, *Poverty in Remote Rural Areas in India – A Review of Evidence and Issues,* CPRC-IIPA Working Paper 21, Chronic Poverty Research Centre University of Manchester and Indian Institute of Public Administration, New Delhi.

Shahabuddin, G. & Bhamidipati, P. L. 2014, 'Conservation-induced displacement, recent perspectives from India', *Environmental Justice,* vol. 7, no. 5, pp. 122–129.

Sharma, N. K. 2009, 'Special economic zones: socio-economic implications', *Economic and Political Weekly,* vol. 44, no. 20, pp. 18–21.

Sharma, R. N. & Singh, S. R. 2009, 'Displacement in Singrauli region: entitlements and rehabilitation', *Economic and Political Weekly,* vol. 44, no. 51, pp. 62–69.

Shrivastava, A. & Kothari, A. 2012, *Churning the Earth: The Making of Global India,* Penguin, New Delhi.

Sigamany, I. 2020, *Forced from the forest; Mobile Indigenous Peoples, Gender Equality and the Forest Rights Act,* Berghahn Books.

Srinivas, A. 2020, 'Why green reform is a grave step back', *Live Mint,* 31 August, viewed 20 December 2023, <www.livemint.com/news/india/why-green-reform-is-a-grave-step-back-11598882280124.html>.

State Agrarian Relations Committee. 2009, Report of committee on State Agrarian Relations and the Unfinished Task of Land Reforms, 8 June, viewed 20 December 2017, <https://dolr.gov.in/en/documents/report-of-committee-on-state-agrarian-relations>.

Subramaniam, T. S. 2013, 'There is no openness', *The Hindu*, 9 August, viewed 20 September 2019, <https://frontline.thehindu.com/the-nation/there-is-no-openness/article4945319.ece>.

Sundar, N. 2007, *Subalterns and Sovereigns: An Anthropological History of Bastar, 1854–2006*, Oxford University Press, New Delhi.

Sundar, N. 2011, 'The rule of law and citizenship in central India: postcolonial dilemmas', *Citizenship Studies*, vol. 15, no. 3–4, pp. 419–432.

Sundar, N. 2023, 'We will teach India democracy: indigenous voices in decision making', *Journal of Imperial and Commonwealth History.*

Sunil, M. & Dungdung, G., 2013. *Jharkhand mein Asmita Sangharsh.* Ranchi Jharkhand Human Rights Movement.

Tatpati, M., Kothari, A. & Mishra, R. 2016, *Challenging the Idea of Growth without Limits*, Vikalp Sangam, Pune, viewed 20 September 2019, <http://vikalpsangam.org/static/media/uploads/Resources/niyamgiricasestudy.pdf>.

Thompson, L., Mackey, B., McNulty, S. & Mosseler, A., 2009. *Forest Resilience, Biodiversity, and Climate Change. A Synthesis of the Biodiversity/Resilience, Stability, Relationship in Forest Ecosystems.* Secretariat of the Convention on Biological Diversity, Montreal, Technical Series no. 43.

Trivedi, N. 2023, 'Chhattisgarh Adivasis rue loss of land', *IndiaSpend*, 28 October, viewed 20 December 2023, <www.indiaspend.com/adivasi/chhattisgarh-adivasis-rue-loss-of-land-879971>.

Tyagi, N. & Das, S. 2018, 'Assessing gender responsiveness of forest policies in India', *Forest Policy and Economics*, vol. 92, July, pp. 160–168.

Udaykumar, S. P. (ed.) 2004, The Koodankulam Handbook, Transcend South Asia, Nagercoli, pp. 300–315.

Vats, S. 2023, 'How MP Govt's VanMitra Portal threatens tribal rights over forest land', *IndiaSpend*, 6 November, viewed 20 December 2023, <www.indiaspend.com/land-rights/how-mp-govts-vanmitra-portal-threatens-tribal-rights-over-forest-land-880848>.

Wahi, N., Bhatia, A., Shukla, P., Gandhi, D., Jain, S. & Chauhan, U. 2018, *Land Acquisition in India: A Review of Supreme Court Cases (1950–2016).*

Worsdell, T. & Shrivastava, K. 2020, *Locating the Breach: Mapping the Nature of Land Conflicts in India*, Land Conflict Watch, February, New Delhi, viewed 20 December 2023, <https://assets-global.website-files.com/5d70c9269b8d7bd25d8b1696/5ecd20dd626f166d67f67461_Locating_the_Breach_Feb_2020.pdf>.

Legislations

Coal Bearing Areas (Acquisition and Development) Act *1957.*
Compensatory Forestation Fund Act *2016.*
Land Acquisition Act 1894.
Land Acquisition, Rehabilitation and Resettlement Act 2013.
Mahatma Gandhi National Rural Employment Guarantee Act 2005.

National Green Tribunal Act 2010.

Panchayats (Extension of Scheduled Areas) Act 1996.

Right to Fair Compensation and Transparency in Land Acquisition, Rehabilitation and Resettlement Act 2013.

Right to Information (RTI) Act 2005.

Scheduled Tribes and Other Traditional Forest Dwellers (Recognition of Forest Rights) Act 2006.

Special Economic Zones Act 2005.

Websites

Environment Justice Atlas https://ejatlas.org
Environment Performance Index https://epi.yale.edu
Human Rights Watch www.hrw.org/world-report/

4 Countering coal in India

Politics of the Mahan coal mine

Coal is king and paramount Lord of industry is an old saying in the industrial world. Industrial greatness has been built upon coal by many countries. In India, coal is the most important indigenous energy resource and remains the dominant fuel for power generation and many industrial applications.

Statement by a judge of the Supreme Court of India (*Deccan Chronicle*, 2014).

Between 2011 and 2015, Greenpeace India mobilised forest-dependent communities in Central India against a private coal mine that had been allocated to a joint venture between Essar Power Ltd. and Hindalco Industries Ltd. Although it has campaign offices in countries in the global South, Greenpeace is largely a global North-based and driven organisation with a centralised structure (Doherty and Doyle, 2006). This was one of the international NGO's very few grassroots campaigns in India, and the only one that involved the mobilisation a rural constituency against coalmining. Greenpeace stated its objective as 'bringing out the true cost of coal, which is not just economic, but also environmental, social and spiritual' (Greenpeace Campaigner quoted in Niyogy, 2015, para 3).

The mobilisation of 11 villages in the Mahan forests, in a region known as India's energy capital, was a central plank in Greenpeace India's Climate and Energy (C&E) Program. Other elements in the Program included policy analysis around phasing out coal and expanding small and medium-scale renewable energy sources, rural electrification through a solar-micro grid project in the eastern state of Bihar (Greenpeace India, 2013), and policy-driven advocacy and urban mobilisation on air pollution.

The coal mine was stopped in March 2015 after the Narendra Modi government acted on a Supreme Court order that cancelled 214 coal blocks including Mahan; the previous Congress-led government had allocated them without following due process. A coal block is an area allocated for coalmining.

However, the Modi government then began a systematic campaign of disrupting the work of civil society groups and people's movements. It alleged

DOI: 10.4324/9781003410416-4

that international environmental and human rights groups were acting at the behest of the foreign hand to stall India's growth. Reports prepared by the government's surveillance agency the Intelligence Bureau (IB) targeted Greenpeace's campaign against coal mines and thermal power plants.

Greenpeace's anti-coal mobilisation and the hostility of the State's response brought out the contradictions of the political economy of coal in India. It agitated a civil society debate on the social and environmental costs of coal. While coal is still India's primary energy source, coalmining and thermal power generation are also responsible for a full spectrum of social and environmental issues. Greenpeace's activism and the State's response offer perspectives on how the script of global climate activism through stopping coal is necessarily transformed in the socio-political context of a developing, coal-dependent economy. It also offers perspectives on the risks of global environmentalism's scrutiny in such a context.

This chapter traces the build up and eventual crackdown on Greenpeace India's anti-coal campaign between 2006 and 2016 during a dynamic time period in the political economy of coal in India. Between 2004 and 2015 India both deepened and privatised its coalmining capacity for the stated purpose of providing electricity to all Indians. Parallelly, India declared ambitious targets to increase its renewable energy capacity and reduce the emissions intensity of its development under the Paris Agreement. The Greenpeace campaign was formed and operated within this contradictory context.

This chapter answers the second question of this book regarding India, about the dialectic between the State and civil society and how coal-extraction is being countered by civil society. The Background traces Greenpeace's activism in India before the Mahan campaign, describing how a global North-oriented environmental organisation situated its activism within a global South political–economic context. In Section 4.1, I outline the political economy of coal in India, the changing context of coal generation since neoliberalisation, and how India's recent coal growth was designed and delivered. In Section 4.2, I trace the various stages of Greenpeace's advocacy against coalmining, its rural grassroots mobilisation in Mahan, and its urban public outreach. The movement operated within the political milieu of India's 'coal rush' that was characterised by corruption, inter-ministerial clashes, and blatant corporate favouritism within the Congress-led United Progressive Alliance government (2004–2014).

In Section 4.3, I trace the crackdown on Greenpeace during the first 2 years of the Bharatiya Janata Party-led National Democratic Alliance government (2014–2016) under the Prime Ministership of Narendra Modi, and Greenpeace's fight back through legal and public forums. I summarise the Indian case study through a timeline of the politics and activism of coal at the end of Section 4.3. In the final section of this chapter, I analyse the debates by which coal was delegitimised in Indian civil society and conclude by arguing why it is critical to include such debates and discourses emerging from a global South political–economic context in a global narrative of environmentalism for climate action.

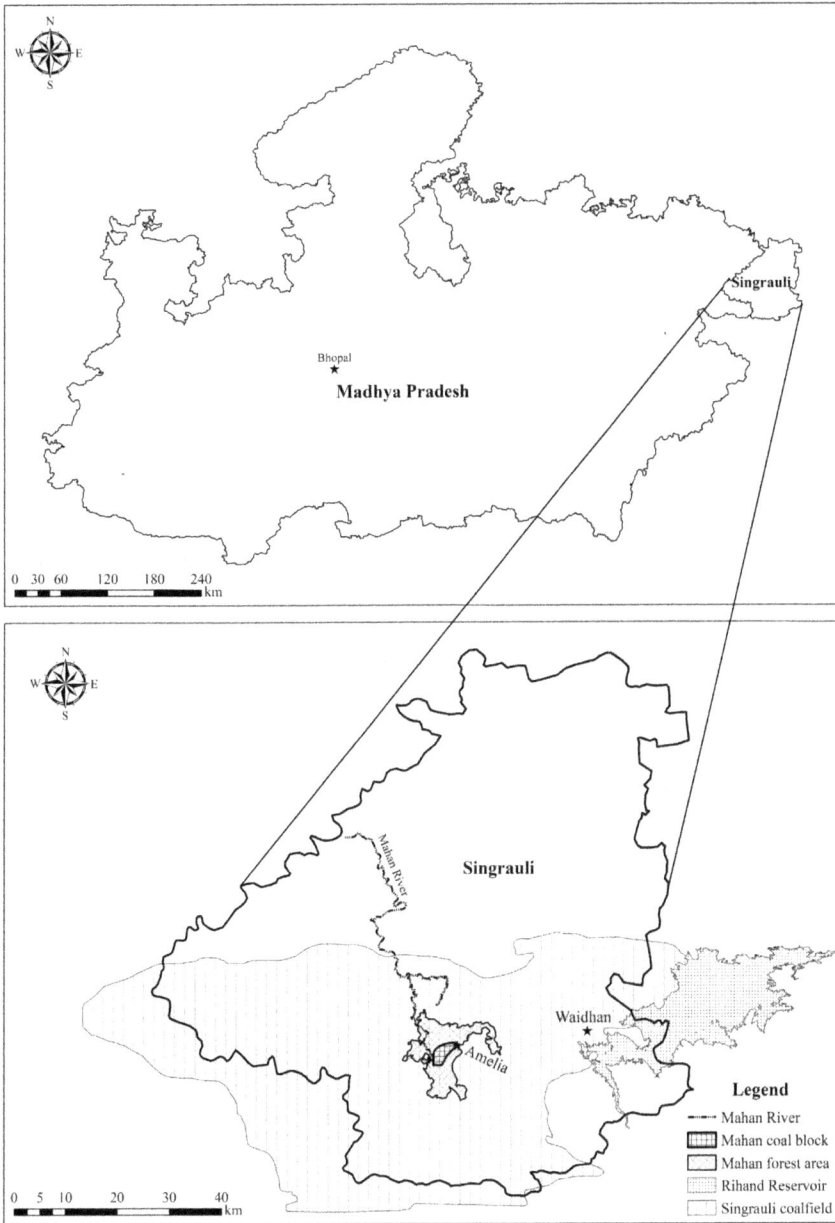

Figure 4.1 Madhya Pradesh with the Singrauli region.

Background: Greenpeace in India

Greenpeace officially registered in India in 2001, although India-based volunteers had staged anti-toxic waste dumping protests in the 1990s. One of the organisation's earliest successes in India came in 2006 when a global Greenpeace campaign forced France to recall the decommissioned warship, *Clemenceau*, bearing high levels of asbestos, from being sent to the world's largest shipbreaking yard in Gujarat (BBC News, 2006). With the Bhopal gas tragedy of 1984 still strong in public memory, Greenpeace strategically inserted itself into India's environmental debate with campaigns against toxic pollution by foreign corporations.[1]

Greenpeace's early successes tracked the international trail of hazardous substances that made third-world countries a waste-dump for the first world. But subsequent campaigns began to directly question India's developmental paradigm. Greenpeace's advocacy helped civil society to reframe debates on contentious issues such as genetically modified (GM) crops through challenging the official rhetoric of food security and scientific progress in agriculture (Mehta, 2014). Against official discourse that presented India's nuclear sector as environmentally benign and economically viable in the face of climate change, Greenpeace, along with other groups, sought accountability for risky and publicly contentious projects such as at Kudankulam in Tamil Nadu (Doshi, 2016). Protests against the Kudankulam project and the questioning of risky nuclear technology by civil society networks were met with repressive tactics by the State as I have outlined in Section 3.4 of the previous chapter.[2]

To align its global climate and energy advocacy within the context of India, a developing country, Greenpeace first set out to establish credibility in the space of India's energy generation and development by focussing on the issue of energy efficiency as a cost-saving measure. Greenpeace India's 'Ban the Bulb' campaign instigated a consumer focussed government subsidy scheme to replace inefficient light globes with compact fluorescents (Commonfloor, 2011). Greenpeace usually frames its campaigns on moral grounds (Doherty and Doyle, 2006), but establishing the moral case against ending coal usage to tackle climate change proved challenging in the Indian context of the need for development, which was anchored in the aim to provide electricity (through coal-based power, India's most abundant source of energy) to all.

In the early period of international climate negotiations, a position by an international group like Greenpeace on India needing to reduce its coal-based emissions would likely not have found sympathy from many Indian civil society groups who stood by the principle of common but differentiated responsibility (CBDR), asking the global North to act first and do more for the climate, than the global South. Therefore, the Greenpeace Report *Hiding Behind the Poor* (Ananthapadmanabhan et al., 2007) attempted to redirect civil society's gaze on climate injustice away from the global North–South divide and toward India's internal economic disparity. The research found that:

> The considerably significant carbon footprint of a relatively small wealthy class (1% of the Indian population) is camouflaged by the 823 million poor who keep the overall per capita emissions below 2 tonnes of CO_2 per year.

(p. 2)

Figure 4.2 Greenpeace nuclear liability bill campaign image.

Source: Photo by Amit Madheshia/Greenpeace.

In a similar vein to environmentalism of the poor, the *Hiding Behind the Poor* report asked for intra-generational equity for the Indian poor, asking India to commit to climate-action for justice toward the Indian poor who are bearing a disproportionate burden of climate change. The report was discredited by the government and proved controversial amongst some civil society groups as anticipated (Subramanian, 2015). Nevertheless, it sparked a debate about the need for domestic climate accountability in India. As in the case of other environmental issues in the Indian context, the case for reducing coal usage gained legitimacy amongst civil society organisations through the argument of social justice for the poor.

4.1 Political economy of coal in India

The plans and policies of India's coalmining and power sectors underwent significant changes between 2004 and 2015, and contrasted with India's historic approach towards 'coal nationalism' that I describe under the first subsection here. This 11-year period coincided with the timelines of India's 10th (2002–2007), 11th (2007–2012), and 12th (2012–2017) Five-Year Plans that aimed for an ambitious growth rate of 8% of the GDP (Planning Commission, 2013).[3] Politically, this period covered two terms of the Congress-led United Progressive Alliance government under Prime Minister Manmohan Singh and the first 2 years of the first term of the

Bharatiya Janata Party-led National Democratic Alliance government under Prime Minister Narendra Modi.

India's coal-fired growth engine was overhauled during this period through privatisation and massive increases in coal production and thermal power generation, engineering what can be called India's 'coal rush'. This period also witnessed government corruption in the allocation of coal blocks to private corporations, clearly indicating favouritism in India's state-corporate nexus. The policies and politics, corruption and favouritism that characterised India's coal rush between 2004 and 2015, brought out deep and persistent contradictions between the stated moral intent – that of poverty eradication and electricity for all Indians – and the effect – windfall gains by private corporations – of coal-led economic growth in neoliberal India.

Coal Nationalism

The approach of resource extraction as a source of State wealth began in the colonial era. This economic pathway was intensified in post-colonial India by making minerals, particularly coal, central to economic development. The development of minerals was prioritised to service the nine major public sector undertakings (PSUs) that produced iron, steel, and heavy electrical equipment, materials that were vital for a newly industrialising nation. As India's most abundant major fuel, coal formed the resource-backbone for post-independence development to build and strengthen public sector industries. India's first Five-Year Plan identified increasing coal production and efficiency as a key requirement for a newly independent nation (Lahiri-Dutt, 2014).

Economic development in turn was deemed crucial for alleviating poverty. State discourse linked industrial development in India to the moral imperative of eradicating poverty through providing electricity to the masses. Electricity use is strongly linked to overall development and the improvement of specific indicators such as child mortality and female life expectancy in the Human Development Index (Ghosh, 2016). Over 300 million Indians lack access to electricity according to India's draft National Energy Policy 2018 (Niti Aayog, 2018a).[4] Coal was deemed synonymous with the public interest owing to this imperative.

India's coal reserves are estimated at 276.81 billion to a depth of 1200 meters, concentrated in coal-bearing areas (CBAs) covering large parts of the states of Jharkhand, Orissa, Chhattisgarh, West Bengal, Madhya Pradesh, Andhra Pradesh and Maharashtra. Coal reserves are distributed across 27 major coalfields (Ministry of Coal (MoC) website, https://coal.nic.in/, n.d.).

Through a framework of laws and policies that gave coal legal eminence, India built a 'national coal economy' (Lahiri-Dutt, 2016, p. 204). Policies and public companies kept the price of coal low with the principle aim of electrifying the nation and power sectors that were critical for growth (Gopal, 2016). The *Mines and Minerals (Development and Regulation) Act 1957* (MMDR) reserved coal and lignite (brown coal) exclusively for the public sector by categorising them as major minerals. The *Coal Bearing Areas (Acquisition and Development) Act 1957*

(CBAA) gave coal greater priority over other legislative land-uses (including the inalienable land-rights of Adivasis in designated Scheduled Areas) to 'establish greater public control over the coal mining industry and its development' (CBAA, 1957, p. 1). Together, the CBAA and the (now repealed) *Land Acquisition Act 1894* (LAA) vested ultimate power in the State to acquire any land for coalmining (Lahiri-Dutt, 2016). In this manner, India's coal nationalism essentially undercut the intent to protect the lands of Adivasis as stated in the Constitution.

The *Coal Mines Nationalisation Act 1973* (CMNA) consolidated the vision of the MMDR by bringing coal mining more systematically within the purview of the public sector and effectively making coal identical to the nation State. The Act stated that:

> No person other than the central government or a government company or a corporation owned, managed or operated by the central government shall carry on coal mining in India, in any form.
>
> (CMNA, 1973, section 3)

Coal mining and electricity production remained solely as State preserves till the 1980s. After the nationalisation of coal mines in 1973, mining was driven through the State-controlled Coal India Limited (CIL) under the MoC that was established in 1975. With seven fully owned coal-producing subsidiaries, CIL is the world's largest coal producer (Lahiri-Dutt et al., 2012). While CIL had a complete monopoly over coal mining till the 1980s, after the liberalisation of the energy sector it still controlled 81% of India's coal production (Lahiri-Dutt, 2016). Electricity generation was largely controlled by the State-run National Thermal Power Corporation (now NTPC Ltd.), India's largest power company that was set up in 1975.

The overlapping layers of coal, forests, and high concentration of Adivasi groups in India's coal-bearing areas made coalmining the chief agent of disruption for forest-dependent communities. A major proportion of such people were not absorbed into the new coal economy. In a poignant account of the unmaking of traditional livelihoods by coalmining, Geographer Kuntala Lahiri-Dutt writes in *The Diverse Worlds of Coal in India: Energising the Nation, Energising Livelihoods* (2016) that with their livelihoods and land gone, Adivasi communities were compelled to subsist by collecting coal 'in lieu' of forest products, creating the 'subsistence coal economy' in India (2016, p. 204).

Central India's coal-rich regions formed the centrepiece of India's coal-led development. The Singrauli region that stretches across the northeastern part of the central Indian state of Madhya Pradesh is called the energy capital on account of producing 10% of the country's thermal power. It is one of South Asia's biggest industrial areas and contains some of India's oldest State-owned thermal power plants and coal mines.[5] The 1960s marked the beginning of Singrauli's saga of displacements for large hydroelectric projects followed by thermal power.[6]

On the basis of a $150 million loan from the World Bank, the State-owned National Thermal Power Corporation set up the Singrauli Super Thermal Power Project (SSTPP) in the region in the early 1980s (Clark, 2003). This was followed by the Vindhyachal Super Thermal Power Project (VSTPP) and the Rihand Super Thermal Power Project (RSTPP). Vindhyachal is India's largest thermal power plant with a current generating capacity of 4760 megawatts (see www.ntpc.co.in/power-generation/). The Uttar Pradesh State Electricity Board (UPSEB) set up its power plant at Anpara, also in Singrauli. The Northern Coalfields Limited (NCL), one of the seven coal-producing subsidiaries of CIL, started nine open coal mines to supply fuel to the power projects. Nigahi, part of NCL's mine cluster in Singrauli, is one of India's largest open-cast coal mines, providing fuel for India's largest thermal power plant (www.ntpc.co.in).

Singrauli's ecology and society bore the brunt of this intensive development. It told a story over decades of human displacements, environmental and social neglect, and underdevelopment (Singh, 2017). The region stands out amongst other displacement-affected landscapes from mega projects in India, both in terms of the intensity and frequency of land acquisition and consequent displacements over five decades (Sharma and Singh, 2009). Adivasis bore the biggest brunt of Singrauli's industrial development, comprising nearly half the number of displaced people from large-scale industrial projects from the 1960s onwards (Pillai et al., 2011).

Coalmining has acted as a double-edged sword in the coalmining regions in major coal-producing states. While it has brought in incomes from work for the coal mines, due to the dominance of coal in the economy, these regions have experienced multidimensional poverty due to the lack of diversity of trades and the inability to absorb Adivasis displaced by coalmining into the new coal economy (Oxford Poverty and Human Development Initiative, 2018). Being strongly dependent on coalmining revenues also makes it challenging for these regions to transition away from coalmining (Lahiri-Dutt, 2016).

'Neoliberal coal'

Policy changes to boost the production of both coal and electricity after economic liberalisation led to the entry of private players. The *National Mineral Policy 1993* (NMP) was announced to encourage private and foreign direct investment and attract state-of-the-art-technology to India's mining sector. In the same year, through an amendment to the CMNA, State-run corporations and private companies were allowed to take captive coal mines for power generation, washing coal, or any other end use notified by the Indian government.[7]

Following this amendment, the World Bank offered India a $20 million loan for technical assistance in the negotiation of purchase agreements and privatisation of power projects (Marston, 2011). But despite legislative changes, few allocations were made to private players till 2004 (Gopal, 2016). The pace of privatisation of power generation remained slow in the first decade of liberalisation, with the bulk of India's thermal power capacity still being produced by NTPC Ltd. (Rosewarne, 2016). As for coal mining, CIL remained the sole entity to commercially mine the fuel.

Since 2004 the Indian government rapidly allocated 194 coal blocks either directly to private corporations or State-private partnerships for captive-coal mining to generate power. Where coal blocks were allocated to State-enterprises, private companies were contracted to undertake mining operations and other onsite technical projects (Gopal, 2016).[8] Policy hurdles were removed to allow foreign-owned engineering companies to enter into coal handling and processing, further demonstrating the prominence given to the private sector in increasing India's coal production (Rosewarne, 2016).

Structural changes within the electricity sector paved the way for the entry of Indian corporate giants – Adani, Reliance Power, Tata Power and Essar Power – to generate electricity and sell it to the national grid. The *Indian Electricity Act 2003* (IEA) transformed the power sector that had been fully state-controlled till 1991. It introduced competition and choice and was passed with the intent of providing complete commercial autonomy to buy and sell power (Thakur et al., 2005). Privatisation and the entry of foreign direct investments in power generation in India led to the setting up of super-sized thermal power plants, often with foreign companies as primary promoters (Ahmed, 2010).

Private companies began to profit through the corporate-owned-and-operated model of mining and electricity generation on account of the constantly rising demand for electricity, producing what Kuntala Lahiri-Dutt calls 'neoliberal coal' (2016, p. 205). Lahiri-Dutt argues that privatisation exposed the contradiction in the political economy of coal in India in two significant ways. As government rhetoric continued to equate coal to development even after economic liberalisation, the Indian State now began to define private corporations as indispensable to India's energy security. And because the regulatory framework that granted legal eminence to coal in the days of complete State-control remained intact after liberalisation, private corporations became beneficiaries of the State's subsuming of Adivasi and peasant lands and livelihoods for coal and thermal power production (Lahiri-Dutt, 2016, p. 206).

Whether thermal power generation in the neoliberal era was able to meet the objective of poverty alleviation through access to electricity stands contested. Electricity distribution in India follows the same pattern as income distribution – that of high inequality. Electricity distribution through the central grid system is largely concentrated in India's urban and industrial regions (IEA, 2015). Owing both to the challenges of extending the structure over vast distances to remote locations, as well as the inability of rural populations to meet the utility's cost of extending the grid. The poverty in such regions stands in sharp contrast to those regions that rapidly flourished in the neoliberal era. In an ironical contrast indicative of spatial injustice between coalmining and thermal power producing regions and states and those that are more prosperous in India, although the former demonstrate high energy use because of industrial activities – coal extraction and thermal power generation – their domestic electricity use remains low on account of poverty (IEA, 2021).

As of 2015, 50% of the population in the state of Madhya Pradesh lived below the poverty line and an equivalent proportion lives without electricity (Singh, 2015b). Singrauli in particular, although it supplies electricity to 16 states, is one of India's least developed districts according to the Indian government's own expert estimate

(Niti Aayog, 2018b). Balanced against structural and economic contradictions, the Indian government's mission of 'Power to all by 2012' that later turned into '24×7 and Power to all by 2022', through the boosting of neoliberal coal-fired electricity generation, appeared misdirected (Shahi, 2003; Waray, 2018).[9]

According to Indian government estimates, 96.7% of households had been 'electrified by 2019' under the Indian Prime Minister's household electrification scheme named *Saubhagya*. However, taking a closer look at the quality of access to electricity, particularly by rural households in India's least electrified states including Madhya Pradesh, an independent study has stated that India is yet to achieve 'access to affordable, reliable, sustainable and modern energy for all' (a subset of Sustainable Development Goal 7 of the United Nations to be achieved by 2030) (CEEW, 2020, p. 6).

Finally, despite providing raw material from their lands and bearing the brunt of the loss of lands, forests, livelihoods, and environmental pollution for coal-fired electricity, several communities in coal-bearing areas in central India still effectively remain without electricity, creating a case of both environmental and energy injustice (Bhushan and Hazra, 2008; Chhotray, 2022).

Engineering India's coal rush

Coal assumed greater significance after 2003 when the Indian government pronounced the 'Power to all by 2012' mission. The Planning Commission made a corresponding increase in coal production and power generation in the 10th and 11th Five-Year Plans (2002–2012) aiming to add 100,000 megawatts of generating capacity to the power grid by 2012.[10] From the 10th Plan onwards, India also set an annual economic growth rate of 8% of the GDP, requiring a capacity increase of at least 8000–10,000 megawatts of annual power generation to bridge the energy deficit (Tongia, 2003).

The increased coal demand for thermal power generation could not be met by CIL alone, leading to a suite of policies to allow private and other State-enterprise-based coalmining for electricity. An Expert Committee Report on India's Energy Policy in 2006 estimated coal production to expand to over 2 billion tonnes per annum to meet the energy deficit created by the policy demand. It recommended structural changes to the coal sector to meet the necessary electricity demand, by allotting coal blocks to other eligible players including other public sector and private companies for thermal power generation (Fernandes, 2012).

After 2007, there followed a rush of approvals for thermal power plants, both new and those doubling their capacities (IEA, 2014). The Indian government incentivised up to 106 mega thermal power plants to be set up with private and foreign direct investment by reducing import duties on equipment (Ministry of Finance, 2012). Many of the new approvals were for ultra-mega power projects, which were super-critical coal plants with a capacity to generate above 4000 megawatts of electricity (Rosewarne, 2016).

Finally, India's 12th Five-Year Plan (2012–2017) set a very ambitious national target of 100,000 megawatts of installed capacity, with coal as its mainstay (Pillai et al., 2011, p. 1). The energy deficit implied an increase in domestic mining as well as coal imports. The chairman of Coal India said that the complete demand could not be met from domestic sources by 2020, despite India's domestic production growing at 8–9%. The target was double of what had been achieved under the 11th Plan, and it made a deep impact on the landscape of Singrauli, India's energy capital (Saikia, 2012).

Around the time of the opening up of India's power sector, NTPC Ltd. received a new $400 million loan from the World Bank for the expansion of the Rihand and Vindhyachal power plants in Singrauli, making it the single largest borrower in the Bank's history (Clark, 2003). Singrauli also has India's largest private capital investment for super thermal power plants and coal mines (singrauli.nic.in). Between 2005 and 2015, privatisation brought massive corporate investments, estimated to the tune of Rs. 1 Lakh Crore (A$20 billion) into Singrauli (Singh, 2015b). Some of India's largest private sector energy producers such as Essar, Hindalco, Reliance, Jaypee, and Dainik Bhaskar, as well as public–private partnerships (PPPs) led by state governments, have been operating mines and super critical thermal power plants in the region (Pillai et al., 2011).

The region's borders were redrawn in 2008 evidently to facilitate private coal mining and thermal power generation in a move critics considered similar to the creation of the new states of Jharkhand and Chhattisgarh (Singh, 2015b, 2017).[11] Singrauli was expected to supply an additional 35,000 megawatts of electricity to the central grid by 2017 toward fulfilling the gargantuan ambition of India's 12th Plan (2012–2017) to add 100,000 megawatts of generation capacity. In addition to increased generation from state-owned plants, this was to be achieved through the addition of five private power projects that would add 13,000 megawatts of electricity (Singh, 2017). Singrauli now has 10 thermal power plants, five of which are State-owned and five are under private ownership. Reliance Power's mega 3960 megawatts private plant in the Sasan village in Singrauli was commissioned in 2013 despite Singrauli being designated a critically polluted area by the Ministry of Environment and Forests (MoEF; Vyawahare, 2018).[12]

In 2014, the Modi government allowed public sector utilities to mine coal commercially. Commentators regarded this as the first step towards opening the coal sector for allowing private commercial coal mining (Singh, 2015a). India sent mixed messages internationally about its ambition to double its coal production by 2020, and its intention to reduce its coal import dependency (Das, 2015; PTI, 2016). Under these scenarios of expanded coal usage laid out by subsequent governments, India's Planning Commission envisaged the dominance of coal in the energy mix to continue into the foreseeable future (Planning Commission, 2015).[13,14]

Adivasis and the environment deprioritised

The scale of policy changes to manufacture India's coal rush led to unprecedented land grabs and land use changes across forested regions. A 2012 report by Greenpeace India highlighted that the government's ambitious plans could destroy over 1 million hectares of forest, an area twice the size of India's five largest cities combined. Dense forests cover nearly three-fourths of the area for potential new coal mines. Assessing the environmental impacts of proposed coal mines across 13 major coalfields, the report said:

> From 2007 to 2011, the coalmine lease area and coal production capacity have approximately doubled compared to pre-2007 levels. Virtually all new coal mining, and most of the planned power plants are located in central India, India's largest contiguous tiger landscape...the forest areas under discussion are also a critical livelihood resource for forest-dependent populations, including Adivasi communities...If India is to continue on its current path of increasing reliance on coal for electricity, it will mean the eventual fragmentation and destruction of large areas of forest habitat, the loss of vital connecting corridors for the tiger and other species, destruction of important watersheds for peninsular India's major rivers and the displacement and further impoverishment of large numbers of forest dependent communities.
>
> (Fernandes, 2012, p. 5).

Between 2002 and 2011, the MoEF had already diverted 400,687 hectares of forestland, out of which mining and power projects contributed 38% (Fernandes, 2012). Coalmining accounted for 65% of all forestland approved for diversion for mining by the Environment Ministry (CSE, 2012). The targets for coal and thermal power generation of the 12th Plan suggested forest clearance on an unprecedented scale. Coal India's 2011–2012 report indicated that clearances still pending for 179 coal blocks and approval to divert 28,771 hectares of forestlands were hampering the government's power generation plans (Kumar and Buchar, 2012).

However, available statistics on clearances for coal mines and thermal power plants during the 11th Plan period proved such allegations to be false (Gopal, 2016).[15] Not only had very few projects been rejected, but also the Environment Ministry had approved double the coal production capacity needed during the 11th Plan period. Clearances for coal mines and power plants under the 11th Plan period stood out as the highest number of projects ever cleared in any Five-Year Plan period in India (CSE, 2012).

However, false narratives about regulatory hurdles from the industry, parts of the government, and the media, continued to dominate. A cover story in a national weekly magazine titled *Green Terror* accused the Indian Environment Minister of jinxing India's development (Kumar and Buchar, 2012). To remove hurdles in coal block allocations in forested areas, an Indian government-appointed high-level committee simplistically recommended that 'coal-bearing blocks should normally be taken up for mining' (Government of India, 2011, p. 11), discounting the costs and consequences on Adivasis and other forest-dependent communities (Gopal, 2016).

The arrival of 'a new flood of companies' during this period in Singrauli triggered a third phase of development-induced displacements (Sharma and Singh, 2009; Dokuzović, 2012).[16] Five new designated super thermal power projects required an estimated 10,000 acres of land and threatened the future of over 4000 families in the district (Sharma and Singh, 2009, p. 65, Table 3). The mining sites for the new thermal projects were mostly in the last remaining intact forests in the region. The total forest area cleared in Singrauli since the days of nationally owned mines and power stations in the 1980s stands at an estimated 4990.450 hectares, with only 18,548 hectares still remaining (Chakravartty, 2011, Table 1). By 2011, 15 mining projects by large private corporations had already begun operations in Singrauli and the region had already lost one-third of its forest cover (Pillai et al., 2011).

Measures to expedite coal and associated infrastructure approvals under the Modi government since 2014 have included a more interventionist role by the Prime Minister's Office (PMO) (PTI, 2015). The Modi government also strengthened actions by the previous government that had exempted major infrastructure projects from having to obtain the consent of communities for forest clearance. Against the backdrop of the UPA government's 'Coalgate' scam that I discuss next, the Narendra Modi government began in 2014 by promising to better manage India's coal sector. However, it has not yet lived up to this promise (Kohli and Menon, 2020).

The Modi government's performance on reforming India's coal sector falls outside the timeframe and scope of this research. But it is worth briefly noting that as compared to the Congress-led UPA government's 'Coalgate' scam that was revealed in detail through the 2012 Comptroller and Auditor General (CAG) report and ruled against through the 2014 Supreme Court judgement, the Narendra Modi-led Bharatiya Janata Party government's special favours to the Adani Enterprises Limited at costs to the Indian exchequer are only now beginning to emerge (Jalihal and Sambhav, 2023).[17] These events indicate the susceptibility of governments under both major Indian parties to special favours to corporations and expose the corruption inherent in the political economy of coal in India. In addition and finally, the Modi government's decision to auction 50 new coal blocks in 2020 undermined India's efforts to transition towards clean energy, and exposed the contradiction in the government's approach of 'talking renewables and walking coal' (Roy and Schaffartzik, 2021).

Corruption in coal block allocation

The MoC had asked for the allotment of coal blocks to be made through a process of competitive bidding to avoid excessive gains to recipients on account of the price difference between coal supplied by CIL and coal mined from captive coal blocks. Competitive bidding would have established transparency and objectivity and ensured that the state benefitted from the coal rush.

A 2012 report by the CAG of India revealed that in reality the government rapidly allocated 194 coal blocks from 2004 without any transparent assessment of financial capacity or technology (CAG, 2012). In many cases coal blocks were provided in excess of the coal required for the captive project or at higher grades. Loopholes in the allocation process made coal a lucrative business for private players. In some cases, companies had acquired coal cheaply from allocated coal blocks and then diverted them to power plants that were sold at market rates. In other cases, companies had taken coal blocks as assets they had no intention of mining. This corruption in the Congress-led United Progressive Alliance government's coal block allocations came to be known as the 'Coalgate scam'. It revealed the extent of the corporate-political nexus, with the State giving land and coal to corporations for free. It is one of the largest scams to be exposed in India to that date (Inamdar, 2013).

The CAG report attributed the real reason for coal shortages and the consequent energy deficit to corruption at the highest level within government. As a result of corruption in the allocation process, out of 86 blocks that were supposed to start production in 2010–2011, only 28 blocks had started by March 2011. Many new thermal plants had been approved without coal linkages being established, causing India's coal imports to increase by 20% to 105.8 million tonnes to cover the coal supply shortage (Gopal, 2016). The Environment Ministry pointed out how private players took advantage of the flawed process:

> MoEF in the five years till August 2011 has granted clearance to 210,000 megawatts of thermal power capacity. However, most of these projects have not been commissioned…(this) looks similar to what we are finding in the case of coal allocation. Proponents have sought and taken environmental clearance as this provides them with land and water allotment as well.
>
> (Excerpt from Environment Minister Natarajan's letter, in Mazoomdar, 2014a).

A report by the Prayas Energy Group showed that another 508,907 megawatts of energy projects – of which coal-based plants accounted for an overwhelming 84% – were at various stages in the environmental clearance cycle (Dharmadhikary and Dixit, 2011). The report found that the pipeline projects could add three times more installed capacity in the 12th Plan period than India needed under a high-efficiency and high-renewable energy pathway. The Central Electricity Authority's (CEA) National Electricity Plan (2017–2022) indicated that owing to massive additions of renewable-based capacity in this period, no more coal-based capacity additions would be required till 2022 (Jai, 2016). Under the circumstances, excess approvals created the risk of many incomplete projects turning into stranded assets that tie up large amounts of financial assets (Dharmadhikary and Dixit, 2011).

New coal projects have significant externalities such as loss of common lands and livelihoods in forested and rural areas, displacement of communities, and destruction of forests and local water systems.[18] The biggest costs of the large-scale and reckless transfers of lands and coal resources to private corporations through the coal block misallocation were borne directly by project-affected Adivasi and peasant communities, and indirectly by the exchequer and Indian taxpayers. The CAG report estimated the loss of revenue to the government, and consequent gain to private companies, at over a $200 billion (Mehdudia, 2012).[19]

4.2 Politics and resistance of the Mahan coal mine (2010–2014)

Reflecting the global shift in climate activism, Greenpeace's international climate strategy prioritised the need to stop coal mines by supporting local and community resistance. Although Greenpeace's policy advocacy in India had previously outlined pathways for phasing out coal, the new global direction lent credence to a direct approach against coalmining.

Speaking to me during an interview in the organisation's New Delhi office, Greenpeace India's Climate and Energy Campaign Manager further clarified climate activism's new approach after the Copenhagen Climate Summit in 2009, and how Greenpeace adapted the global approach to an Indian context. 'Campaign frames' changed from climate change to that of local issues, and 'keep coal in the ground' became the new principle for designing campaign action after Copenhagen. In India, this implied focussing on all the interconnected effects from coal on forests, water, and Indigenous peoples. Taking a local approach meant that Greenpeace India needed to acknowledge this complexity at the ground level. Greenpeace's new strategy in India became about talking about the various benefits of stopping coal (interview 3 April 2017).

A report by Greenpeace India *How Coal Mining is Trashing Tigerland* (Fernandes, 2012) set the context for the anti-coal campaign through a focus on coal mining's impacts in the central Indian forests, and their implications for India's international climate commitments:

> At international climate negotiations the government has put forward the role played by India's forests as a CO_2 sink, and the potential for further increasing the carbon stock in forests, in an attempt to tap the REDD+ funds. As with the forest cover statistics, estimates of ongoing carbon sequestration by Indian government agencies are at odds with those of independent researchers. Independent estimates (in 2011) indicate that the carbon stock in India's forest biomass decreased continuously from 2003 onwards, despite a slight increase in forest cover…A loss in carbon stock makes sense when looked at in conjunction with increased rates of forest land diversion (almost always natural forest) for industrial use over the last decade…mining and power projects account for bulk of (38%) of this diversion.
>
> (p. 18)

The report focussed on the bigger risk of an extensive loss of forests from new coal projects and down to the specific and critical need to preserve forest corridors that offered vital connectivity between existing wildlife reserves in central India:

> Taken together with a rapid acceleration in the pace of forest clearances for coal mining over the last decade, this implies the rapid destruction and fragmentation of large areas of forests within high priority landscapes that have been scientifically identified as crucial for sustaining tiger blocks…As fragmentation

increases, corridors take on an even greater importance. Corridors and forest patches provide 'stepping stones' and continuity between larger forests…(but) lacking the higher degree of legal protection afforded by the Wildlife Protection Act of 1972…these areas are easier to 'sacrifice'.

(Fernandes, 2012, p. 15)

A large section of the extensive Singrauli coalfields lie within two major tiger reserves that connect to sanctuaries further north through two wildlife corridors. These wildlife corridors contain intact forests running parallel to each other. Governments had allocated coal blocks both within the 10 kilometres stipulated buffer zone of the Protected Areas and in a major wildlife corridor through Dongraital–Mahan–Chatrasal–Amelia. The report recommended excluding protected areas from the purview of coal mining and not granting forest clearance to the coal blocks in the major corridors.

This focus on forests and coal in central India brought Greenpeace India face-to-face with challenges it had only tangentially or partially engaged with earlier. The first was working with local communities and grassroots movements. The Climate and Energy Campaign Manager provided insights about this challenge for Greenpeace India, and the strategic approach their climate campaign took to address it. Grassroots movements often saw Greenpeace as an international NGO that was prone to 'doing some drama', stealing the limelight, and then leaving. That had to change. After choosing Mahan as their campaign site based on a ground analysis, Greenpeace had to decide what they could leave the community with. The climate campaign team agreed that community forest rights (CFRs) should be in place for the villages around the coal block. And that CFR education would be their entry point to talk about coalmining in the local forest-village communities (interview 03 April 2017).

The other challenge was negotiating the vortex of India's coal politics during the two terms of the Congress-led United Progressive Alliance government (2004–2009, 2009–2014). The Mahan coal block became the bone of contention between the Environment and Coal Ministries and emblematic of India's environment versus development debate. Framed and pushed by strong corporate interests that influenced the highest levels of the Indian government, this debate leaned heavily on the narrative of development, even though the gaps between the proposed model of India's coal rush and its benefits for communities had clearly begun to show. In the face of egregious administrative mismanagement of coal resources, the development mantra was operationalised by parts of the Indian State in nexus with private corporations to legitimise painting the processes of environmental approvals, community consent, and consultation for diversions of forestlands as blockages.

The following five subsections trace the build up of this coal politics between 2006 and 2014, the shaping of the Greenpeace anti-coal movement along its contours, the extent of corporate influence in government decision-making and attempts to silence the environmental NGO and the eventual Supreme Court judgment that called for the allocation of the Mahan coal block to be cancelled.

Tug of war between Environment and Coal Ministries over Mahan

In 2006, the Indian Ministry of Coal allocated the Mahan coal block in the Singrauli Coalfields in northeastern Madhya Pradesh to a private corporation for captive coal mining. Mahan Coal Ltd, a Rs. 5000 crore (A$1billion) joint venture between the London Stock Exchange-listed Essar Power Ltd. and the Indian aluminium manufacturing company Hindalco Ltd. (Kohli, 2012a). Essar Power is one of India's largest private thermal power producers with plants in three states. It is part of the Essar Group, a multinational conglomerate with investments across steel, infrastructure and energy in more than 29 countries across five continents (Ramanathan, 2015). The Essar conglomerate is also highly leveraged, with a large debt mostly owed to state-owned banks in India (Kaushik, 2015). Hindalco, the biggest producer of primary aluminium in Asia, is a subsidiary of the Aditya Birla Industrial Group, also a multinational conglomerate with operations in 34 countries across diverse portfolios ranging from aluminium, cement and telecom to insurance and lifestyle (www.hindalco.com).

The proposed Mahan coal block was to supply Essar's 1000 megawatts and Hindalco's 650 megawatts captive power plants, respectively. It was expected to produce 8.5 million tonnes of coal per year through a 1000-hectare open-cut coal mine over a lifetime of 14 years (Fernandes, 2012). It was granted environmental clearance in 2008 by the Environment Ministry (Chakravartty, 2011). But forest clearance for the coal block, a separate process, proved contentious.

In the developed northern parts of the Singrauli coalfield forest cover had already been fragmented by coalmining. But Mahan was located in the coalfield's southern part amidst undulating hills covered by thick deciduous forests. Mahan is one of Asia's oldest contiguous Sal tree forests and supports the livelihood needs of largely Adivasi communities (Kohli, 2013). The Mahan forests also serve as a prominent tiger habitat and elephant corridor (Chakravartty, 2011). The coal block also fell within the catchment of two major perennial rivers. A 1000-hectare open-cut coal mine at Mahan would have cleared an estimated 500,000 trees, destabilised the watershed of the Rihand reservoir and jeopardised the livelihoods of 50,000 people across 54 villages (Padel, 2016).

Three years after the allocation of the coal block at Mahan and a year after its environmental clearance, the MoEF and the MoC established a joint criterion to designate which forest zones across nine major coalfields in central India were to be deemed 'inviolate' on account of their outstanding biodiversity. Inviolate forests would be exempt from coalmining.[20] In 2010, the joint process designated Mahan and six other coal blocks in the Singrauli coalfield as 'no-go' zones for coalmining in acknowledgment of the high quality of forests there. The 'go-no-go' exercise was part of a six-point agenda prepared by the two ministries to remove delays from the clearance process for coalmining while keeping environmental interests in mind (Kohli et al., 2012).

But once the 'go-no-go' exercise was completed, upon seeing the extent of coal blocks rendered inviolate, both CIL and the Coal Ministry began stepping back from the idea. The initial exercise identified 396 coal blocks as go areas and the rest 206 as no-go; the no-go coal blocks had a combined production potential of 660 million tonnes of coal per year. Within a year, following interventions from the Prime Minister's Office, and objections from the Coal, Power and Steel Ministries, the Environment Ministry had appended 70 no-go coal blocks to the go list (Down to Earth, 2015). It has been pointed out that the MoEF's time frame for lifting Singrauli's moratorium on industrial development – only a year after imposing it – coincided with several coal blocks in the region initially deemed as no-go areas being re-categorised as go areas.[21]

The Coal Ministry moved a cabinet note to make it mandatory for the Environment Ministry to approve every allocated coal block without taking into account the effects on forests, wildlife and the environment (Ramesh, 2015). Publicly, the Environment Minister spoke of a compromise and not the abandonment of the go/no-go principle for the government to increase its coal production, particularly in light of nuclear power becoming an area of concern and hydropower facing public opposition owing to its displacement issues (Press Trust of India (PTI, 2011).

Environment Minister Jairam Ramesh confessed to constant pressure from the MoC and the PMO, as well as demands from the Essar and Aditya Birla chief executives, for granting forest clearance to the Mahan coal block. In March 2010, the chairperson of the Essar Group Shashi Ruia wrote to Prime Minister Manmohan Singh that 65% of the Essar Thermal Power Plant near Mahan had already been constructed. The correspondence emphasised that project delays would result in 'a huge loss to us as well as the country' (Chakravartty, 2011, para 3).[22] Meetings between the Chief Executives and the Environment Minister followed in May 2010. Essar was also pressuring the Indian government to clear two more coal blocks in a thickly forested no-go area in the state of Jharkhand, because 'the projects are in an advanced stage of execution' (para 5).

Responding to the Prime Minister regarding Essar's concerns, the Environment Minister objected to 'fait accompli' arguments for approvals, asking why power plants were being built pre-emptively before securing forest clearances, creating a situation and compelling Ministries to push 'done deals' (Ramesh, 2015). The Environment Minister was concerned that approving Mahan 'will open a Pandora's box which we should avoid at all costs' (MoEF, 2011, p. 1).

In the interest of clearing Essar and Hindalco's coal block, an inter-ministerial exercise conducted by the Prime Minister agreed on six points for allowing coal blocks in no-go areas to be cleared: whether land and water allocations had been secured, environmental clearance obtained, orders for plant equipment placed, work at a site begun, substantial expenditure made or committed, and, finally, whether the plant was likely to be commissioned in under 3 years (Kaushik, 2015). The extent of Essar's involvement in steering the government toward clearing Mahan was exposed in 2014 through a series of emails leaked by a company employee turned whistle-blower.[23] These emails formed the basis for a Public Interest Litigation against Essar in the Supreme Court (Suresh and Sarin, 2015).

The Environment Minister still declined forest clearance for Mahan. On the other hand, based on these revisions, the Environment Ministry granted the first state of forest clearances to coal blocks in the vicinity of Mahan that had also initially been placed on the no-go list.[24] It is mandatory under the Forest Conservation Act (FCA) 1980 to obtain a two-stage forest clearance for mining and other major projects on forestland. The Environment Minister justified the revision by clarifying that these coal blocks were either situated in already-fragmented forests in the already-developed northern part of the Singrauli coalfield, or their mining involved felling fewer trees. They were also situated outside the periphery of the undisturbed 20,000 hectares of ancient Sal forests within which the Mahan coal block fell. The MoEF had not issued any clearances within this area (Ramesh, 2015).

Fait accompli approval

The Environment Ministry's Forest Advisory Committee (FAC), a statutory body set up under the Forest Conservation Act (FCA) 1980, had found on a visit to Mahan that the quality of its forests was far superior to what project proponents and the state government of Madhya Pradesh had disclosed (Sethi, 2015). The Environment Minister said in a statement that the MoEF had rejected forest clearance for the Mahan coal block based on the exceptional quality of the local forests (MoEF, 2011).

The site visit by the FAC also highlighted other risks that project proponents had not reported. The Mahan coal block was located on a hilly terrain where mining would produce excessive mining overburden. Storing the overburden on the hill slopes would prove hazardous; it would risk silting the Rihand Reservoir and the two rivers on which the region's farming and industrial sectors depend (Chakravartty, 2011). The MoEF recommended the Sohagpur coalfield in the neighbouring Shahdol district where 70% of the coal blocks were on the 'go' list as an alternative fuel source for the Essar power plant (Ramesh, 2015, p. 33).

Correspondence released by the Environment Minister showed that the project proponents approached the Indian Finance Minister asking for a quick decision on Mahan. In the meantime, the chief minister of Madhya Pradesh added to the chorus for Mahan's development by announcing an indefinite fast to protest the central government's continued discrimination toward a 'backward' state by delaying the coal block approval, even though two companies had invested millions of dollars in the Singrauli region (Chakravartty, 2011).

In 2011, the matter of the approval of the Mahan coal block was taken out of the hands of the MoEF and placed with a high-level Empowered Group of Ministries (EGoM) headed by the Finance Minister. The EGoM's mandate was to suggest solutions for regulatory hurdles to mining and industrial projects. The Environment Minister issued a statement on the MoEF website stating the Ministry's position on the controversy surrounding the Mahan coal block approval:

The power plants do not have the redeeming feature of being super critical units that generate 5–8% lesser emissions of carbon dioxide. By Essar's own admission, the Mahan coal block will meet the coal requirements of the two 600 megawatts units for fourteen years only. There is no coal linkage for the

Figure 4.3 Mahan Sangharsh Samiti members walking across Mahan River and forest.

Source: Photo by Reuters.

balance of its life, which could extend for another ten–fifteen years at least. I am not entirely clear why such a good quality forest area should be broken up for such a partial requirement. A third 600 megawatts unit is planned as part of Phase II of the project for which the coal linkage has yet to be firmed up. I am unable to agree to consider the Mahan coal block for State 1 clearance. The Mahan coal block is therefore being submitted to the GoM for its consideration with a recommendation that an alternative coal linkage be provided for the two power plants.

(MoEF, 2011, p. 4)

The EGoM formed a high-level committee to suggest clearance measures for coal blocks in no-go areas. The committee's report declared the go/no-go exercise as uncomprehensive and without legal basis as it mapped only 37.5% of India's CBAs. The report concluded that the go/no-go concept should not be applied while giving forest clearance to coal projects (Chakravartty, 2011). Minutes of an EGoM meeting in May 2011 indicated an intergovernmental tussle between Ministries on the issue of the Mahan coal block; while the Environment Minister argued against its approval, the Power and Home (internal affairs) Ministers pressed for clearance (Kohli, 2013).

The next day's media reports announced that the EGoM had approved coal mining in Mahan (PTI, 2012). The decision demonstrated how easily the government's empowered approval bodies undermined expert and scientific evidence and the position of one of its most crucial Ministries. Following this decision, Stage 1 Forest Clearance was granted to Mahan in October 2011, based on 36 conditions including the completion of a range of studies and compliance with the processes under the Forest Rights Act (Greenpeace India, 2014j).

The new Environment Minister Jayanthi Natarajan who had replaced Jairam Ramesh by then reluctantly accepted the decision, noting that:

Despite reservations against the diversion of the dense forest land expressed strongly by the environment ministry at the GoM, and the fact that the entire civil work and construction of the plant is already complete after procurement of environmental clearance – and resulting inter alia in huge exposure to nationalised banks – Forest Clearance may be granted to the Mahan coal block..

(Quoted in Sethi, 2015)

The Environment Minister's concerns on the approval process were expressed in notes acquired through Right to Information:

Needless to say, it is crucial to avoid such classic fait accompli situations in the future in order to preserve the integrity of our forests.

(Quoted in Kohli, 2013)

Leaked emails exposed the underhand means through which Essar had secured the Stage 1 clearance for Mahan from the EGoM, including personal favours by providing employment to close associates and relatives of the finance minister (Kaushik, 2015).[25]

The tussle between the economic and environmental priorities within the Indian government was further increased after the National Investment Board (NIB) was established in 2012, to clear large infrastructure projects above Rs. 1000 crore that were being held back by concerns surrounding land acquisition and environmental and forest clearance processes. The concept of the NIB was formally established through the setting up of the Cabinet Committee on Investments (CCI), a standing committee with overarching powers that could bypass the approvals process of the MoEF (Kohli, 2012b).

With the granting of a conditional, first-stage forest clearance for the contested coal block, the setting for the playing out of Mahan's environment versus development conflict shifted from the upper echelons of power in India located in the national capital New Delhi to the villages surrounding the proposed coal mine in the Mahan forests. In collaboration with Greenpeace, a local grassroots mobilisation came together under the banner of the Mahan Sangharsh Samiti (MSS) or the Mahan Resistance Front in 2012 (Pillai, 2017). From the nexus between the central government's Ministers and the Essar and Hindalco CEOs, the focus now moved to the State–corporate nexus in Mahan, and how this affected the daily lives of local communities, particularly the activities of MSS. Finally, the spotlight now shone on the politics of the Forest Rights Act, its adoption by the communities, and its abuse by the State–corporate nexus.

Greenpeace and Mahan Sangharsh Samiti

Greenpeace used the Forest Rights Act as an organising tool to engage people in the 11 villages surrounding the Mahan forests. The community started coming together to express their concerns about the potential harm to their livelihoods. They asserted what the law under the FRA stipulates – that forest clearances should not be granted till the process of recognising their individual forest rights (IFRs) and community forest rights (CFRs) had been completed (Kohli et al., 2012).

The people of Mahan asserted their legal right to decide on the matter of mining in their local forests. The local administration and local company officials colluded to disrupt the process stipulated under the FRA. In 2012, company-hired-goons disrupted village council proceedings in Amelia, the largest village in Mahan, to stop people from registering their CFRs under the FRA (*Economic Times*, 2013). In March 2013, a village council meeting in Amelia to determine people's consent for mining was followed by a large-scale forgery of signatures that was orchestrated by the local administration and agents of the company. The incident of forgery was revealed much later through documents obtained by Greenpeace under the right to information (Pioneer, 2014).

Greenpeace and MSS took the matter to the Minister for Tribal Affairs in the central government; the Minister directed the state government of Madhya Pradesh to conduct new village council meetings to determine people's consent and expressed

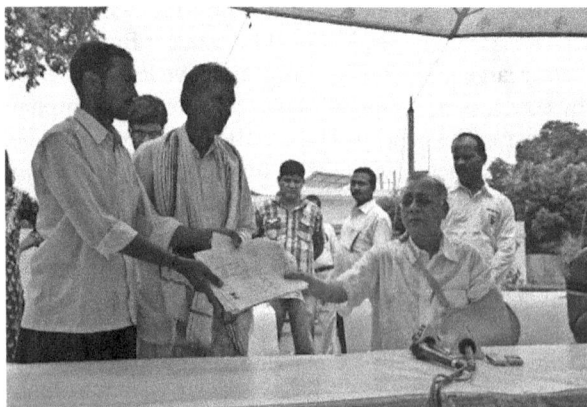

Figure 4.4 MSS members hand over documents showing the forgery of signatures during the mining referendum in Mahan to India's Tribal Affairs Minister.

Source: Photo by Greenpeace.

concerns about the large-scale violations (Sethi, 2015). While the state government had committed fraud at the Amelia village meeting, it had altogether avoided the community consent process in all other 53 potentially affected villages around the Mahan forests, indicating the extent of violation of the FRA in coalmining approval processes (Greenpeace India, 2014j).

Even as the Tribal Affairs Minister asked for a new determination of people's consent for mining in Mahan, a new environment minister, Veerappa Moily, issued a final forest clearance for the Mahan coal block on the basis of the fraudulent village council resolution. The conflict between the Tribal Affairs and Environment Ministries in the Indian government reflected yet another dimension of the inter-governmental conflict over Mahan. The Environment Minister Veerappa Moily's special favours towards the Essar Corporation stood exposed when he allocated an oilfield to the energy conglomerate at 'dirt cheap' prices around the same time period of issuing the forest clearance for Mahan (PTI, 2014).[26]

MSS filed a petition in the National Green Tribunal (NGT) challenging the validity of the final forest clearance for Mahan. The petitioners stated that forestlands had been diverted without any scrutiny, the findings of the Forest Advisory Committee had not been taken into consideration, and the coal mine had been approved without undertaking a comprehensive wildlife study and cumulative impact assessment for the region.[27] Essar Energy was already on the NGT's radar due to environmental hazards from the collapse of a mud wall of the fly-ash-dyke at the company's power plant, near the villages of Mahan (Greenpeace India, 2014f). The Madhya Pradesh government responded to the NGT with an undertaking not to fell trees in the Mahan forests till October 2014.

However, the company had already begun preparations to log the Mahan forests. The local movement responded by forming human chains around the trees and sitting in the forests from dawn to dusk. Women led the blockades in a manner common to people's movements in India. Women are likely to be the most affected by the loss of forest produce and firewood for domestic use, and the loss of common lands for cattle grazing. They are also disproportionately impacted by coalmining-induced displacements and rehabilitation (Ahmad and Lahiri-Dutt, 2006). The administration and the police sided with the company. The police raided the Greenpeace office located in Waidhan, the administrative headquarters of the Singrauli district, and arrested Greenpeace and MSS members on the basis of false allegations (Greenpeace India, 2014h). In the face of these setbacks for MSS and Greenpeace, the NGT's order and the state government's undertaking brought temporary relief to one of the most fraught phases of the movement.

Targeting Essar, getting 'SLAPP'ed

The Essar Group is regarded as better at 'environmental management', a code word for lobbying by corporations (of politicians, bureaucrats, and senior journalists) than financial management. Essar is known to have extended special treatment across party lines (Kaushik, 2015; Suresh and Sarin, 2015).[28] By challenging the Mahan coal mine, Greenpeace had challenged the State-corporate nexus in India, formed of deeply embedded personal gains and favours and business interests that extended to the highest levels of the government. As a result, it became a target of both the State and the corporation.

The Greenpeace report *Trashing Tigerland* had named Essar Power and Hindalco, amongst others, as 'companies involved in forest destruction' in central India. Side by side with stepping up the community mobilisations in Mahan, Greenpeace also started an urban-focussed public campaign to expose Essar's role in destroying central India's old forests and tiger habitats.

The tiger character Sheroo was created as the face of Greenpeace's urban outreach campaign titled 'Junglistan' ('forest kingdom'). The campaign raised awareness across major cities about the risks of coalmining in central India, and generated public support for a moratorium on coal block allocations in forests (Greenpeace India, 2011). Greenpeace volunteers appeared at forums attended by Essar CEO Shashi Ruia and Environment Minister Verappa Moily dressed in tiger suits, aiming to 'embarrass' and highlight the State–corporate entanglement (Greenpeace India, 2014d). A Greenpeace activist spent a month in a tree in the central Indian forests to promote the campaign issue through social media (Lakshmi, 2013).

In January 2014, Greenpeace pulled a characteristic 'stunt action' by unfurling a 72-foot-long and 36-foot wide banner with the message 'We Kill Forests: Essar' across the face of Essar Energy's corporate headquarters building in Mumbai. The banner also contained images of Prime Minister Manmohan Singh and Environment Minister Moily (Greenpeace India, 2014a). Greenpeace volunteers and trained climbers approached the building management disguised as a cleaning team for a fake company, climbed to the roof, and hung down with the banner. The banner and the message were visible for kilometres around (Kaushik, 2015).

Twenty-seven MSS members, men and women, had travelled from Mahan to participate in the Greenpeace action, with many of them travelling out of their regions for the very first time. A total of 67 protestors that included Greenpeace staff members, volunteers and the MSS, were arrested after the action (Greenpeace India, 2014c). The trial for the charges pressed by Essar continued in the Mumbai High Court for 6 years. Essar also pressed separate defamation charges against Greenpeace for Rs. 500 crores (A$100 million), alleging that the ENGO had displayed false and malicious content on the banner and in pamphlets handed out at the protest (Shrivasava, 2014).[29]

Reacting to a Greenpeace campaign video on the Mahan forests that had been running in cinemas across India, Essar sought a blanket ban on any Greenpeace criticism of the project (Greenpeace India, 2014g). It appealed for the removal of all content related to the company from Greenpeace's website and campaign materials such as posters, leaflets and pamphlets. Essar justified its targetting of Greenpeace's activism on the basis of the IB reports that had accused Greenpeace of slowing down India's development (Greenpeace India, 2014f). The Mumbai High Court even accepted the reports as critical evidence of Greenpeace's malicious

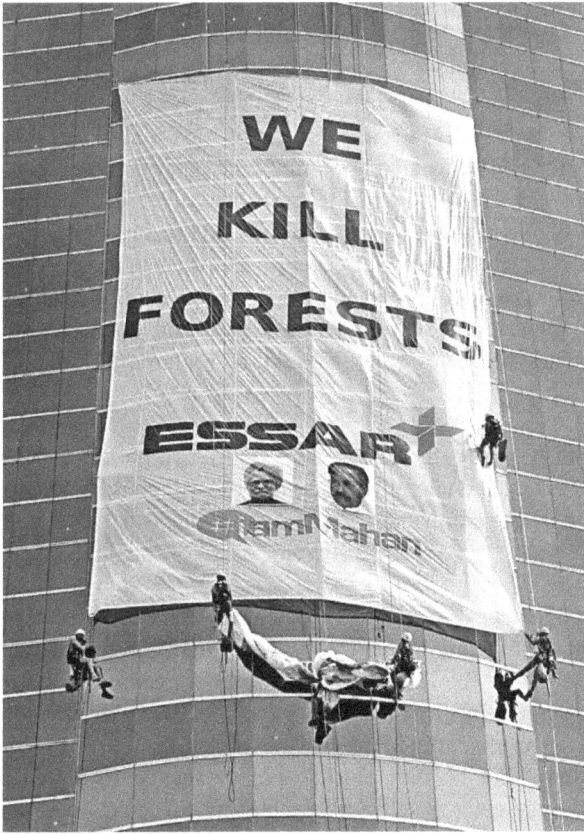

Figure 4.5 Greenpeace activists hang a banner on the Essar Headquarters in Mumbai.

Source: Photo by Sandeep Mahankal/IANS.

intent towards the corporation in the high-stakes defamation case (Greenpeace India, 2014b).

Essar attempted to extend a 'gag and restrain' strategy towards Greenpeace's activities in Mahan, by appealing to the Singrauli District Court to prevent the ENGO from approaching within a 100 meters of the thermal power plant or the resettlement colony (of peoples displaced by the plant). The District Court however rejected Essar's SLAPP (Strategic Lawsuit against Public Participation) suit against Greenpeace (Greenpeace India, 2015a).

De-allocation of the Mahan coal block

For Greenpeace and Mahan, the fight to stop mining took a fortuitous turn after the Supreme Court of India ruled against the Coalgate scam, and a new Indian government under Narendra Modi subsequently de-allocated Mahan (and other) coal blocks from the auction list statedly on account of the 'inviolate' quality of their forests.

In September 2014, the Supreme Court of India delivered a judgement on the Congress-led United Progressive Alliance government's coal block allocation scam, ordering the cancellation of 214 (out of a total of 218) coal blocks that had been 'illegally' and 'arbitrarily" allocated to private companies between 1993 and 2011 without following the due process of competitive auctioning (Rajagopal, 2014). The judgement highlighted the findings of the CAG report, that instead of auctioning coal blocks to corporations and generating revenues, the government had issued several leases to companies owned by or connected to politicians. Even when coal blocks were allocated to State-enterprises, politically connected private companies still emerged as winners by entering into joint ventures (Nileena, 2018). The ruling said:

> This modus operandi has virtually defeated the legislative policy in the Coal Mines Nationalisation Act, and winning and mining of coalmines has resultantly gone in the hands of private companies.
>
> (Supreme Court of India, 2014, p. 159)

The Supreme Court judgement provided an opportunity to fix the coal scam by reforming legislation and prosecuting violators (Nileena, 2018). The newly elected BJP-led National Democratic Alliance government under Narendra Modi brought in a new legislation, the *Coal Mines (Special Provisions) Act 2015,* which mandated auctions for all captive blocks offered to private companies. It still retained the power for the government to allot blocks without auction to State-owned enterprises, but banned State-companies from forming any new joint ventures with private firms for new coal block allocations, or from bringing them on as partners via mining-services agreements, though it permitted private firms to be brought in as contractors or subcontractors.

All extant joint venture agreements were dissolved after the Supreme Court ruling. Based on the new rules, the Modi government auctioned the first set of coal blocks to private companies in 2015. The auctions were characterised by aggressive biddings and the auction amount and royalty payable to six mineral-rich states has been estimated to be to the tune of Rs. 1 lakh crore (A$20 billion) over the next 30 years (Jai, 2015).

The risks involved in this new approach need to be understood in light of trans-formations brought to India's coal mining sector since 2015 under the Modi govern-ment, and its consequences on Adivasi and forest-dwelling communities. Although the system of auction in the 2015 Act appears as a transparent process, it essen-tially means that money will determine everything, and other social determinants including the consent of landowners, will become dispensable. India's coal mining sector has undergone a massive overhaul from the era of public sector monopoly, to allowing captive mining by private corporations for their thermal plants, to the most recent policy changes by the Modi government in 2018 allowing private and foreign corporations to dig and sell Indian coal (Bharadwaj, 2018).[30]

The Narendra Modi government's actions since 2016 to transform the coal sector stand outside the timeline and scope of this research. But it is worth noting that, to a large extent, promises of reforming India's coal sector through transparency in coal block allocations and making commercial auctions profitable for the Indian exchequer remain unmet. Investigative reportings on India's coal mine auctions by the independent Reporter's Collective show that the Narendra Modi Indian government 'gives away coal blocks to private companies through a discretionary government allotment route when only a single bidder turns up for the auctions' (Jalihal, 2023b). Through these actions, once again corporations stand to make a windfall gain at considerable costs to the Indian exchequer.

In 2022, the Indian government said forests with green cover of more than 40% would not be mined and then abandoned the decision. Yet again it showed a lack of consistency in decision-making. As a result, the Mahan coal block, at the epicentre of the previous UPA government's 'go-no-go' controversy and cancelled by the Modi government, has been reinstated in India's coal mine auction list in 2023. It has also been reported that an adjoining coal block also located in thick forests is likely to be allocated to a subsidiary of Adani Power Limited that acquired the nearby Mahan Thermal Power Plant from Essar in 2022 for underground coalmining (Jalihal, 2023b). The investigative reports conclude that with these actions by the Modi government, India has come full circle since the Supreme Court ruling in 2014 (Jalihal, 2023a, 2023b).

But in early 2015, a time frame central to the scope of this research, the circumstances were slightly different. For Greenpeace and the MSS, the 4-year struggle ended when the Modi government dropped the Mahan coal block from the first round of fresh auctions (Times News Network, 2015). The decision to keep the Mahan coal block (one of the few to be kept off the auction list), announced in March 2015, came as a big relief to the movement that was hoping the 'new government would throw the old rules out of the window' (Greenpeace India Climate Campaign Manager interview 3 April 2017).[31]

The Narendra Modi government's intervention brought the political dispute over Mahan and the immediate threat of mining to one of Asia's oldest stretches of Sal forests to a halt. Greenpeace India's Climate Campaign Manager told me during the interview that the movement in Mahan and its eventual success against coal mining stood out as 'a model campaign for Greenpeace internationally, most certainly a first-of-its-kind for Greenpeace in terms of organising at the grassroots in the global South' (interview 3 April 2017). But the government's intervention instigated another phase in the cycle of coal politics in India. This time, instead of the Mahan forests, Greenpeace would find itself at the heart of the conflict between the environment and development debate as the Narendra Modi government began to crack down on civil society organisations.

4.3 State crackdown and fight back by Greenpeace (2014–2016)

Anti-coal activism is spearheaded by US-based 'green' organisations and Greenpeace, which have formed a 'Coal Network' to take down India's 455 proposed CFPPS (520 GW) amongst 999 globally…Since 2013, through front entities, Greenpeace has initiated protests in the Singrauli region (Madhya Pradesh) which produces 15,000 megawatts (projected to double to 30,000 megawatts).

(Intelligence Bureau of India, 2014a)

A report submitted to Prime Minister Modi's office by India's primary intelligence agency the IB titled *Impact of NGOs on Development* (2014a) accused foreign-funded NGOs of 'serving as tools for foreign policy interests of Western governments' by running campaigns to support human rights and environmental issues in the country (Ranjan, 2014). Over 22 listed organisations and networks, including international NGOs Action Aid, Amnesty and Greenpeace, were alleged to have collectively brought down India's GDP by 2–3%. Activities ranging from anti-nuclear and anti-coal protests to campaigns against GM food, agitations against hydroelectric projects and mega industrial projects in the eastern state of Odisha were labelled as anti-national.

Although its main concern was cited as the misuse of foreign funds by NGOs in violation of India's *Foreign Contribution (Regulation) Act 2010* (FCRA), the report failed to demonstrate how organisations actually violated the law (Sarma, 2014).[32] The allegations appeared motivated by Narendra Modi's criticism of foreign NGOs and overlooked genuine evidence of financial violations in India's voluntary sector (Mazoomdar, 2014b).[33]

The IB reports made an example of Greenpeace. The first report alleged that the organisation was a 'threat to national economic security' and was contravening laws in an attempt to change the dynamics of India's energy mix, and that Greenpeace only attracted domestic donations to mask its actual sources of funding. A second exclusive dossier titled *Greenpeace Spearheading a Concerted Campaign against India's Energy Expansion Plans* (Intelligence Bureau, 2014b) mapped out Greenpeace's organisational network all the way from the global down to the grassroots, even mentioning prominent locals within the MSS. It alleged that Greenpeace's activities in Singrauli channelled the pro-environmental policies of European governments and recommended its FCRA licence for receiving international funding be cancelled.

Created during the British Raj to spy on freedom fighters, the IB has continued in post-independence India without a statutory basis or parliamentary and judicial oversight, helping governments to crush political insurgencies through intelligence gathering from the shadows (Nigar, 2014). The Bureau's reports served as tools for the State's crackdown on NGOs and legitimised the impunity with which corporations like Essar Energy acted against local resistances like the MSS.[34] The Modi government restricted international funding for 11 out of the 22 listed organisations on grounds of non-compliance with the FCRA (Yadav, 2014).[35]

The witch-hunt and the fight back by Greenpeace

The Government of India attempted three strikes on Greenpeace India. Within weeks of the IB report leaks to the media in June 2014, the Ministry of Home Affairs (MHA) froze funding from Greenpeace International and the US-based Climate Works Foundation into the organisation's Indian bank accounts. Greenpeace India obtained an order for the unblocking of international funding from the Delhi High Court in January 2015, which declared that the Home Ministry's actions were illegal and unconstitutional (Burke, 2015).[36]

In January 2015, the Greenpeace activist who led the anti-coal mobilisations in Singrauli was offloaded from a London-bound flight in New Delhi, and a look-out-circular was issued in the activist's name. The Greenpeace activist was scheduled to present on the treatment of Adivasis in Mahan by the London Stock Exchange-listed Essar Energy before a UK cross-Parliamentary Committee. The Indian government's stated concerns were about Greenpeace presenting India in a poor light. Greenpeace yet again challenged and won against the government order (Mathur, 2015a). The Delhi High Court's landmark judgement in March 2015 upheld the need for dissent in a vibrant democracy, ruling that 'contrarian views held by a group of people (who form a nation) do not make them anti-national' (Justice Sakhdher, 2015, para 12.3).

The Indian government froze all of Greenpeace's bank accounts, involving both foreign and domestic funds. It blocked the organisation's online donation page, leaving no room to seek financial support. The reasons cited were those of a discrepancy in Greenpeace's records in demonstrating proper usage of foreign funds. A statement by Greenpeace India's Executive Director warned that the organisation would have to close within a month. The statement said that the actions of the MHA:

> Could lead to not only the loss of 340 employees but a sudden death for its campaigns which strove to represent the voice of the poor on issues of sustainable development, environmental justice and clean, affordable energy.
>
> (Greenpeace India, 2015c)

The Delhi High Court ruled in Greenpeace's favour for a third time in May 2015, directing the MHA to unblock the domestic bank accounts (Mathur, 2015b).

Greenpeace's FCRA licence and its registration to operate in India were next in line for an attack by the Indian government later in 2015 (Gopal, 2015).[37] Greenpeace took the case to the High Court in Chennai where Greenpeace is registered in India and secured an unconditional stay on the cancellation of its registration, thereby avoiding having to completely shut down (Greenpeace India, 2015d). The Court ruled in November that the cancellation of the FCRA licence was 'unwarranted'.

The series of attacks severely affected Greenpeace's campaigns. Staff worked without pay when all its accounts were frozen (Bhalla, 2015). For the first time in 14 years of campaigning in India, its survival looked doubtful (Greenpeace India,

2015c). A smear campaign that paralleled the attacks portrayed Greenpeace as a foreign agent. A government affidavit to the Delhi High Court labelled the Greenpeace activist who led the mobilisation in Mahan as a 'bad activist' who testified before foreign committees against their own country (Narayanan, 2015). To discredit Greenpeace's anti-coal activism, it alleged that Greenpeace was channelling foreign funds worth crores (tens of millions) of rupees into MSS to mobilise the locals against coal mining (Nair, 2015).

Greenpeace declared its donation sources and financial figures as a demonstration of transparency. It countered the 'foreign-agent' allegation by showing that 70% of its funds came from Indian citizens (Sreenivas, 2015). Despite budget cutbacks, Greenpeace was able to keep some significant campaigns afloat, and garner strong civil society support through these projects.

The 'victory' for the Mahan forests in March 2015 coincided with the Delhi High Court judgment against the removal of the Greenpeace activist from the flight to London. Greenpeace and its allies called these outcomes a victory for democracy wherein the right to dissent has been upheld as a fundamental right for citizens under the Indian Constitution (Greenpeace India, 2015b). Just on its own, the incident of the removal of the Greenpeace activist from the flight had attracted strong criticism from various sections of civil society as a 'brazen assault on a citizen's

Figure 4.6 Greenpeace members hold a banner saying 'You cannot muzzle dissent in a democracy' in a park in Bangalore after its bank accounts are frozen.

Source: Photo by Dipti Desai/Greenpeace.

right to liberty and free expression that India has seen in decades' (Varadarajan, 2015b, para 2.). The high-level media attention that followed the flight off-loading created a sympathetic platform for Greenpeace's activism, especially its anti-coal protests in Singrauli.

The crackdown in the context of neoliberalism

The crackdown on Greenpeace came at a time when the social and environmental effects of two decades of rapid, resource extraction-driven development, had begun triggering civil society debates about the deep corporate bias, and erosion of public interest, in economic growth. The government's attempt to label Greenpeace a foreign agent to disrupt the organisation's work was therefore met with criticism from civil society groups and people's networks (Vaidyanathan et al., 2015). Against the broader context of a crackdown by the State on civil society's dissent against industrial projects that were failing to serve the public interest, the Courts upheld Greenpeace's right to protect the environment and livelihoods of vulnerable communities as an assertion of a democratic right. Various High Court judgements rescued Greenpeace from the brink of collapse a total of six times in the span of 1 year (Greenpeace India, 2015d).

However, since Greenpeace's legal success, further attacks on civil society organisations have shrunk the space for democratic dissent in India. In December 2016, in a step deemed illegal by the National Human Rights Commission, the Modi government refused to renew the FCRA licences of 25 prominent rights-based organisations on the grounds that the organisations had undertaken activities that were detrimental to the national interest, effectively negating their chance to continue functioning (Sampath, 2016).[38] Beyond the year 2016, which corresponds with the timescales of this research, the Narendra Modi Indian government has continued to invoke the FCRA 'in a systematic way' to shut down civil society organisations, research institutions, think tanks and international non-profit organisations operating in India for alleged violations (Telegraph, 2024).

Timeline of the Indian anti-coal resistance

In Table 4.1, I summarise a timeline of the major actions of the Indian government, the Essar Corporation, Greenpeace, and MSS. This timeline of the politics and resistance of the Mahan coal mine highlights the patterns of power in India's coal sector and the dialectic between the State and ENGOs working with a forest-rights-based agenda. It also draws on events and milestones discussed in the previous Chapter 3 and following Chapter 5.

Table 4.1 Timeline of the Indian anti-coal resistance

Year	Event
Congress-led United Progressive Alliance government	
2006	Forest Rights Act 2006 (FRA) enacted to redress historic wrongs towards forest-dwelling communities by returning individual and community rights.
	Mahan coal block allocated to Mahan Coal Ltd., a private joint venture between Essar Power and aluminium manufacturer Hindalco.
2009	Greenpeace prioritises 'keeping coal in the ground' by supporting local resistances; in India it focuses on the interconnected effects from coalmining on forests, water, and Indigenous peoples.
	An environment ministry circular to state governments emphasises that forest rights claims must be settled before allowing forests to be 'diverted' for mining.
2010	Joint 'go-no-go exercise' by ministries of environment and coal designates Mahan as 'inviolate' for mining on account of high-quality forests, but the Coal Ministry capitulates on the decision.
	Greenpeace organises an awareness camp on the effects of coalmining for youth in the Singrauli region.
	Greenpeace activists begin visiting Mahan to build awareness about forest rights.
2011	Inter-ministerial tussle over Mahan. Essar Power – whose thermal power plant is nearing completion in the vicinity – plays a strong role in steering the government towards approving Mahan.
	Environment Ministry is compelled to grant a first stage forest clearance for Mahan despite suggesting alternative fuel sources for Essar. It asks why are power plants being built before securing the coal and warns about 'fait accompli' approvals.
	Coal Curse, a fact-finding report on Singrauli by Greenpeace finds that legislations to improve lives in India's mineral rich areas have failed in Singrauli due to chronic violations and neglect by a State–corporate nexus.
2012	*Countering coal*, a discussion paper by Greenpeace and Kalpavriskh concludes that awareness followed by claiming legal rights under the FRA would empower forest-dwelling communities to assert their rights and protect their forests and forest produce.
	'Coalgate' scam is exposed through a Comptroller and Auditor General (CAG) report, that the government allocated coal blocks without any transparent assessment, leading to windfall gains for private corporations at considerable costs to the exchequer.
	The Greenpeace report *How Coal Mining is Trashing Tigerland* shows that forests in central India that serve as vital wildlife corridors and are the cultural and economic lifeline of forest-dependent communities are at risk from expanded coalmining, concluding that forest rights are essential.
	Mahan Sangharsh Samiti is formed in collaboration with Greenpeace.
	Village council proceedings to register community forest rights in Amelia disrupted by company-hired goons.

Table 4.1 (Continued)

Year	Event
2013	Forgery of signatures after a *gram sabha* in Amelia village by company agents and the local administration turns a referendum that said no to mining into consent.
	Ministry of Tribal Affairs orders a fresh village council resolution to be conducted after Greenpeace and MSS take the matter of fraudulent signatures to them.
2014	Greenpeace unfurls a giant banner saying 'We kill forests: Essar' from the company's headquarters in Mumbai. MSS members join Greenpeace in the protest.
	Essar presses Rs. 500 crore (A$100 million) defamation against Greenpeace and brings a SLAPP suit.
	A new environment minister issues a final forest clearance based on the fraudulent consent; the minister is known for his special favours to Essar.
	Greenpeace organises mahua-collection camps in the Mahan forests for urban youth to understand and support the Mahan struggle.
	Greenpeace-MSS challenge the fraudulent forest clearance in the NGT; based on a NGT stay order the state government guarantees not to fell trees in Mahan for a few months. Yet, the company prepares to fell trees.
	MSS members form human chains around trees, sitting in the forest from dawn to dusk; women often lead these blockades.
	Arbitrary arrests without warrant of MSS members and Greenpeace staff by the local police devised to thwart the movement when people began protecting trees from being felled.
	MSS and Greenpeace form a giant human sign 'Essar Quit Mahan' in a field in Mahan.
	Company agents bribe and intimidate villagers ahead of a new village council meeting in Amelia to coerce a yes vote on mining.

Narendra Modi's Bharatiya Janata Party-led National Democratic Alliance Government

Year	Event
2014	Supreme Court orders 214 'illegal' coal block allocations including Mahan to be cancelled in response to the CAG report.
	Intelligence Bureau Reports alleges that Greenpeace is risking India's economic security by 'spearheading a concerted campaign against India's energy expansion plans'. The report also names prominent MSS members.
	Modi government blocks major international funding for Greenpeace.
2015	Delhi High Court orders Greenpeace's international funds to be unblocked and calls the government's actions 'unconstitutional'.
	A Greenpeace activist is offloaded from a flight to London; the government issues a lookout circular issued in her name and labels her a 'bad activist' who wants to testify before foreign committees against her own country – she was going to talk about the treatment of Adivasis in Mahan by Essar before a UK cross-parliamentary committee.
	Greenpeace challenges the government action and wins in the Delhi High Court, which upholds the need for dissent in a vibrant democracy.
	Coal Mines (Special Provisions) Act 2015 mandates auctions for captive coal blocks for private companies; the first set of coal blocks are auctioned based on the new rules.

(Continued)

Table 4.1 (Continued)

Year	Event
	The movement celebrates as the government designates Mahan 'inviolate' and removes it from the mining auction list. The people of Mahan had joined with other local people's movements to mark the victory as 'democracy day'.
	The government blocks all of Greenpeace's bank accounts citing discrepancy in the ENGO's records in using foreign funds.
	The Delhi High Court directs the government to unblock Greenpeace's domestic bank accounts, and rules in favour of Greenpeace and against the government's actions a third time in a row.
	The government cancels Greenpeace's FCRA licence to receive international funding and registration to operate in India.
	The Chennai High Court orders an unconditional stay on the cancellation of Greenpeace's registration and calls the government's cancellation of its FCRA licence 'unwarranted'.
	The government alleges Greenpeace is channelling crores (tens of millions) of rupees of foreign funds into Mahan Sangharsh Samiti to mobilise the locals against coal mining.
	Government attacks affect Greenpeace's work in India, but despite budget cutbacks it is able to keep some significant campaigns afloat.
	Indian civil society supports Greenpeace for raising concerns on human rights and democracy around coal. And supporting the Mahan community's struggle for justice and rights.
2016	Amelia North coal block also in the Mahan forests allocated to a public sector company, risking forests around Pidarwah village approximately 15 kilometres from Amelia. Yet again the government failed to apply a common rationale in making decisions that are critical for the environment and communities.
	Amelia, Budher and Suhira villages eventually manage to file for community forest rights (CFRs) despite threats and disruptions to the process by company agents in collusion with the local administration.
	India dilutes provisions of the FRA through relaxing the need for community consent before destroying forests.
2018	Greenpeace shuts down its Singrauli campaign office and does not attempt further anti-coal community mobilisations; its anti-coal activism takes the less risky approach of exposing the economic implausibility, and ecological impacts with a major focus on air pollution, of new coal-fired investments. It also focuses on decentralised solar projects.
	Another government attack on funding forces Greenpeace to close several long-running campaigns and shrink to a fifth of its capacity after two decades of environmental activism. It now operates on a significantly reduced capacity in India.
	Read more at: https://newmatilda.com/2019/03/29/ profit-before-people-why-india-has-silenced-greenpeace/

Outside research timeline: Post-pandemic coal production

2020	Modi government's decision to auction 50 new coal blocks statedly for economic recovery from the COVID pandemic exposes contradiction in the government's approach of 'talking renewables and walking coal.
	India loosens forest clearance processes that compromises the autonomy of *Gram Sabhas* to boost coalmining

Table 4.1 (Continued)

Year	Event
2021	Indian government writes to state governments, which are responsible for recognising forest rights, that 'despite a considerable lapse of time since it (FRA) came into force, the process of recognition of forest rights is yet to be completed.
2022	The government said forests with more than 40% green cover will not be mined and then abandoned the decision, showing a lack of consistency in decision-making yet again.
	A subsidiary of Adani Power Ltd. acquires the nearby Mahan Thermal Power Plant from Essar Power.
	Changes to India's Forest Conservation Rules allow forest clearance for industrial projects (including coal mines) on Adivasi lands even before securing consent from *Gram Sabhas* contradicting the FRA's provisions.
2023	The Mahan coal block is reinstated in India's coal mine auction list after the government capitulates on its decision to not mine areas with 40% or more forest cover.
	An adjoining coal block also in thick forests is likely to be allocated to Adani Power's subsidiary for underground coalmining for the Mahan thermal plant.
	An investigative report on coal auctions in the Modi era concludes that with these (amongst other) actions India has come full circle since the Supreme Court ruling in 2014: www.reporters-collective.in/trc/coal-forests-part-1 www.reporters-collective.in/trc/coal-forest-part-2

Analysis: countering coal through asserting democratic rights

The power vested in coal in the Indian political economy makes it a risky proposition to raise the difficult questions that need to be raised about the relationship between climate action and coal-dependent economic growth. Internationally, India sends signals about being a good climate player based on an ambitious and rapid-scale development of renewable energy. But these actions belie the underlying realities of India's ongoing coal developments. Coalmining in India today disproportionately benefits private corporations while disenfranchising vulnerable Adivasi and peasant communities and destroying the last remaining forests. India's renewable energy development pathway also remains disconnected from India's coal trajectory and energy policies do not reflect an interconnected energy transition pathway (Roy and Schaffartzik, 2021).

The severity of the crackdown on Greenpeace exposed the contradiction between India's articulations about a renewables transition on a global platform and its internal priorities of attracting foreign investment and continuing coal production for energy security. A New Delhi-based independent senior journalist and board member of Greenpeace India told me during an interview that:

Greenpeace and the government want the same thing, less coal, more renewables, so then why was it cracked down on? One obvious reason is that the previous government but more so this government has a phobia of civil society organisations mobilising on specific projects and not just advocating on policy, especially when the work is connected with foreign funding. There is a hugely different implication of commenting on policy and mobilising on the ground.

(Interview 12 April 2018)

As opposed to Northern environmentalism, which is characterised by professional ENGOs and movement networks that strategically target mineral resource sectors such as coal or uranium, sectoral targeting as an activist strategy largely does not apply to India's people's movements and livelihood struggles. Livelihood movements are mostly constituted of an interconnected web of local struggles to protect lands, livelihoods and forests from displacement and destruction. Where sectoral targeting has been applied, such as also in the case of the People's Movement against Nuclear Power (PMANE), the severity of the State's crackdown has demonstrated the risk of strategic mobilisations against energy sources. The State's tactic of labelling movements as foreign agents demonstrates a politics of anxiety around India's neoliberal economic growth. The senior environment law and policy researcher said during the interview:

A direct anti-coal approach might have led to it being sensed by governments as 'sectoral targeting'. There has been financial and technical support to groups, environmental movements, and particularly to transition away from coal: But not necessarily to target coal or to directly and mobilise against coal.

(Interview 05 May 2017)

With the State facilitating private interests in coalmining and power generation against the interests of vulnerable communities, the contradiction inherent in India's extractive economic growth has been clearly revealed: India talks about inclusive growth and still states poverty alleviation as the foremost objective of economic development, but under neoliberalism gains to private corporations at the expense of communities are evident in industrial projects. Economic growth is motivated by a bond between business and political elites or what is understood as crony capitalism.

The high-level corruption and mismanagement in coal block allocations can be understood as a manifestation of this contradiction through crony capitalism in coal's political economy. Greenpeace's anti-coal activism – through the grassroots alliance with MSS and other tactics – negotiated the contradictions within coal's political–economic structures, and the subsequent crackdown further exposed the contradictions. An energy and water analyst and Greenpeace India board member told me during an interview:

The context is not the same as before. Coal is now piling in the mies and coal plants are shutting down; it's partly because of the slowing down of the global economy but largely because they recklessly built power plants in 2006. So we now have a huge spectacle of coal capabilities lying idle, coal lying piled at coalmines, while on the other side, there's a solar boom. You think they would have thought twice about targeting the NGO in such a high-profile manner from such a weak position?

(Interview 12 April 2018)

To many civil society actors, the crackdown indicated the state of Indian democracy today, when economic development is driven by high-stakes foreign investments in major industrial projects. The State regards the activism of an international group like Greenpeace as an economic risk because it has the ability to influence global institutions that are financing industrial projects and can bring international scrutiny to energy developments. Globally, divestment campaigns, where activists have appealed to investors not to fund new coal mines that are economically unviable and are violating Indigenous consent, have been effective in 'keep(ing) coal in the ground'. We will see this in the Australian case in Chapters 6, 7 and 8. However, the Indian government's crackdown on Greenpeace serves as a reminder of the precariousness of global anti-coal alliances and the challenges for global NGOs to operate within a southern economic context of growing anxieties around energy security.

High stakes in industrial projects owing to large foreign investments have made the context for today's environmental conflicts highly charged. Environmental protests unfolding within the current context of favouritism (of corporations) versus dispossession (of communities) see their actions as critical assertions of democracy. From the beginning of its activism in India which roughly corresponds with the time frame of India's neoliberal reforms, Greenpeace critically inserted itself into India's environment versus development debate. Against this context of neoliberalism's effect on democracy, particularly through the State's suppression of dissent, the Delhi High Court judgement mentioned that I have mentioned earlier found it necessary to remind about democratic freedoms enshrined in the Constitution. It interpreted the significance of the Greenpeace-MSS resistance as a constitutional right. In this way, Greenpeace's confrontational anti-coal activism and the political and legal debate that it generated helped to further the significance of today's environmental protests and livelihood struggles in India.

In India, the issue of climate change has largely remained at the level of policy advocacy (Yla-Anttila and Swarnakar, 2017). Activisms of the environmentalism of the poor have largely not drawn the links between coal and climate change. Their experience of environmental injustice has continued to be dominated by the risk of loss of lands and livelihoods from industrial development. The Kalpavriksh member said during the interview that:

Climate change is not necessarily perceived as a grassroots issue, communities in their struggles are not articulating them as part of climate movement. Greenpeace's focus is less on grassroots livelihood issues, than on making the connection between the big and the small picture; on its own the climate change narrative would not have won the Mahan campaign. People talk about climate change in environmental circles, but groups helping to make changes and provide support on the ground do not necessarily do so.

(Interview 05 May 2017)

In the absence of a mass climate movement in India's southern and postcolonial context of environmentalism, the strategic role played by an international NGO helped to elevate dissent against coal mining's socio-ecological impacts, and anti-coal mining resistance through forest rights, to a national debate and towards international attention. However, it is probable that the Greenpeace-MSS movement would not have received such critical global awareness without the extent of the crackdown on Greenpeace; offering a sobering perspective on how environmental conflicts related to coal mining are a normalised affair in India.

Concern from some sections of the Indian and global media about an attack on the 'fundamental rights of an Indian citizen' after the removal of a Greenpeace activist from an international flight scaled up the significance of the resistance to coalmining at Mahan beyond the habitual environment versus development frame toward fundamental freedoms in a democracy. As a senior journalist remarked:

If left unchecked, it (the removal) will become the thin end of the edge, where the government will show its intolerance to dissent in this manner...Tomorrow they will say 'I don't like an article, so let's get rid of freedom of press.'.

(Varadarajan, 2015a, quote from TV interview)

Owing to the series of government crackdowns that followed the de-boarding incident, civil society came to see the Greenpeace–MSS resistance as a fight for democratic rights in India's current political–economic landscape. The dominance of coal, India's most abundant energy source that built the postcolonial economy, was questioned on the grounds of civil and democratic rights. The Kalpavriksh member says:

On the contrary, even if they had not used the climate change argument at all in the movement along with the grassroots groups, they (Greenpeace) would have still won the fight in the real and in the moral sense.

(Interview 05 May 2017)

Allies, however, have regarded Greenpeace's narrative frames on coal, popularised through catchy slogans such as 'keep coal in the ground' and 'End Coal', as unrepresentative of a global-South complexity of economic development. Greenpeace's two decades in India have not been without ideological and practical tensions with the broader civil society network on account of its perceived inability to nuance messages to the intricacies of the Indian context (Talukdar, 2019). The comments related to Greenpeace's 'sectoral targeting of coal' also point to a widely held view within the Indian civil society network about the organisation's apparent intransigence. Collaborators have chafed at Greenpeace's black-and-white positions, which at face value have betrayed a lack of reason, even a disregard, for the intricate contexts of economic and environmental justice in the global South.

Paradoxically, Greenpeace's stand-out activism in India, due to its foreign status and confrontational politics and tactics, and its grassroots alliance with the MSS, served to expose the contradictions in coal's power structure. The involvement of a foreign NGO in targeting coal and its leading role in mobilising a grassroots resistance against coal extraction revealed the double movement of the neoliberal Indian state, in exacerbating environmental injustices and then moving to repress dissent. The participation of an international environmental activist group in an alliance with an Indian livelihood movement against coal in central India, and the consequent state crackdown, challenged coalmining in various ways and at various levels – from local democratic assertions of forest rights to national issues of human rights and the democratic right to challenge mainstream development, to international climate change.

Conclusion

For scholars of global climate justice and global North climate activists, this chapter will offer insights into the political–economic context of coal in India and the possibilities and challenges of countering coal within the Indian context. The Indian context is one of many political-economic contexts that frame environmental injustices in the global South today.

In India, the contradictory actions of the State towards energy security and climate action create an inherent conflict. Historically, India's developmental goals have linked coal-powered electricity generation with the moral imperative of poverty eradication. India's postcolonial economy was built off the back of 'King Coal' through a centralised and State-run structure of coal and thermal power production. Economic policies elevated coal to the 'commanding heights' of the Indian economy while laws gave coal legal eminence, making coal synonymous with the national interest. This prioritisation continued even after the coal mining and thermal power generation sectors began to be privatised, paradoxically making the profit of private corporations from coalmining and electricity generation central to the national interest. In the neoliberal era, the State has driven massive land acquisitions for private coal and thermal power generation, often in violation of its own laws on land rights and the consent of Adivasi and forest-dependent communities. The moral logic of 'greater common good' from India's postcolonial development has been strongly tested against the reality of the State blatantly acting in favour of corporations and against communities, for coalmining and power generation.

Activisms that not only challenge coalmining's grounded issues – loss of land, livelihood, and forest – but also ask the difficult questions around climate action and development are deemed risky to India's national interest. The risk borne by the Greenpeace-MSS activism is exacerbated because of its stand-out politics. Links between climate change and coal extraction are not readily made by grassroots environmentalisms of the poor, and Indian civil society groups largely remain cautious about directly targeting the coal sector. Today's political climate of intolerance toward dissent in India makes this activism a relevant human rights struggle and offers insights from the South about the significance of multi-scalar anti-coal resistances.

Coal retains its historic significance to both Australian and Indian States. But the Indian case of anti-coal activism and subsequent crackdown by the State raises pertinent points about North–South differences in political economic contexts, environmentalisms, and the state of democracy under neoliberalism today. It raises questions about the future of North-South civil society environmental solidarities, and highlights the need for global climate activism's new approach to advocate for the strengthening of Adivasi and land rights in the Global South. I discuss these further in Chapter 9.

Notes

1 In another early campaign in India, along with local groups and the worker's union, Greenpeace exposed mercury contamination by Unilever's thermometer factory that had relocated to Tamil Nadu from New York after failing to comply with US regulations, leading to its closure (Hiddleston, 2010).

2 Repressive tactics by the State also extended to cancellation of the foreign-donations licences of organisations, travel bans on activists, and prohibitions on journalists from visiting protest sites (IANS, 2013).

3 The 10th Plan achieved a growth of 7.6% and the 11th Plan achieved a growth rate of 8%. The 12th Plan set a growth target of 8.2% (Planning Commission, 2013, p. V).

4 Further, although 99.6% of Indian households are shown to now have electricity in various international surveys, these can take only a very basic definition of usage, such as one light bulb and one charging outlet into account. Access, not just electrification is crucial for households in India (Rockerfeller Foundation, 2018).

5 Coal was discovered in Singrauli in 1840. Singrauli's first open cast coal mine was set up in 1857 during colonial rule (Chakravartty, 2018). But the region's coal deposits largely remained unexplored till the late 1960s owing to its relatively inaccessible terrain. Thereafter, abundant water supplies from the Rihand Reservoir made the prospect of mining coal in Singrauli, with its long stretch of open coalfields extending over 200 kilometres, lucrative.

6 The construction of India's largest artificial lake the Gobind Sagar Reservoir, and the Rihand Dam, originally displaced 200,000–300,000 people from 146 villages in Singrauli (Clark, 2003). With no rehabilitation policy in place, many families did not receive any compensation (Dokuzović, 2012). An estimated 60% of dam oustees resettled close to the reservoir site only to be displaced again during the setting up of the Super Thermal Power projects of the National Thermal Power Corporation (NTPC) and coal mines of the NCL in the 1980s (Sharma and Singh, 2009).

7 The CMNA has been amended before to allow captive coalmining by private corporations: an initial amendment to the CMNA in 1977 allowed captive mining by private companies engaged in the production of iron and steel and sub-leasing to private parties for coal mining; a subsequent amendment in 1996 allowed the mining of coal for captive use in cement production by private companies.

8 The government justified the allotment system of coal blocks on the grounds that allowing state-owned enterprises to exploit captive blocks would allow them to reduce their costs on coal, with the end result of passing the benefits to consumers through lower electricity costs. But it was found that the hefty payments that needed to be made to private contractors drove up expenses, making the prospects of consumer benefits negligible at best (Nileena, 2018).

9 Former BJP Prime Minister Atal Bihari Vajpayee first launched a programme 'Power for All by 2007' and introduced a new scheme for rural electrification and another scheme for modernising the power distribution systems in 56 urban circles. Next, the Congress-led government set the new goal of 'Power for All by 2010', which subsequently changed to 'Power for All by 2012'. In 2014, Narendra Modi's BJP government set the new target of '24×7 Power for All' by 2019. But the target date for '24×7 Power for All' later stood shifted to 2022, indicating recurrent problems faced by governments in inducing last mile connectivity and constant power supply to remote locations through the (primarily) coal-fired central grid, instead of prioritising investments in and establishing decentralised power-systems of renewables to reach all corners of India.

10 India added 21,180 megawatts of generating capacity in the 10th Plan period, and 52,000 megawatts in the 11th Plan period (Ministry of Power, 2012).

11 The Singrauli region covers 2200 square kilometres and is shared between the states of Uttar Pradesh (UP) and Madhya Pradesh (MP) in central India. A district bearing the same name was carved out as an independent entity from the Sidhi District in 2008 (Singh, 2017).

12 In January 2010, the MoEF placed a temporary ban on the expansion of industries in the Singrauli industrial cluster. No new coal mines were to be approved for environmental clearance on account of the region having been identified as a critically polluted area (CPA) with a pollution index of 81.73. But the moratorium was lifted only a year later upon assurances from central and state pollution control board that they had started preparing action plans for controlling pollution (Pillai et al., 2011). All pending environmental and forest clearances, particularly for the new proposed coal and thermal power projects, could now proceed with safeguards attached.

13 The Planning Commission was an institution within the Government of India that formulated India's Five-Year Plans. It was suspended in 2014 after Narendra Modi's BJP government came to the helm. The last two years of the 12th Plan, India's last Five-Year Plan, was administered by the newly formed National Institute for Transforming India, widely known as the NITI Aayog (Planning Commission, 2014).

14 As of the latest UN Climate Summit COP 28 in Dubai in December 2023, India continues to emphasise the prominence of coal in the energy mix into the foreseeable future while committing to Net Zero by 2070 (Press Trust of India, 2023).

15 A Centre for Science and Environment factsheet showed that 181 coal mines (including those applying for capacity expansion) with a combined capacity of 583 million tons per annum had been granted environment clearance during the 11th Plan period (CSE, 2012).

16 Estimates places the total displacements in Singrauli since the district was formed in 2008 at 10,000 families from the acquisition of 15,000 acres (Singh, 2015b).

17 The Modi government allowed Adani to continue mining from a coal block holding over 450 million tonnes of coal in one of India's densest forests in central India, that had first been allocated during the Coalgate scam. As a private player, Adani operated as the mine developer for a state-government entity that owned the coal block. Acting on the

Supreme Court judgement, the Modi government cancelled all coal-scam-era contracts between government coal block owners and private coal mine developers. It asked that fresh allocation of coal blocks to government entities be followed by fresh auctions for private entities that will mine them. Adani Enterprises was allowed to be reinstated as the mine developer for this coal block. While the *Coal Mines (Special Provisions) Act 2015*, brought in with the intended purpose of starting transparent auctions and allocations of coal blocks, was changed to disallow other private players from continuing mining their coal-scam-era coal blocks as mine developers (Jalihal and Sambhav, 2023).

18 A study mapping the dependency of the poor on common property for their livelihoods indicates that 70–80% of non-timber forest produce that constitutes a substantial part of forest-dependent people's household incomes, comes from common resources (Beck and Ghosh, 2000).

19 The CAG Report named 25 companies including some of India's largest power and resources corporations – Essar Power, Hindalco, Tata Steel and Power, Adani Group, Lanco, Vedanta group, Arcelor Mittal, Jindal Steel and Power – as beneficiaries in the private allocation process (CAG, 2012).

20 The original suggestion to clearly demarcate forest regions that should not be opened up to coalmining came from the chairman of CIL, following which a joint declaration was made by the Coal and Environment Ministers that approvals would not be granted for mining in the no-go areas (Ramesh, 2015).

21 A list obtained by Greenpeace through the Right to Information Act 2005 indicated that the initial exercise had deemed 222 coal blocks in the central and eastern coalfields, which amounted to 48% of the area being classified, as no-go areas for mining approvals. But within a year, following inter-ministerial negotiations, the number of no-go coal blocks had come down to 153 (Pillai et al., 2011).

22 Investments of around Rs 3600 crore (A$720 million) had already been made in the two power plants linked with the Mahan coal mine (Telegraph, 2011).

23 One email exchange between two senior Essar officials included the summary note of the inter-ministerial meeting. Shashi Ruia wrote to the PM after the meeting asking to instruct the Environment Minister to hasten forest clearance (Kaushik, 2015).

24 Adjoining blocks from Mahan included Chhatrasal, Moher Amlohri and Moher coal blocks allocated to Reliance Industries' Sasan Ultra Mega Power Project, Amelia North and Amelia coal blocks allocated to Jaypee Nigrie's thermal power project, and the Dongri Tal 2 coal block allocated to the state-owned Madhya Pradesh Mining Corporation (Chakravartty, 2011). Excepting Chhatrasal, all other adjoining coal blocks from Mahan were cleared under the above exercise. The MoEF refused clearances for Mahan and Chhatrasal on grounds of their high-quality forests. But they were subsequently cleared by an Empowered Group of Minsiters set up to clear roadblocks in coal block allocations (Kohli, 2013).

25 The Director of Corporate Relations at Essar had met the Finance Minister asking for an early date for the EGoM meeting on Mahan. The leaked emails showed that the Finance Minister had expressed annoyance during the meeting at the company's delay in employing his favoured candidates. The Office of Pranab Mukherjee, the UPA Government's Finance minister and later the President of India, had asked Essar to provide jobs to three people, one of them his granddaughter who had been offered an internship in Essar's London headquarters. Essar resolved the matter within days. And the EGoM meeting approved Stage 1 Forest Clearance for Mahan.

26 As per the leaked Essar emails, in his previous portfolio as Minister for Petroleum and Natural Gas, Moily had kept close contact with the Essar Group's executives throughout 2013 (Kaushik, 2015).

27 The petition stated that apart from the Forest Rights Act 2006, the clearance for Mahan also violated the provisions of the *National Forest Policy 1988*, the *Forest (Conservation) Act 1980*, and the *Biological Diversity Act 2002* (Greenpeace India, 2014e).

28 Leaked emails from Essar revealed the BJP leader Nitin Gadkari who became Minister for Roads and Transport and in the Modi Cabinet, as one of Essar's biggest beneficiaries. The emails revealed that Gadkari and his family had stayed in the private yacht of the Essar CEO Shashi Ruia during a trip to France (Suresh and Sarin, 2015).

29 The Essar Group also pressed defamation charges worth Rs. 250 crores (A$50 million) on *Caravan*, a political and current affairs magazine, after it published an investigate report on the group's influence on politicians based on the Essar company staff turned whistle-blower's leaked emails (Ramanathan, 2015). The article 'Doing the Needful: Essar's Industry of Influence' (Kaushik, 2015) chronicled the underhand means through which Essar wielded influence throughout the Mahan coal block clearance process.

30 Since 2016, the Modi government has paved the way for commercial mining of coal and allowed 100% foreign direct investments in coalmining. Between 2015 and 2018, the Modi government allocated 84 (24 auctioned to private companies, 60 allocated to PSUs) coal blocks. By 2019, 29 of these coal blocks had begun operations, while others were awaiting a range of approvals or tied in conflict and litigation. The Modi government proposed sweeping dilutions to tackle environmental, forest clearance and land acquisition related hurdles, most critically, the de-linking of the FRA from the forest clearance process for coal mines to bypass the process of community consent (Chatterjee, 2019).

31 In the lead up to the 2014 general elections the BJP had criticised the scam-ridden Congress government and promised action against corruption.

32 The FCRA was first passed in 1976 under the twin conditions of a national emergency and the Cold War to curb the foreign hand in domestic politics by preventing political parties from receiving international funding (Sampath, 2016). It was made more draconian in 2010 through the inclusion (amongst other amendments) of organisations of a political nature. The UN Special Rapporteur on the Rights to Freedom of Peaceful Assembly and of Association concluded that the FCRA 2010 is 'not in conformity with international law, principles and standards' (UNHR, 2016). Further amendments in 2014 legitimised foreign funds to political parties, as long as they were routed through Indian subsidiaries, exposing double standards in the Law. Amendments in 2015 legitimised scrutiny over NGOs on account of national interest and required organisations to provide details of social media accounts. The 2020 amendments, by far the most stringent, that commentators have said can 'choke' India's civil society sector, curtailed the ability of bigger NGOs that receive foreign funds to sub-grant to smaller, grassroots entities working at the ground-level; capped how much foreign donations organisations can spend on administration at 20% from a previous 50%; and mandated that foreign funds can only be received through a New Delhi-based account with India's public sector bank State Bank of India (Saraogi, 2022).

33 The 2013 report 'India's funds to NGOs squandered' by the Asian Centre for Human Rights (ACHR) exposed the favouritism and corruption surrounding the allocation of government funds to NGOs (ACHR, 2013).

34 The allegations in the IB report acted to strengthen the arm of corporations facing resistances from people's movements. Essar Energy submitted the IB reports as evidence against Greenpeace at hearings in the Mumbai High Court (Greenpeace India, 2014i).

35 This included international NGOs Greenpeace, Oxfam and Action Aid, smaller national groups like Indian Social Action Forum (INSAF), anti-GM advocacy group Gene Campaign, anti-Monsanto activist Dr Vandana Shiva's organisation Navdanya, four regional groups protesting the nuclear power at Kudankulam in Tamil Nadu, and one opposing the 'Gujarat model of economic development'.

146

36 The Court observed, 'Non-Governmental Organisations often take positions, which are contrary to the policies formulated by the government of the day. That by itself…cannot be used to portray the petitioner's action as being detrimental to national interest'. The judgement is available at: https://elaw.org/in.gpindia.15.

37 The cancellation of Greenpeace's FCRA licence was part of a mass cancellation by the Modi government on grounds of violation of the FCRA. This move affected 9000 of the 40,000 odd FCRA registered organisations in the country (Kaushal, 2015).

38 The most prominent amongst these include the human rights advocacy and legal advisory group Lawyers Collective whose members have spoken out against the violation of human rights by state deployed forces in India's Maoist insurgency affected tribal geography of Bastar in the mineral rich state of Chhattisgarh.

References

Ahmad, N. & Lahiri-Dutt, K. 2006, 'Engendering Mining Communities: examining the missing gender concerns in coalmining displacement and rehabilitation in India', *Gender, Technology and Development*, vol. 10, no. 3, pp. 313–339.

Ahmed, W. 2010, 'Neoliberalism, corporations, and power: Enron in India', *Annals of the Association of American Geographers*, vol. 100, no. 3, pp. 621–639.

Ananthapadmanabhan, G., Srinivas, K. & Gopal, V. 2007, *Hiding Behind the Poor*, Greenpeace India, Bangalore, viewed 20 March 2020, <www.greenpeace.org/india/Global/india/report/2007/11/hiding-behind-the-poor.pdf>.

Asian Centre for Human Rights. 2013, *India's Fund to NGOs Squandered*, 31 January.

Beck, T. and Ghosh, G. M. 2000, 'Common Property Resources and the Poor: Findings from West Bengal', *Economic and Political Weekly,* vol. 35, no. 3, pp. 147–153.

Bhalla, N. 2015, 'With funds and people power, charities globally face government crackdown', *Reuters*, 9 July, viewed 14 August 2019, <www.reuters.com/article/us-global-charities-crackdown-insight-idUSKCN0PJ1AX20150709>.

Bharadwaj, S. 2018, 'The New Land Acquisition Act, does it help the tribal people?', *Journal of Resource, Energy and Development,* vol. 15, no. 1–2, pp. 53–61.

Bhushan, C. & Hazra, M. Z. 2008, *Rich Lands, Poor People: Is Sustainable Mining Possible?* Centre for Science and Environment, New Delhi.

British Broadcasting Corporation. 2006, *Chirac Orders 'Toxic' Ship Home*, 16 February, viewed 20 September 2019, <http://news.bbc.co.uk/2/hi/europe/4716472.stm>.

Burke, J. 2015, 'Indian government ordered to unblock Greenpeace funds', *Guardian*, 21 January, viewed 20 September 2019, <www.theguardian.com/world/2015/jan/21/indian-government-ordered-unblock-greenpeace-funds>.

Centre for Science and Environment. 2012, 'Overview (Fact Sheet on Environment and Forest Clearances)', *Public Watch,* viewed 14 August 2019, < http://cdn.cseindia.org/userfiles/fsheet1.pdf>.

Chakravartty, A. 2011, 'Mahan at all costs', *Down to Earth*, New Delhi, 31 October, viewed 14 August 2019, <www.downtoearth.org.in/news/mahan-at-all-costs-34230>.

Chakravartty, A. 2018, 'Returning home: a story of displacement, dispossession and homecoming', *Dispossession and Resistance in India and Mexico*; Ritimo, 14 May, viewed 30 September 2019, <www.ritimo.org/Returning-Home-A-story-of-displacement-dispossession-and-homecoming>.

Chatterjee, A. 2019, 'Eye on coal: Govt to ease forest, land usage guidelines to boost mining,' *Financial Express*, 27 September, viewed 27 September 2019, <www.financial express.com/economy/eye-on-coal-govt-to-ease-forest-land-usage-guidelines-to-boost-mining/1718906/>.

Chhotray, V. 2022, 'Extractive regimes in the coal heartlands of India: difficult questions for a just transition', in P. Kashwan (ed.), *Climate Justice in India*, Cambridge University Press, Cambridge, pp. 74–97.

Clark, D. 2003, 'Singrauli: an unfulfilled struggle for justice', in D. Clark, J. Fox & K. Treakle (eds), *Demanding Accountability: Civil Society Claims and the World Bank Inspection*, Rowman and Littlefield, pp. 167–190.

Commonfloor. 2011, *'Ban the Bulb' – Campaign in India*, 19 January, viewed 20 September 2019, <www.commonfloor.com/guide/ban-the-bulb-campaign-in-india-4178.html>.

Comptroller and Auditor General of India. 2012, *Performance Audit of Allocation of Coal Blocks and Augmentation of Coal Production*, Ministry of Coal, Report No. 7, 2012–2013.

Council on Energy, Environment and Water. 2020, *State of Electricity Access in India: Insights from the India Residential Energy Survey (IRES) 2020*, October, viewed 20 January 2024, <www.ceew.in/sites/default/files/ceew-research-on-state-of-electricty-access-and-cover age-in-india.pdf>.

Das, A. K. 2015, 'Coal India aiming to double output to one billion tons by 2020', *Mining Weekly*, 17 July.

The Deccan Chronicle. 2014, *Coal is King and extremely important element in developing India: SC*, 26 August, viewed 30 April 2016, <www.deccanchronicle.com/140826/nation-current-affairs/article/coal-king-and-extremely-important-element-developing-india-sc>.

Dharmadhikary, S. & Dixit, S. 2011, *Thermal Power Plants on the Anvil: Implications and Need for Rationalisation*, Prayas Energy Group, August 2011, viewed 14 August 2019, <www.prayaspune.org/peg/publications/item/164-thermal-power-plants-on-the-anvil-implications-and-need-for-rationalisation.html>.

Doherty, B. & Doyle, T. 2006, 'Beyond borders: transnational politics, social movements, an modern environmentalisms', *Environmental Politics,* vol. 15, no. 5, pp. 697–712.

Dokuzović, L. 2012, *Bhikharipore Singrauli: A Case for Just Development [1]*, 30 September, viewed 27 September 2019, <http://sanhati.com/excerpted/5621/>.

Doshi, V. 2016, 'The lonely struggle of India's anti-nuclear protesters', *The Guardian in Idinthakarai*, 6 June, viewed 20 September 2019, <www.theguardian.com/global-deve lopment/2016/jun/06/lonely-struggle-india-anti-nuclear-protesters-tamil-nadu-kudanku lam-idinthakarai>.

Down to Earth. 2015, *'Auto Yes to Coal: High Powered Panel Recommends Freeing Coal Blocks in Forests'*, 27 June.

Economic Times. 2013, *Tribal Affairs Ministry Orders Probe into Mahan Coal Block Allocation*, New Delhi, 22 July, viewed 30 September 2019, <https://economictimes. indiatimes.com/industry/indl-goods/svs/metals-mining/tribals-affairs-ministry-orders-probe-into-mahan-coal-block-allocation/articleshow/21229133.cms?from=mdr>.

Fernandes, A. 2012, *How Coal Mining is Trashing Tigerland*, Greenpeace India Society, Bangalore, viewed 14 August 2019, <www.greenpeace.org/india/en/publication/984/how-coal-mining-is-trashing-tigerland/>.

Ghosh, D. 2016, 'We don't want to eat coal: development and its discontents in a Chhattisgarh district in India', *Energy Policy*, vol. 99, pp. 252–260.

Gopal, V. 2015, *Today Is Another Bad Day for Indian Democracy'*, Greenpeace India, 6 November, viewed 14 August 2019, <www.greenpeace.org/india/en/Blog/Campaign_bl ogs/today-is-another-bad-day-for-indian-democracy/blog/54682/>.

Gopal, V. 2016, 'Coal Accounting: "The story of a fuel kept cheap" ', in K. Kohli & M. Menon (eds), *Business Interests and the Environmental Crisis*, Sage Publications India Ltd, New Delhi.

Government of India. 2011, *Report of the Committee Constituted by Group of Ministers to Suggest Solutions on Coal and Other Development Issues*, New Delhi, 22 July, viewed 14 August 2019, <www.indiaenvironmentportal.org.in/content/341083/report-of-the-committee-constituted-by-group-of-ministers-to-suggest-solutions-on-coal-and-other-development-issues/>.

Greenpeace India. 2011, 'Greenpeace launches Junglistan campaign to garner public support for forests; Challenges government stance on coalmining in forests', *Press Release*, 22 November, viewed 20 September 2019, <www.greenpeace.org/india/en/press/328/gre enpeace-launches-junglistan-campaign-to-garner-public-support-for-forests-challenges-government-stance-on-coal-mining-in-forests/>.

Greenpeace India. 2013c, 'Greenpeace announces revolutionary electricity model in Bihar', *Press Release*, 4 September, viewed 27 September 2019, <https://www.greenpeace.org/india/en/press/3086/greenpeace-announces-revolutionary-electricity-model-in-bihar/>

Greenpeace India. 2014a, 'Greenpeace forest activists scale Essar HQ', *Press Release*, 22 January, viewed 20 September 2019, <www.greenpeace.org/india/en/press/2804/greenpe ace-forest-activists-scale-essar-hq/>.

Greenpeace India. 2014b, 'Greenpeace activists granted bail after being detained by Mumbai police', *Press Release*, 23 January, viewed 20 September 2019, <www.greenpeace.org/india/en/press/2800/greenpeace-activists-granted-bail-after-being-detained-by-mumbai-police/>.

Greenpeace India. 2014c, 'Greenpeace will prove defamation charges wrong. Billionaire corporation sues environment watchdog for Rs. 500 crore and demands a gag order on contents against the company', *Press Release*, 29 January, viewed 20 September 2019, <www.greenpeace.org/india/en/press/2789/greenpeace-will-prove-defamation-charges-wrong-billionaire-corporation-sues-environment-watchdog-for-rs-500-crore-and-dema nds-a-gag-order-on-contents-against-the-company/>.

Greenpeace India. 2014d, 'Greenpeace activists embarrass Moily and Essar', *Press Release*, 14 February, viewed 20 September 2019, <www.greenpeace.org/india/en/press/2746/gre enpeace-activists-embarrass-moily-and-essar/>.

Greenpeace India. 2014e, ' "No tree felling in Mahan till October": MP Govt to NGT', *Press Release*, 26 May, viewed 27 September 2019, <www.greenpeace.org/india/en/press/2596/no-tree-felling-in-mahan-till-october-mp-govt-to-ngt/>.

Greenpeace India. 2014f, 'Essar tries to use "leaked" IB report to clamp down on people's protests in Mahan forest, MP', *Press Release*, 17 June, viewed 20 September 2019, <www.greenpeace.org/india/en/press/2565/essar-tries-to-use-leaked-ib-report-to-clamp-down-on-peoples-protests-in-mahan-forest-mp/>.

Greenpeace India. 2014g, 'Essar seeks a blanket ban on all criticism of their proposed coal mine in Mahan', *Press Release*, 14 July, viewed 20 September 2019, <www.greenpeace.org/india/en/press/2556/essar-seeks-a-blanket-ban-on-all-criticism-of-their-proposed-coal-mine-in-mahan/>.

Greenpeace India. 2014h, 'Crackdown on Forest Rights Activists Intensifies. Police arrests two Greenpeace activists in the middle of the night in bid to "intimidate people" before Gram Sabha', *Press Release*, 30 July, viewed 20 September 2019, <www.greenpeace.org/india/en/press/2543/crackdown-on-forest-rights-activists-intensifies-police-arrests-two-greenpeace-activists-in-the-middle-of-the-night-in-bid-to-intimidate-people-before-gram-sabha/>.

Greenpeace India. 2014i, 'Bombay HC accepts Essar's affidavit to include malicious IB report that was "leaked" last month', *Press Release*, 31 July, viewed 20 September 2019, <www.greenpeace.org/india/en/Press/Bombay-HC-accepts-Essars-affidavit-to-include-malicious-IB-report-that-was-leaked-last-month/>.

Greenpeace India. 2014j, 'NGT renders forest clearance to Mahan Coal Ltd. invalid after SC verdict; Greenpeace, MSS celebrate but vow to oppose any attempt to re-allocate Mahan', *Press Release*, 26 September, viewed 20 September 2019, <www.greenpeace.org/india/en/press/2481/ngt-renders-forest-clearance-to-mahan-coal-ltd-invalid-after-sc-verdict-greenpeace-mss-celebrate-but-vow-to-oppose-any-attempt-to-re-allocate-mahan/>.

Greenpeace India. 2015a, 'Waidhan district court dismissed Essar's plea against Greenpeace', *Press Release*, 20 February, viewed 27 September 2019, <www.greenpeace.org/india/en/press/2357/waidhan-district-court-dismisses-essars-plea-against-greenpeace/>.

Greenpeace India. 2015b, 'High Court overturns Priya Pillai offloading, declares government move undemocratic', *Press Release*, 12 March, viewed 20 August 2019, <www.greenpeace.org/international/en/press/releases/2015/High-Court-overturns-Priya-Pillai-offloading-declares-government-move-undemocratic/>.

Greenpeace India. 2015c, 'Greenpeace India Director tells staff: "one month left to fight" before shutdown', *Press Release*, 5 May, viewed 14 August 2019, <www.greenpeace.org/india/en/Press/Greenpeace-India-Director-tells-staff-one-month-left-to-fight-before-shutdown/>.

Greenpeace India. 2015d, 'Madras High Court stays Greenpeace cancellation', *Press Release*, 20 November, viewed 14 August 2019, <www.greenpeace.org/india/en/Press/Madras-High-Court-stays-Greenpeace-cancellation/>.

Hiddleston, S. 2010, 'Poisoned ground', *Frontline Magazine*, vol. 27, no. 19, 24 September.

IANS. 2013, 'Anti-Kudankulam activists to raise protest to next level', *The Hindu*, 17 August, viewed 20 September 2019, <www.thehindu.com/news/national/tamil-nadu/antikudankulam-activists-to-raise-protest-to-next-level/article5032106.ece>.

Inamdar, N. 2013, '7 things you wanted to know about "Coalgate": a quick summary of the coal allocation scam', *Business Standard*, 15 October, viewed 14 August 2019, <www.business-standard.com/article/companies/7-things-you-wanted-to-know-about-coalgate-113101500366_1.html>.

Intelligence Bureau. 2014a, *Subject: Concerted Efforts by Select Foreign Funded NGOs to "Take Down" Indian Development Projects*.

Intelligence Bureau. 2014b, *Subject: FCRA Registered Greenpeace Spearheading a Concerted Campaign against India's Energy Expansion Plans*, viewed 14 August 2019, <www.documentcloud.org/documents/1201267-greenpeace-ib-report.html>.

International Energy Agency (IEA). 2014, *World Energy Outlook 2014*, Paris.

International Energy Agency (IEA). 2015, *India Energy Outlook: World Energy Outlook's Special Report*, Paris.

International Energy Agency (IEA). 2021, *India Energy Outlook 2021; World Energy Outlook Special Report*, viewed 20 January 2024, <www.iea.org/reports/india-energy-outlook-2021/energy-in-india-today>.

Jai, S. 2015, 'First coal block e-auction earns Rs 1 lakh cr. for States', *Business Standard*, 23 February, viewed 14 August 2019, <www.business-standard.com/article/economy-policy/first-coal-block-e-auction-earns-rs-1-lakh-cr-for-states-115022300039_1.html>.

Jai, S. 2016, 'India does not need any more coal-based capacity addition till 2022: Central Electricity Authority', *Business Standard*, 13 December, viewed 15 September 2019, <www.business-standard.com/article/economy-policy/india-does-not-need-more-coal-based-capacity-addition-till-2022-central-electricity-authority-116121300042_1.html>.

Jalihal, S. 2023a, 'Advantage Adani: power industry lobbies, coal ministry unlocks dense forests for mining', *The Reporters Collective*, 9 October, viewed 20 November 2022, <www.reporters-collective.in/trc/coal-forests-part-1>.

Jalihal, S. 2023b, *Coal Reform Overturned: After Failed Auctions, Centre Hands Out Coal Blocks at Discretion'*, 11 October, viewed 20 December 2023, <www.reporters-collect ive.in/trc/coal-forest-part-2>.

Jalihal, S. & Sambhav K. 2023, 'Modi government allowed Adani coal deals it knew were inappropriate', *Al Jazeera*, 1 March, viewed 20 January 2024, <www.aljazeera.com/econ omy/2023/3/1/modi-govt-allowed-adani-coal-deals-it-knew-were-inappropriate#:~:text= The%20Modi%20govt%20brought%20in,coal%2Dscam%20era'%20deals.&text= New%20Delhi%2C%20India%20%E2%80%93%20The%20Indian,his%20coal%20b usiness%2C%20documents%20reveal>.

Kaushal, A. 2015, 'Government cancels registration of 9,000 NGOs', *Business Standard*, 29 April, viewed 14 August 2019 <www.business-standard.com/article/current-affairs/gov ernment-cancels-registration-of-9-000-ngos-115042800367_1.html>.

Kaushik, K. 2015, 'Doing the needful: Essar's industry of influence', *Caravan Magazine*, 1 August, viewed 27 September 2019, <https://caravanmagazine.in/reportage/doing-need ful-essar-industry-influence>.

Kohli, K. 2012a, 'Mera Bharat Aur Mahan', *India Together*, 31 March, viewed 15 September 2019, <http://indiatogether.org/mahan-environment>.

Kohli, K. 2012b, 'Who needs CCI?', *Civil Society*, January.

Kohli, K. 2013, 'Two coal blocks and a political story', *Economic & Political Weekly*, vol. 48, no. 41, p. 13.

Kohli, K., Kothari, A. & Pillai, P. 2012, *Countering Coal?*, Discussion paper by Kalpavriksh and Greenpeace India, New Delhi/Pune and Bangalore, viewed 15 September 2019, <www.greenpeace.org/india/en/publication/989/countering-coal-community-forest-rig hts-and-coal-mining-regions-of-india/>.

Kohli, K. & Menon, M. 2020, *India's U-Turn on "Clean" Energy is a Bad Move*, 16 June, viewed 20 September 2020, <https://thewire.in/environment/coal-washing-environment-ministry-changing-rules>.

Kumar, D. and Bhuchar, P. 2012, 'Green Terror: outdated environmental laws and inflexible ministers strangle Indian economy', *India Today*, 5 October, viewed 14 August 2019, <www.indiatoday.in/india/story/green-terror-outdated-environmental-laws-and-inflexi ble-ministers-strangle-indian-economy-117900-2012-10-05>.

Lahiri-Dutt, K. 2014, 'Introducing coal in India: energising the nation', in K. Lahiri-Dutt (ed.), *The Coal Nation: Histories, Politics and Ecologies of Coal in India*, Ashgate, Aldershot, pp. 1–37.

Lahiri-Dutt, K. 2016, 'The diverse worlds of coal in India: energising the nation, energising livelihoods', *Energy Policy*, vol. 99, pp. 203–13.

Lahiri-Dutt, K., Balakrishnan, R. & Ahmad, N. 2012, 'Land acquisition and dispossession: private coal companies in Jharkhand', *SSRN*, 3 March, viewed 15 September 2019, <https://ssrn.com/abstract=2015125>.

Lakshmi R. V. 2013, *Junglistan Diaries – TEDx with Brikesh Singh*, Greenpeace India, 24 May, viewed 27 September 2019, <www.greenpeace.org/india/en/story/294/junglistan-diaries-tedx-with-brikesh-singh/>.

Marston, A. 2011, *No Fairy Tale: Singrauli, India, Still Suffering Years After World Bank Coal Investments*, Bretton Woods Project, London, November, viewed 20 September 2019, <http://old.brettonwoodsproject.org/doc/env/singrauli.pdf>.

Mathur, A. 2015a, 'HC rules in favour of Greenpeace Pillai: State can't muzzle dissent', *Indian Express*, 15 March, viewed 14 August 2019, <http://indianexpress.com/article/india/high-court-quashes-look-out-circular-against-greenpeace-activist-priya-pillai/>.

Mathur, A. 2015b, 'Delhi High Court allows Greenpeace to operate 2 domestic accounts for donations', *Indian Express*, 28 May, viewed 14 August 2019, <http://indianexpress.com/article/india/india-others/delhi-hc-allows-greenpeace-to-use-two-accounts-for-domestic-donations/>.

Mazoomdar, J. 2014a, 'Delhi pollution higher than Beijing's report denied', *BBC*, 31 January, viewed 20 September 2019, <http://mazoomdaar.blogspot.com/2014/01/>.

Mazoomdar, J. 2014b, 'IB's NGO-scare report to Modi plagiarises from old Modi speech', *Indian Express*, 13 June, viewed 14 August 2019, <http://indianexpress.com/article/india/politics/ibs-ngo-scare-report-to-modi-plagiarises-from-old-modi-speech/>.

Mehdudia, S. 2012, 'Congress will take on CAG over Coalgate report before PAC', *The Hindu*, 21 September, viewed 15 September 2019, <www.thehindu.com/todays-paper/tp-national/congress-will-take-on-cag-over-coalgate-report-before-pac/article3920568.ece>.

Mehta, N. 2014, 'Don't succumb to unscientific prejudices against Bt. crops: Manmohan Singh', *Live Mint*, 3 February, viewed 20 September 2019, <www.livemint.com/Politics/jUYRmnHqK5WEX7p55VWlPK/Dont-succumb-to-unscientific-prejudices-against-Bt-crops.html>.

Ministry of Environment and Forests. 2011, *Allocation of Mahan Coalblock in Madhya Pradesh to Essar and Hindalco*, Government of India, 8 July, viewed 20 September 2019, <www.moef.nic.in/downloads/public-information/Mahan.pdf>.

Ministry of Finance. 2012, *Budget Notification No. 12/2012 – Customs*, Government of India, 17 March, viewed 20 September 2019, <http://indiabudget.nic.in/ub2012-13/cen/cus1212.pdf>.

Ministry of Power. 2012, *Report of the Working Group on Power for Twelfth Plan (2012–17)*, Government of India, New Delhi, January 2012, viewed 20 September 2019, <http://planningcommission.gov.in/aboutus/committee/wrkgrp12/wg_power1904.pdf>.

Nair, R. 2015, 'Priya Pillai case: the shadows lengthen', *Economic and Political Weekly*, vol. 50, no. 10, pp. 12–14.

Narayanan, N. 2015, 'As government denounces "bad activist" Priya Pillai in court, "good activists" come to her defence', *Scroll.in*, 17 February, viewed 20 September 2019, <https://scroll.in/article/707224/as-government-denounces-bad-activist-priya-pillai-in-court-good-activists-come-to-her-defence>.

Nigar, S. 2014, 'The Case for an Intelligible Agency', *HARDNEWS*, 12 November, viewed 20 August 2019, <www.hardnewsmedia.com/2014/11/6443>.

Nileena, M. S. 2018, 'Coalgate 2.0: The Adani Group reaps benefits worth thousands of crores of rupees as the coal scam continues under the Modi government', *Caravan Magazine*, 1 March, viewed 20 September 2019, <https://caravanmagazine.in/reportage/coalgate-2-0>.

Niti Aayog, Government of India. 2018a, *Draft National Energy Policy*, version as of 27 June 2017, viewed 20 January 2024, <www.niti.gov.in/sites/default/files/2022-12/NEP-ID_27.06.2017.pdf.pdf>.

Niti Aayog, Government of India. 2018b, *Transformation of Aspirational Districts: Baseline Ranking and Real-time Monitoring Dashboard 2018*, viewed 20 January 2024, <www.niti.gov.in/sites/default/files/2018-12/AspirationalDistrictsBaselineRankingMarch2018.pdf>.

Niyogy, K. 2015, 'A war to save our forests', *Live Mint,* 7 November, viewed 15 September 2019, <www.livemint.com/Leisure/fAB8jZp3H6OUkFxdAUVbXK/A-war-to-save-our-forests.html>.

Oxford Poverty and Human Development Initiative. 2018, *Multidimensional Poverty Index,* University of Oxford, viewed 20 January 2024, <https://ophi.org.uk/multidimensional-poverty-index/>.

Padel, F. 2016, 'Investment induced displacement and the ecological basis of India's economy', in S. Venkateshwar & S. Bandyopadhyay (eds), *Globalisation and the Challenges of Development in Contemporary India,* Springer, Singapore, pp. 147–169.

Pillai, P. 2017, 'An ongoing battle', *Frontline Magazine,* 30 August, viewed 20 September 2018, <https://frontline.thehindu.com/social-issues/an-ongoing-battle/article9831547.ece>.

Pillai, P., Gopal, V. & Kohli, K. 2011, *Singrauli: The Coal Curse – A Fact Finding Report on the Impact of Coalmining on the People and Environment of Singrauli,* Greenpeace India, September 2011, viewed 27 September 2019, <www.greenpeace.org/india/en/publication/1006/singrauli-the-coal-curse/>.

Pioneer. 2014, *HC Asks SP to Probe Gram Sabha Forgery',* 16 July, viewed 27 September 2019, <www.dailypioneer.com/2014/state-editions/hc-asks-sp-to-probe-gram-sabha-forgery.html>.

Planning Commission. 2013, *Twelfth Five Year Plan (2012–2017): Faster, More Inclusive and Sustainable Growth,* Government of India, vol. 1, viewed 20 September 2019, <http://planningcommission.gov.in/plans/planrel/12thplan/pdf/12fyp_vol1.pdf>.

Planning Commission. 2014, 'Government establishes Niti Ayog to replace Planning Commission', *Press Release,* Government of India, 15 August, viewed 20 September 2019, <http://planningcommission.gov.in/press-release.pdf>.

Planning Commission of India. n.d, *Archived Website,* viewed 15 November 2015, <http://planningcommissionarchive.nic.in>.

Press Trust of India. 2011, 'Compromise, not abandonment needed in "Go, No-Go" policy: Ramesh', *The Hindu,* New Delhi, 9 April, viewed 27 September 2019, <www.thehindu.com/news/national/Compromise-not-abandonment-needed-in-lsquoGo-No-Gorsquo-policy-Ramesh/article14677028.ece>.

Press Trust of India. 2012, 'Cab nod for panel to clear mega projects, renames NIB as CCI', *Business Standard,* 13 December, viewed 20 August 2019, <www.business-standard.com/article/economy-policy/cab-nod-for-panel-to-clear-mega-projects-renames-nib-as-cci-112121303016_1.html>.

Press Trust of India. 2014, 'Moily planning to give oil field to Essar cheaply, alleges AAP', *Economic Times,* New Delhi, 17 April, viewed 27 September 2019, <https://economictimes.indiatimes.com/news/politics-and-nation/moily-planning-to-give-oil-field-to-essar-cheaply-alleges-aap/articleshow/33828233.cms>.

Press Trust of India. 2015, 'Parliamentary panel welcomes PMO monitoring coal sector issues', *Press Trust of India,* 16 December, viewed 19 August 2021, <https://economictimes.indiatimes.com/industry/indl-goods/svs/metals-mining/parliamentary-panel-welcomes-pmo-monitoring-coal-sector-issues/articleshow/50206613.cms?from=mdr>

Press Trust of India. 2016, 'India to stop thermal coal imports; save Rs 400,000 crore: Piyush Goyal', *Business Standard,* 16 April, viewed 20 September 2019, <www.business-standard.com/article/pti-stories/india-to-stop-thermal-coal-imports-save-rs-40-000-cr-116041500319_1.html>.

Press Trust of India. 2023, 'India will continue to rely on coal power until it becomes developed country, says Bhupinder Yadav', *The Hindu,* 19 December, viewed 20 January

2024, <www.thehindu.com/sci-tech/energy-and-environment/india-resisted-pressure-from-developed-nations-for-phase-out-of-fossil-fuels-says-bhupender-yadav/article6 7653184.ece>.

Rajagopal, K. 2014, 'Supreme court quashes allocation of 2014 coal blocks', *The Hindu*, 24 September, viewed 15 September 2019, <www.thehindu.com/news/national/supreme-court-quashes-allocation-of-all-but-four-of-218-coal-blocks/article6441855.ece>.

Ramanathan, S. 2015, 'Essar goes after the Caravan with lawsuit for damning article, magazine gives it right back', *The News Minute*, 25 August, viewed 27 September 2019, <www.thenewsminute.com/article/essar-goes-after-caravan-lawsuit-damning-article-magazine-gives-it-right-back-33667>.

Ramesh, J. 2015, *Green Signals; Ecology, Growth and Democracy in India*, Oxford University Press.

Ranjan, A. 2014, 'Foreign-aided NGOs are actively stalling development, IB tells PMO in a report', *Indian Express*, 7 June, viewed 20 August, 2019, <http://indianexpress.com/article/india/india-others/foreign-aided-ngos-are-actively-stalling-development-ib-tells-pmo-in-a-report/>.

The Rockerfeller Foundation. 2018. *24X7 Power is About Access, Not Electrification*, 22 January, viewed 20 January 2024, <www.rockefellerfoundation.org/insights/perspective/24x7-power-access-not-electrification/>.

Rosewarne, S. 2016, 'The transnationalisation of the Indian coal economy and the Australian Political economy: the fusion of regimes of accumulation?', *Energy Policy*, viewed 20 June 2020, <https://doi.org/10.1016/j.enpol.2016.05.022>.

Roy, B. & Schaffartzik, A. 2021, 'Talk renewables, walk coal: the paradox of India's energy transition', *Ecological Economics*, vol. 180, no. 106871.

Saikia, S. 2012, 'CIL foresees coal demand-supply gap till 13th Plan', *The Hindu Business Online*, 19 November, viewed 20 September 2019, <www.thehindubusinessline.com/economy/cil-foresees-coal-demandsupply-gap-till-13th-plan/article4110865.ece>.

Sakhdher, Justice R. 2015, *Priya Parameshwaran Pillai vs Union of India and Ors.*, Delhi High Court, 12 March, viewed 20 September 2019, <https://indiankanoon.org/doc/64486 862/>.

Sampath, G. 2016, 'Time to repeal the FCRA', *The Hindu*, 27 December, viewed 20 September 2019, <www.thehindu.com/opinion/lead/Time-to-repeal-the-FCRA/article1 6946222.ece>.

Saraogi, A. 2022, *FCRA Judgement: Assessing the impact on NGOs'*, Supreme Court Observer, 6 June, viewed 20 January 2024, <www.scobserver.in/journal/fcra-judgment-assessing-the-impact-on-ngos-noel-harper/#:~:text=First%2C%20the%20Amendm ent%20bars%20recipient,from%2050%25%20to%2020%25>.

Sarma, *E.A.S. 2014, quoted in 'BJ*P, Congress guilty of taking foreign funds Delhi HC', *Association for Democratic Reforms*, March, viewed on 20 June 2021, <www.businesss tandard.com/article/economy-policy/nda-govt-s-grouse-with-greenpeace-mahan-coal-block-protests-115022000023_1.html>.

Sethi, N. 2015, 'NDA govt's grouse with Greenpeace: Mahan coal block protests', *Business Standard*, New Delhi, 20 February, viewed 20 September 2019, <www.businessstand ard.com/article/economy-policy/nda-govt-s-grouse-with-greenpeace-mahan-coal-block-protests-115022000023_1.html>.

Shahi, R. V. 2003, 'Mission 2012: Power for all', *ELECTRICAL INDIA*, vol. 43, no. 10, pp. 61–64.

Sharma, R. N. & Singh, S. R. 2009, 'Displacement in Singrauli region: entitlements and rehabilitation', *Economic and Political Weekly*, vol. 44, no. 51, pp. 62–69.

Shrivastava, P. 2014, 'Essar v Greenpeace case files', *Legally India,* 23 June, viewed 20 September 2021, <https://www.legallyindia.com/the-bench-and-the-bar/case-files-essar-v-greenpeace-20140623-4810>.

Singh, V. 2015a, 'Foreign Contribution Regulation Act: new crackdown on NGO foreign funds', *Indian Express,* 15 June, viewed 14 August 2019, <http://indianexpress.com/arti cle/india/india-others/foreign-contribution-regulation-act-new-crackdown-on-ngo-fore ign-funds/>.

Singh, S. R. 2015b, 'Neither Switzerland nor Singapore but a site of daily struggle', *The Wire,* 24 July, viewed 30 September 2019, <https://thewire.in/economy/neither-switzerl and-nor-singapore-but-a-site-of-daily-struggle>.

Singh, S. R. 2017, 'Political economy of land acquisition and resource development in India', *Political Economy of Contemporary India,* January, pp. 279–306.

Sreenivas, S. 2015, *How Greenpeace India Raises Its Funds,* Greenpeace India, 20 May 20, viewed 20 September 2019, <www.greenpeace.org/india/en/Blog/Campaign_blogs/how-greenpeace-india-raises-its-funds/blog/52929/>.

Subramanian, S. 2015, 'India's war on Greenpeace', *The Guardian,* 11 August, viewed 15 September 2019, <www.theguardian.com/world/2015/aug/11/indias-war-on-gre enpeace>.

Suresh, A. E., & Sarin, R. 2015, 'Essar Leaks: French cruise for Nitin Gadkari, favours to UPA Minister, journalists', *Indian Express,* 27 February, viewed 27 September 2019, <https://indianexpress.com/article/india/india-others/essar-leaks-french-cruise-for-gadk ari-favours-to-upa-minister-journalists/>.

Talukdar, R. 2019, 'Profit Before People: Why India Has Silenced Greenpeace', *New Matilda,* 29 March, viewed 20 June 2020, <https://newmatilda.com/2019/03/29/pro fit-before-people-why-india-has-silenced-greenpeace/?fbclid=IwAR2NwczXun5q1Ug-g3kSxvaOJLInTfwNlnS6Up3Xxts7A7XZF_wbvOsHKYc>.

The Telegraph. 2011, *Jairam Locks Coal Blocks – GOM to Decide Allotments to Essar, Hindalco Power Plants,* New Delhi, 9 July, viewed 27 September 2019, <www.telegra phindia.com/india/jairam-locks-coal-block-gom-to-decide-allotments-to-essar-hindalco-power-plants/cid/369687>.

The Telegraph. 2024, *New Policy: Editorial on Modi Government Weaponising FCRA to Shut Down Civil Society Outfits,* 26 January, viewed 26 January 2024, <www.telegraphin dia.com/opinion/new-policy-editorial-on-modi-government-weaponising-fcra-to-shut-down-civil-society-outfits/cid/1996222>.

Thakur, T., Deshmukh, S. G., Kaushik, S. C., & Kulshrestha, M. 2005, 'Impact assessment of the Electricity Act 2003 on the Indian power sector', *Energy Policy,* vol. 33, no. 9, pp. 1187–1198.

Times News Network. 2015, 'Mahan coal block won't be auctioned, coal min says in RTI reply', *The Times of India,* 21 March, viewed 20 September 2019, <http://timesofindia.ind iatimes.com/city/bhopal/Mahan-coal-block-wont-be-auctioned-coal-min-says-in-RTI-reply/articleshow/46642702.cms>.

Tongia, R. 2003, *The Political Economy of Indian Power Sector Reforms',* Working Paper No. 4, Program on Energy and Sustainable Development, Centre for Environmental Science and Policy, Stanford Institute for International Studies, Stanford University, Standford, CA, December, viewed 20 September 2019, <https://pdfs.semanticscholar.org/ 9639/07d0425aa8a5d222707cd1feaacc74b75597.pdf?_ga=2.20463096.1196817498.157 6502584-1094641479.1560696696>.

United Nations Human Rights. 2016, 'Analysis of International Law, Standards and Principles – applicable to the – Foreign Contributions Regulation Act 2010', *Information*

Note, 20 April, viewed 14 August 2019, <www.ohchr.org/Documents/Issues/FAssociat ion/InfoNoteIndia.pdf>.

Vaidyanathan, A., Vaidyanathan, A, Vanaik, A. *et al*. 2015, 'Greenpeace Ban: Violation of Rights', *Economic and Political Weekly*, 25 April, vol. 50, no. 17, p. 2.

Varadarajan, S. 2015a, 'To the Point', *Headlines Today*, 14 January, viewed 20 September 2019, <www.youtube.com/watch?v=4cO52Ao_eAI>.

Varadarajan, S. 2015b, 'An attack on the right to liberty', *Live Mint,* 22 January, viewed 14 August 2019, <www.livemint.com/Opinion/l7ThUHOzcRIVNeEl636BwJ/An-attack-on-the-right-to-liberty.html>.

Vyawahare, M. 2018, 'State apathy continues to choke Singrauli,' *Hindustan Times,* Singrauli, 20 August, viewed 20 September 2019, <www.hindustantimes.com/india-news/state-apa thy-continues-to-choke-singrauli/story-xrsGiqftFRnhRjgvNGobgO.html>.

Waray, S. 2018, 'Bringing 24X7 power to all by 2022', *Hindu Business Line,* viewed 20 September 2019, <www.thehindubusinessline.com/opinion/bringing-24x7-power-to-all-by-2022/article22136414.ece1>.

Yadav, S. 2014, 'Only 11 of 22 NGOs in IB report are FCRA compliant, none filed '13–'14 statement', *Indian Express*, 20 June, viewed 14 August 2019, <http://indianexpress.com/ article/india/india-others/only-11-of-22-ngos-in-ib-report-are-fcra-compliant-none-filed-13-14-statement/>.

Yla-Anttila, T. & Swarnakar, P. 2017, 'Crowding-in: how Indian civil society organizations began mobilizing around climate change', *British Journal of Sociology*, vol. 68, no. 2, pp. 273–292.

Legislations

Coal Bearing Areas (Acquisition and Development) Act 1957.
Coal Mines Nationalisation Act 1973.
Coal Mines (Special Provisions) Act 2015.
Foreign Contributions (Regulation) Act 2010.
Indian Electricity Act 2003.
Land Acquisition Act 1894.
Mines and Minerals (Development and Regulation) Act 1957.
National Mineral Policy 1993
Scheduled Tribes and Other Traditional Forest Dwellers (Recognition of Forest Rights) Act 2006.

Websites

Ministry of Coal https://coal.nic.in/
www.hindalco.com
www.ntpc.co.in
https://singrauli.nic.in

5   An anti-coal movement in India's energy capital

'Gaon Chodab Nahi,
Jangal Chodab Nahi,
Maye Mati Chodab Nahi
Ladai Chodab Nahi'

We will not leave our village
Our Forest
Our Land – our Earth Mother
We will not stop fighting.
Song of resistance written by Adivasi leader Bhagwan Majhi from Odisha

In March 2015, a grassroots mobilisation in the Singrauli district of the central Indian state of Madhya Pradesh saved the Mahan forests from the imminent risk of coal mining. The four-year struggle came to an end when a newly elected Indian government under Narendra Modi cancelled the Mahan coal block from the auction list. The cancellation followed a 2014 Supreme Court order that deemed the previous government's allocation of coal blocks to be arbitrary and corrupt. The local resistance Mahan Sangharsh Samiti (MSS) or the Mahan Resistance Front was formed of people from eleven villages fringing the Mahan forests – Amelia, Budher, Bandhaura, Suhira, Barwantola, Nagwa, Khairahi, Karsua Lal, Piderwah, Bandha and Jamgadhi. The largest village Amelia had a population of 2200 as of 2014. Overall, the coal mine threatened to disrupt the forest-dependent livelihoods of 50,000 people from 54 villages in the region (Greenpeace India 2014e).[1]

The local movement was mobilised by the Indian arm of the international environmental non-governmental organisation (ENGO) Greenpeace. Greenpeace activists began visiting Mahan in 2010 to start building awareness about forest rights amongst village communities. Greenpeace established an office in Waidhan, the administrative headquarters of the Singrauli district, to work in a sustained manner in the region. An alliance eventually formed between the previously non-politicised community at Mahan and Greenpeace. Together they ran a movement that ended with the cancellation of the Mahan coal block.

DOI: 10.4324/9781003410416-5

Communities in Mahan, particularly families from the villages of Khairahi, Bandhaura, Karsualal and Nagwa, first faced the prospect of displacement in 2007 when the state government of Madhya Pradesh acquired land for the Essar thermal power plant. Although on paper Essar appeared to be offering a generous rehabilitation and resettlement package for the project affected, in reality project affected people alleged irregularities in the process and cases of compensations not having been granted (Sharma and Singh 2009).[2] The incident that finally shaped peoples' resolve to resist the coal mine occurred in 2013 when the local government and company officials forged villagers' signatures on a referendum on mining.

During my ethnographic research in Mahan, I found that the mobilisation of this rural constituency was motivated by a growing discontent with State and company interference in their everyday lives. Their resolve to fight was shaped by a newfound understanding of forest rights. This chapter answers a central question of the book from the Indian perspective by taking a look at the ground level: what are the discourses, tactics, and relations of anti-coal activism? What is its significance for environmentalism and its context in India? The chapter highlights the process of politicisation of the community and the subsequent formation of the local movement, through biographical accounts of six movement leaders whom I had repeated interactions with and interviewed during my four field trips to Mahan. All names have been changed to protect their identities. I also trace the movement's progress through analyses of resources, tactics, events, campaign moments and protest narratives.

Beyond the secondary sources of news articles, reports and publications, and primary sources from interviews and field notes, the research for this chapter has also been informed by postings by MSS members on Radio Sangharsh (resistance radio), a community portal, and diary entries chronicling the resistance by one of the MSS leaders that were made available to me. The Greenpeace Mahan team referred to in this chapter consisted of the Greenpeace activist who led the local mobilisation, another campaigner, a communications officer, and two community engagement officers. This Greenpeace team was based out of Waidhan for several months in the year during the active stages of the movement.

The Background summarises the history of mass displacements and ecological destruction in Singrauli through three waves of industrial developments from the 1960s. Sections 5.1, 5.2, 5.3, 5.4 and 5.5 describe the build up and the various dimensions of the anti-coal resistance in Mahan based on my ethnographic research. The Analysis discusses the significance of the Greenpeace-MSS anti-coal resistance against a local context of historic marginalisation of forest-dependent communities in Singrauli and a global climate imperative to 'End Coal!'. The Conclusion highlights insights from this resistance for global climate activism's new approach.

Background: discontent and displacement in Singrauli

The success of the grassroots movement in Mahan in getting the voices of village communities heard against the State–corporate apparatus needs to be understood alongside the Singrauli region's history of human displacement and ecological destruction through five decades of concentrated industrial development. The

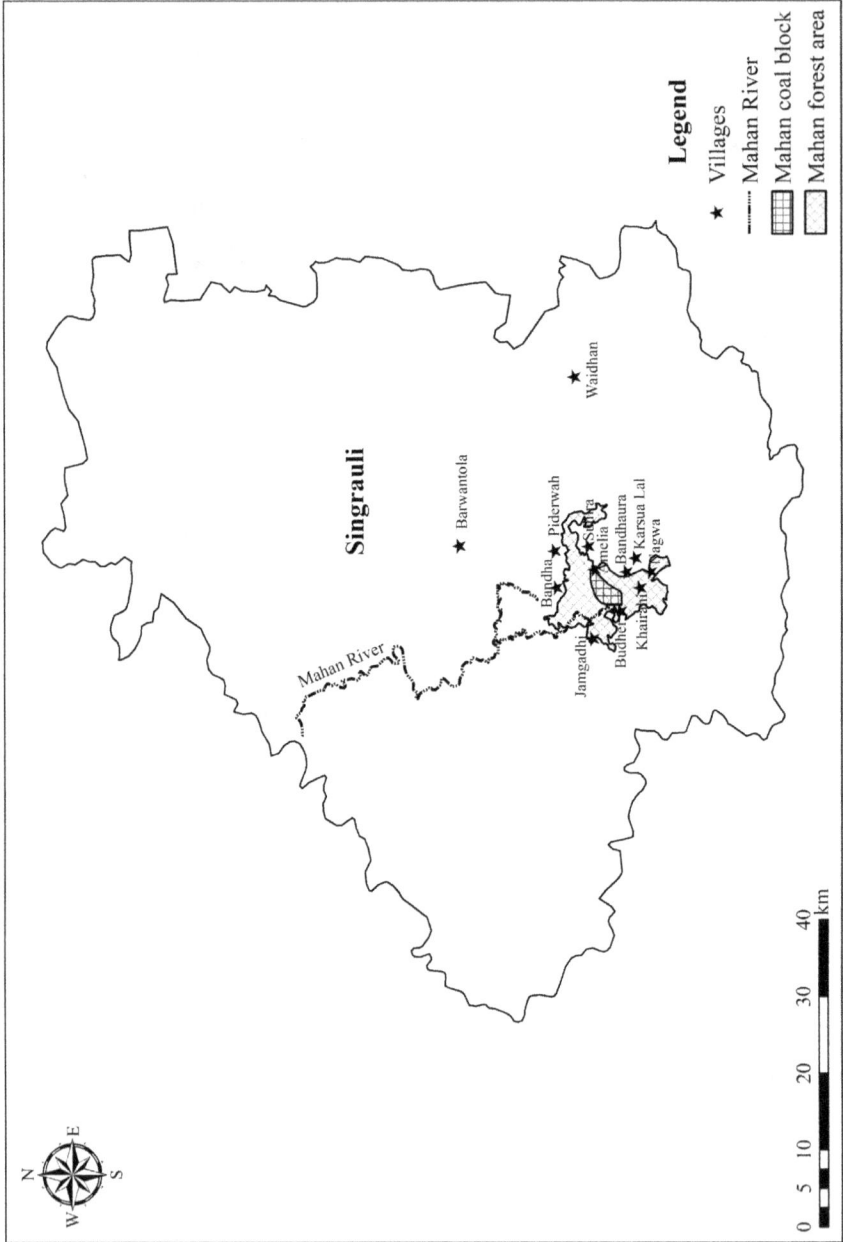

Figure 5.1 Singrauli and Mahan Forest area.

multiple waves of displacements in Singrauli can be partly attributed to inadequate developmental planning, particularly the failure to demarcate coal-bearing and non coal-bearing areas so that people could plan their post-displacement settlement and access alternative lands and livelihoods (Singh 2015).

In many cases, their predicament was exacerbated by an inability to claim compensation or jobs. Most rehabilitation policies only acknowledged officially recognised landowners. Failure to produce proof of ownership even though they had lived on the land for generations made Adivasis particularly vulnerable to dislocation without compensation. Coal India Limited's resettlement policy only considered persons with two or more acres of land as eligible for employment after displacement (Singh 2015). Such policies exacerbated injustices from industrial developments.

In the 1980s, a protest movement of project-affected communities was formed under the banner of the Srijan Lok Hit Samiti (community welfare committee) to fight for the entitlements of the displaced (Dokuzovic 2012). A series of early protests against the Northern Coalfields Limited and National Thermal Power Corporation (that later became NTPC Limited) were followed by a massive demonstration in 1988 of mostly Adivasis who were able to attract the attention of the media, civil society groups, and the World Bank (that funded the thermal projects). The State and State-owned companies responded with remedial measures that were nevertheless meagre and were ridden with corruption (Sharma and Singh 2009).[3] The structural divide between responsibility for the environment and minority rights that belonged to the State and resettlement and rehabilitation that belonged to the companies and investors resulted in making affected communities particularly vulnerable to the process of industrial expansion. Historical evidence shows that they often fell through the gap between the State–corporate divide in responsibilities for their futures.

Despite NTPC Ltd routinely violating the World Bank's policy guidelines for rehabilitation, the latter 'turned the other way' on the company's violations by ignoring local and non-governmental organisation (NGO) testimonials (Dokuzovic 2012).[4,5] Other deprivations of the rights of displaced locals included the company employing a majority of migrant workers at lower wages from even poorer areas such as Bihar, Jharkhand and Chhattisgarh at project construction sites (Pillai et al. 2011).

Since the days of large dams and State-owned coal projects, resistances like the Narmada Bachao Aandolan and later the National Alliance of Peoples Movement, and legal measures like the *National Rehabilitation and Resettlement Policy 1997*, and later the *Right to Fair Compensation and Transparency in Land Acquisition, Rehabilitation and Resettlement Act 2013*, have created rights for displaced people that did not exist before. In India's energy capital Singrauli, such efforts have however been undercut by a pervasive culture of favouritism by the State towards industry and treating ecological sustainability and social justice with disdain.

Systematic violations of communities' rights and entitlements through decades of intensive industrialisation had created acute and chronic unemployment and a lack of future prospects. Landless project-displaced communities, largely Adivasi, received no compensation and were left with no prospect of livelihood after the loss of shared natural resources such as forests, ponds, grazing lands, river beds and fisheries. *Coal Curse,* a Greenpeace fact-finding report on Singrauli, concluded that the lofty aim of regulations such as the *National Mineral Policy (NMP) 2008* to improve the lives of communities in mineral-rich industrial areas had not been able to improve the circumstances of Singrauli's marginalised residents (Pillai et al. 2011).[6] On the implementation of the *Forest Rights Act 2006* (FRA), although the Greenpeace fact-finding team was assured by the Collector of Singrauli that the new administration would ensure compliance, the report expressed concerns about the fundamental lack of awareness on the part of communities about their rights over forests.[7]

Industrial pollution from three decades of concentrated mining and burning of coal had a telling effect on Singrauli's landscape. Dokuzovic (2012) provides a sordid account:

> Upon entering Singrauli, one is immediately struck by the overwhelming environmental damage, from the hardly breathable air to the blackened, ash-covered landscape. Both land and water have been fiercely and systematically polluted … Coal mining has lead to dangerous coal fires, overheating of the land, clogging of water sources, soil erosion, and loss of soil fertility. Some of the toxins released from coal mining are arsenic, lead, mercury, and radium, as well as uranium and thorium, which lead to radioactive contamination. The direct dumping of toxins has polluted the land, water, and air with mercury and heavy metals. Deforestation has released greenhouse gases, aside from destroying the resources of local people and wildlife.
>
> (para 8)

Toxic fly ash from Singrauli's thermal power plants flowed into the Rihand dam, clogging and contaminating the land and water. High concentrations of toxic acidity turned some water sources into what locals called 'death water' that could corrode human and animal flesh on contact. However, the companies disregarded the pleas of locals to build bridges across the toxic waters (Dokuzovic 2012).

Events such as these do not only belong in the past for Singrauli. As the walls of the fly ash dam of Reliance Power's Sasan mega thermal power plant, one of the world's largest, collapsed in 2020, toxic sludge spilled into Singrauli's villages and agricultural land, contaminated water reservoirs, and even killed six people. While the company had ignored notices from the State Pollution Control Board about the ash dams, the district administration had ignored citizens' appeals to enforce proper guidelines for the building and maintenance of ash dams (Dutta 2020). Such events indicate that an everyday story of corruption, violations and neglect by the State–corporate nexus continues to unfold in India's largest industrial zone and the second most polluted (Talukdar and Pillai 2022).

Since 2006, private energy projects with captive coal blocks allocated in Singrauli's last remaining intact forests portended a new complexity in the region's displacement saga. Forests are home to forest-dependent families that have lived on the land for generations and often without formal property titles (Sharma and Singh 2009). Although Singrauli's present landscape appears overcome by an energy juggernaut that makes any other vision impossible to conceive, just two generations ago the region's small landholders and Adivasis were practising forest-dependent and farming livelihoods. Singrauli was carved out as a separate district in 2008 to facilitate private coal mining. The scale of industrialisation, particularly the extent and concentration of coal mining and burning, had destroyed Singrauli's social and ecological balance, making it neither the Switzerland nor Singapore that governments had promised, but a site of daily struggles for survival and a symbol of dystopia (Singh 2015).[8]

5.1 Use and abuse of the Forest Rights Act

Greenpeace's first step towards engaging the community at Mahan was about building awareness about people's forest rights, about legal provisions that require the State and companies to follow due process in seeking their consent over mining.

The *Panchayat Extension to Schedule Areas Act 1996* (PESA) requires the consent of village councils 'before making the acquisition of land in the Scheduled Areas for development projects and before re-settling or rehabilitating persons affected by such projects in the Scheduled Areas'. The *Forest Rights Act 2006* requires community consent and both individual and community claims over forestland to be settled before allowing the diversion of forestlands for mining to go ahead. This was made explicitly clear through a circular issued by the Ministry of Environment and Forests to state governments (MOEF 2009). The FRA also extends the right to claim forest rights to other forest-dependent communities apart from scheduled tribes.

While the grassroots mobilisation realised their rights to contest the loss of livelihoods in the face of impending coal mining in the region through the FRA, the state government administration that strongly supported the project reverted to what has become a common practice of violation of the Act, by forging consent at village council meetings. How the State used and abused the *Forest Rights Act 2006* made a strong impact on the local movement, its genesis, and its tactics.

Forest rights as an organising plank

In the initial meetings organised in the villages around Mahan, the Greenpeace team engaged the community in a dialogue to understand how they understood their ownership over the forests. A survey of the community's relationship with the Mahan forests demonstrated their indispensability to local livelihoods, as well as a sense of disempowerment due to a lack of awareness of newly instituted legal rights that they can claim over their local forests:

> Every year during the mahua season I shut down my house in Amelia village and come to stay in the Mahan forests for about a month to collect mahua, which I sell at Rs. 17-20 (A$ 0.31-0.36) /kilo. We also collect other forest

produce like tendu leaves, chironji, harra, bamboo, mushroom etc. but we are now hearing that these forests will be given to the company for mining coal. If the government gives away these forests we have no other means to live and we will not even get any compensation because we have no rights over these forests.

Respondent from Amelia, in Kohli et al. (2012, p. 14)

Given the potential of forest rights to empower communities in coal-rich areas to protect their livelihoods from mining-related disruptions, the paper suggested that:

Even though tribal and other forest dwelling communities have been residing on their lands for generations, the lack of recognition of rights and historical oppression at the hands of the State or forest mafia has meant that many communities do not believe that they can assert their rightful claim on the forests and forest produce...The process of recognition of rights which includes getting informed, engaging in collective discussions and evidence-gathering, mapping and so on, would enable a change in this reality. It would allow for communities to believe in their rightful claims over forests. This presumably can translate into forest dwelling communities feeling the need to hold on to what they are able to recognise as theirs.

Kohli et al. (2012, p. 8)

Forging of mining consent by the State and the company

A strong State–corporate nexus and eagerness by governments to secure mining revenues play a significant role in causing violations of processes under the PESA and FRA for community consent (for mining) and forest rights. The forgery of community consent is one of the most common ways in which the local administration violates these processes to favour companies. The incident of the forgery of signatures at the *Gram Sabha* in Amelia village on 6 March 2013 proved to be a turning point in the resistance.

MSS members described the forgery to me. Only a small crowd of 182 residents from Amelia village attended the village council meeting for a referendum on the Mahan coal mine. The meeting was overseen by the head of Amelia village who supported the project, and the local administrative officer. People's request to see the signed *Gram Sabha* resolution was denied. The signatures of 1125 people, including two deceased and one jailed (during the referendum) were subsequently forged on to the resolution document, turning a referendum that had said no to mining into a yes in the final document (Greenpeace India 2013a).

The episode became public knowledge only after Greenpeace obtained documents on the *Gram Sabha* proceedings through the right to information (RTI) process (Pioneer 2014). But MSS members' suspicions had been raised that very evening after the *Gram Sabha,* on seeing the conduct of the local administrator and village chief. After the meeting, the administrator and the police went around the village forcing people to sign the resolution. But that still did not give them enough signatures. MSS members suspect that they forged signatures on a mass scale later at night when company officials were seen at the village chief's residence. Testimonials on Radio Sangharsh indicated MSS members' shock and betrayal at seeing their own names on the resolution document once it had been obtained through the RTI. MSS leaders Dayanath and Ramlal from Amelia travelled to New Delhi, and along with Greenpeace, testified before the central government's Tribal Affairs Minister, about the forgery of the *Gram Sabha* resolution. The Union Minister criticised the state government of Madhya Pradesh for holding a referendum without the entire community's knowledge, and for the local administration's role in forging consent (Ghatwai 2015).

Mahan claims community forest rights

The Mahan Coal Limited (MCL) claimed that the majority of the locals supported the project. The CEO of MCL alleged that the coal mine was being delayed on the pretext of Adivasi rights, using the ploy of claiming forest rights:

> There is not a single person or family residing directly inside the forest area of the Mahan coal block. Thus, in a strict sense, claims being made by vested interests are outside the preview of Forest Rights Act…this is perhaps one of the unique cases in the country where in spite of there not being any displacement, resettlement, benefits amounting to significant sums, will accrue to villagers.
>
> MCL CEO Ramakant Tiwari quoted in Trivedi (2014, para 9)

Tactics of delegitimising the community's rights over the Mahan forests were coupled with attempts at physically disrupting the actual process of claiming forest rights. The village of Amelia first attempted to pass a resolution declaring their rights over Mahan forests in a *Gram Sabha* on the day of Indian Independence, 15 August, in 2012. Local goons hired by the company threatened people and disrupted the meeting. Despite threats and intimidations, the three villages of Amelia, Budher and Suhira eventually managed to file for community forest rights (CFRs) by 2016.

The company's tactics to disrupt forest rights

Bribery is a common tactic deployed by mining corporations to manufacture consent (Chowdhury and Aga 2020). The Indian Minister for Tribal Affairs had ordered fresh *Gram Sabha*s to be held in Mahan. The district administration set the new village council dates for between 16 and 23 August 2014. MSS testimonials on the movement's web-based community Radio Sangharsh demonstrate how company agents bribed and intimidated villagers ahead of the meeting, including by distributing chicken and alcohol amongst the men and saris amongst women and visiting people's homes to compel a vote for mining. One testimony said:

People from the company come to us in groups of three to four, and they tell us to participate in the *Gram Sabha*. They tell us that they will give us money, clothes and later, when the coalmines open, they will give us jobs

Radio Sangharsh testimonial (28/08/2014)

These everyday acts of sabotage, violence, deceit and misinformation by the company and local administration demonstrate how the State–corporate apparatus dilutes and violates the FRA's objective of redressing the historic injustices towards Adivasi and other forest-dependent communities.

5.2 Formation of the Mahan Sangharsh Samiti

The transformation of the people of Mahan into an organised resistance against a coal mine, and their motivation to challenge the State and the corporation, was steered by various imperatives. While threats and intimidations had strengthened their resolve, the forgery of signatures had broken their complacency. Six leaders from the movement presented overlapping yet distinct accounts of why they had joined a fight to stop the Mahan coal mine, as well as decided to trust an international NGO that had never worked in the region before. My interviews with Narayan, Dayanath, Sunita, Deendayal, Ramlal and Raj made it evident that their personal journeys towards politicisation had been gradual and incremental. The willingness to trust an external organisation had grown over time and in direct response to the growing threat of mining.

Narayan's first reaction to Greenpeace had been whether they would last in Mahan: *Yeh log tikenge kya? Pehle bhi bohuto ko dekha tha*, which, translated from Hindi means 'will they last? I have seen other NGOs come and go, none of them stuck around', he had wondered. Aged 30 and with a family of three school-going children, Narayan is one of the few people from Amelia to have received a master's degree. In 2008 he had worked as a supervisor at the Essar thermal power plant, located four kilometres from Amelia. He had adjusted to the red-and-white-banded smokestacks and the fly ash from the power plant as a permanent presence in a previously unpolluted landscape. But the forgery proved to be a turning point. He quit his job after the incident, deciding not to be a *naukar* anymore. The Hindi word means servant or one who gives service, conveying a pejorative meaning for a jobholder.

Dayanath, two years older than Narayan and also from Amelia, used to harbour a deeper cynicism about big NGOs. He told me about an incident with an NGO representative from Bhopal, the state capital of Madhya Pradesh, which had left him wary. Although the representative had won the community's trust and had been running a community development program in Amelia for six months, the person suddenly departed and never contacted any of the locals again. At first, Dayanath ignored Greenpeace's requests to join in the meetings. But an incident in which company agents inflicted violence on him tipped him over to the other side.

MCL officials felt threatened by Greenpeace's presence in Mahan and attempted to sabotage the NGO's reputation. Company agents and family members of

Amelia's village head took Dayanath and four other locals by force to the Hindalco office in Waidhan. Once there, they were threatened at gunpoint to sign a letter with baseless allegations against Greenpeace. However, the five locals managed to escape from the building when an internal scuffle broke out between the agents and family members. After returning to Amelia, Dayanath asked Greenpeace to come and work with his community, promising to get others involved in the fight.

One of the more senior MSS leaders, Ramlal, a grandfather, had experienced three waves of industrial projects-induced displacements in Singrauli, twice for the Rihand Reservoir, and a third time from the setting up of NTPC's thermal power plant in Shaktinagar.[9] The family had finally settled in Amelia. Thinking about a fourth potential risk of displacement, he said:

> There is nowhere else to escape from this menace! Why can't they dig coal out of the ground from the Minister's bungalow in Delhi? Why my home, my forests? For the sake of mining for just 15 years, they want to destroy places where our gods and goddesses have dwelt for centuries?

Sunita, aged 25, from Budher village, is one of the very few women leaders in the MSS. She first learnt about the effects of coal mining in central India at an awareness camp for local youth in 2010 that Greenpeace had organised. She told me that she had been sent to the camp against her will by her father. She struggled through the three-day programme. When Greenpeace started mobilising in Mahan, her father once again forced her to attend the meetings. But she started taking her association with the MSS seriously after the company agents started paying regular visits to Budher village, attempting to convince people about the coal mining project.

Deendayal from Amelia, also 25, is a high school teacher. He belongs to the Gond Scheduled Tribe. Deendayal also attended the Greenpeace camp in 2010. He was motivated to join the MSS from the beginning, on account of the historic experience of dispossession of Adivasis from the forests. Deendayal confirmed that people used to be ignorant of the FRA, and variously believed that the State, the Forestry Department, and even forest guards, were the rightful owners of the Mahan forests. The first meeting with Greenpeace in Amelia helped to clarify his own knowledge of the issue of forest ownership and the rights enshrined in the FRA. This understanding, combined with the knowledge of coal mining's effects in Singrauli from the educational camp, gave him the motivation to fight.

Greenpeace activists had been raising awareness about forest rights in Mahan since 2010. But the mobilisation of the community into a resistance took longer, and was facilitated simultaneously through their growing trust in Greenpeace and distrust in the *sarkar* or government. People from the five villages Amelia, Budher, Bandhaura, Suhira and Barwantola formally came together as a resistance in February 2012 under the banner of the Mahan Sangharsh Samiti (MSS) which means the Mahan resistance front, and joined Greenpeace in stopping the coal mine (Pillai 2017). The other six villages in the movement – Nagwa, Khairahi, Karsua Lal, Piderwah, Bandha and Jamgadhi – joined a few months after the forgery

Figure 5.2 Women form a protective ring around a tree in the Mahan Forest to prevent company workers from felling trees. Photo by Greenpeace.

incident in 2013 (Greenpeace India 2014a). The logo of the *aandolan* or resistance consisted of a ring of dancing people surrounding a circle of peacock feathers. The MSS logo contains a mahua tree at its centre, drawn in the *Warli* Adivasi art style on a bright green background. The logo symbolised the protection of the forest by the community and was inspired by their understanding of forest ownership.

Rajji, a village elder from Amelia and the inspiration behind the MSS logo, confirmed what Deendayal said about the local understanding about forest rights. They used to think that the forests belonged to the *sarkar* or State who could take them away from the locals whenever they chose. At the peak of the conflict in 2014, when the company's contractors were preparing to log the forest, people in Amelia and Budher started putting up triangle-shaped green flags with the MSS logo outside their homes, showing their collective resistance to the coal mine.

Women risked losing even more from the coal mine, including grazing grounds for their cattle and collecting mahua flowers and tendu leaves in the forest to sell in the market. But most of the women in Mahan were unable to join the movement because of the lack of respite from a grinding schedule of daily work at home and in the fields. They also felt withdrawn in public, and hesitant to speak in front of men at gatherings. The Greenpeace activist who led the mobilisation in Mahan had to try a different approach to enlist women. She told me that in the initial months, she organised separate gatherings for women.

Once their confidence grew, women raised a variety of issues concerning their daily lives, including domestic violence, asking that MSS address the issue amongst

Figure 5.3 A Mahan Sangharsh Samiti member holding an MSS flag showing the movement's logo at a meeting. Photo by Greenpeace.

its male members. Speaking against domestic violence thus became a regular part of MSS meetings. Over a time span of four years of addressing domestic violence within the MSS, two core male members transformed from being perpetrators to champions against domestic violence.

The caste divide proved a contentious issue to tackle in the beginning. Although caste reservations have persisted, I gathered from the accounts of the Greenpeace team that their rigidity, where upper caste members refused to eat at the homes of Dalit MSS members, has dissolved over time, as a sense of solidarity has grown from a shared struggle. Testimonials on Radio Sangharsh indicated how Amelia's upper caste *sarpanch* (village head) and his family members harassed lower caste MSS movement members and supported local company agents who threatened MSS members with murder and false charges. Such menaces strengthened solidarities across the social divisions within the movement, and built a shared discontent against the local authority that acted against the interests of the majority of the people. When a Dalit member who was hosting a movement meeting in his courtyard was kicked and abused by the *sarpanch's* brother, all MSS members stood up against the violence.

The company set up a pro-mining front known as the Mahan Vikas Manch (MVM) or the Mahan Development Front, to thwart the MSS's mobilisation efforts against the coal mine. Locals known to have been recruited as *dalals* or agents by the company claimed allegiance to the Mahan Vikas Manch. The MVM accused the MSS of holding up development in the region. The district administration empathised with the pro-mining agenda. It argued that mining and thermal power projects also brought schools and hospitals to 'backward areas' like Singrauli through the corporate social responsibility of companies (Chakravartty 2011).

Figure 5.4 The women of Mahan tie ceremonial threads around trees for their protection on the occasion of the Hindu festival of 'Raksha Bandhan'. Photo by Udit Kulshreshtha/Greenpeace India.

I experienced an interaction between the pro-mining front and MSS members during one of my field trips. A man on a motorbike blocked the path of the Greenpeace activist and local women as they were returning from the weekly market, claiming: '*Aap bahar se aate hai aur hamara vikas bandh kar dete hai*' ('you outsiders come into the region and threaten our development'). Following a well-practised strategy of tackling such incidents of intimidation, the MSS women took over the conversation at this point, telling the *dalal* that they opposed the coal mine since it posed a risk to their livelihoods and their grazing grounds.

Women confronting company agents is a well-worn tactic in grassroots people's movements. One of the frequently repeated jokes within the MSS is about Sunita intimidating the company *dalal* who used to come to Budher village to convince people about the benefits of the coal mine. The account of Sunita standing on the dirt track outside her village, shouting at the *dalal* to leave her village alone, used to generate banter at the MSS gatherings.

Despite the different degrees of awareness of forest rights, and of the willingness to claim them, the realisation that the State was acting against their interests while favouring the corporation, acted as a common catalyst to politicise the village community in Mahan against the coal mine. Testimonials on Radio Sangharsh indicate that the State and company's interference in their daily lives grew significantly over a period of four years. Mutual trust and the willingness to collaborate with Greenpeace strengthened during this time, as a result of their growing disaffection with the State.

5.3 State–corporate nexus in Mahan

In Chapter 4, I discussed the questionable process of allocation and mining approval for the Mahan coal block that demonstrated a high-level nexus between Essar Power and the Indian government. The everyday interventions of the local

Figure 5.5 Women chant slogans and hold movement flags in an MSS meeting. Photo by
Sudhanshu Malhotra/Greenpeace India.

administration, forest department officials, and the local police, in the life of the
MSS members, demonstrate how the State–corporate compact operates on the
ground in a mining region in India. While accounts from the ground rarely reach
the large media establishments in the big cities, in the case of Mahan, the presence
of an international NGO lent visibility to such dealings.

The Greenpeace team and MSS members told me about how the police acted
to disrupt the movement at the behest of the company officials. Radio Sangharsh
testimonies also confirmed that the most common form of disruption involved
police intimidations against organising meetings. Arrests became a common feature during the movement's most critical phase in 2014. The Indian Environment
Minister granted a final clearance for the Mahan coal mine in February 2014, despite the Minister for Tribal Affairs having nullified the fraudulent mining resolution
from Amelia. The locals, a majority of them women, had spread themselves out in
the forest, to stop the company's contractors from marking and felling the trees.

The Greenpeace activist who led the Mahan movement recounted an incident
when an MCL official accosted her in Amelia. He threatened her to stop mobilising
the locals, boasting that he was the 'biggest hooligan in the area'. Another written
account about the risks faced by women in the movement from company agents
and local government officials said:

> Forest officials and company agents tried to manhandle women from the MSS
> who had moved into the forest to peacefully stop the numbering of trees for
> felling. They were threatened and abused (including with sexual innuendos)
> and questions were raised regarding their character. The company agents
> morphed photographs of women in bikinis and threatened to publish it in the
> local papers. When these threats did not work, they went around showing those
> photographs to community members on their laptops, calling them *Mayamohini*

(seductresses). The local member of the legislative assembly threatened women working in Mahan and challenged men to rape them. When they approached the police with complaints and evidence, not only were they sent back but also charged with false cases!

Pillai (2019, p. 152)

I had a first-hand experience of how the local administration and the police disrupted the MSS's activities during my first field visit. On 4 April 2017, MSS organised a public hearing to redress the grievances of people who had been displaced by the Essar Thermal Plant in 2008. People from the four villages Nagwah, Khairahi, Bandhora and Karsua Lal had been sitting on an indefinite *dharna* (protests sit) in demand for their grievances to be heard. Acting at the behest of Essar Power, the administration resorted to desperate tactics to sabotage the event. The Greenpeace team received a phone call from the local police late at night before the next day's hearing, and were informed that a curfew had been imposed within a five-kilometre radius of the Essar Power Plant in anticipation of a risk to property. This meant that the hearing could not be held in the affected villages as they fell within the curfew zone. Police arrived at the designated venue for the hearing late at night, forced MSS members to pull down the tent that had been set up for the next day, and threatened them with arrest.

The public hearing finally went ahead owing to the resourcefulness of the locals and a high level of cooperation and trust between MSS and Greenpeace that made it possible to respond to the challenge quickly. MSS members carried the tent material to a new location on foot and under the cover of dark to avoid being detected. The tent was ready before dawn. The convoy of cars carrying the panellists took a rarely travelled back route to the new venue to avoid being detected. The panel, containing people's movement leaders and environmental lawyers, heard a range of grievances from land-displaced people of Mahan, and prepared an action report. Three weeks after the public hearing, three men and three women from Khairahi (one of the displacement-affected villages from where people had testified at the hearing), one of whom had an eight-month-old baby, were arrested and detained for five days. The grounds provided for their arrests, similar to previous such incidents of baseless arrests, were the obstruction of duty for public servants.

The incidents reveal an entrenched entitlement by local administrations and mining companies in an extractive hotspot to intimidate communities with impunity. They show that governments wield the power to shape the destinies of entire landscapes as per their choosing, and the collusion between various levels of the State to seal the fate of communities residing in resource-rich or rather resource-cursed regions.

5.4 An unusual alliance and its resistance

Greenpeace has previously been perceived as beating a hasty retreat from grassroots issues, leaving communities vulnerable, making networks like the National Association of Peoples' Movements (NAPM) wary of trusting this international

NGO. India's grassroots movements have largely remained ideologically opposed to professionalised and foreign-funded NGOs (Talukdar 2018). But after a visit to Mahan, the leader of the Madhya Pradesh-based People's Resistance Movement acknowledged that Greenpeace's work in Mahan told a different story from its previous work in India since this time it had attempted to empower and organise affected people (Greenpeace India 2013b). This is part of the reason why civil society overwhelmingly supported Greenpeace during the Indian government's crackdown on the organisation in 2015 (Talukdar 2018).

Owing to their alliance with Greenpeace, MSS members participated in activities that could be considered unusual from a grassroots perspective. Twenty-seven men and women from Mahan participated in a Greenpeace banner-drop action outside the Essar Headquarters in Mumbai in January 2014 (Greenpeace India 2014b). The fact that none had travelled to India's financial capital before and that most had never left the boundaries of their own state demonstrated that MSS members had taken a massive leap of faith to work in alliance with Greenpeace.

Narayan and Deendayal broke into peals of laughter recollecting standing outside the Essar headquarters in Mumbai, while 12 activists dressed in tiger suits unfurled a giant banner from the top of the building with the message 'We kill forests: Essar'. Sunita was standing inside the Essar corporate building. When an Essar official asked her why she had illegally entered their building, she turned the question back, asking, 'why does this company enter our village without our permission?'

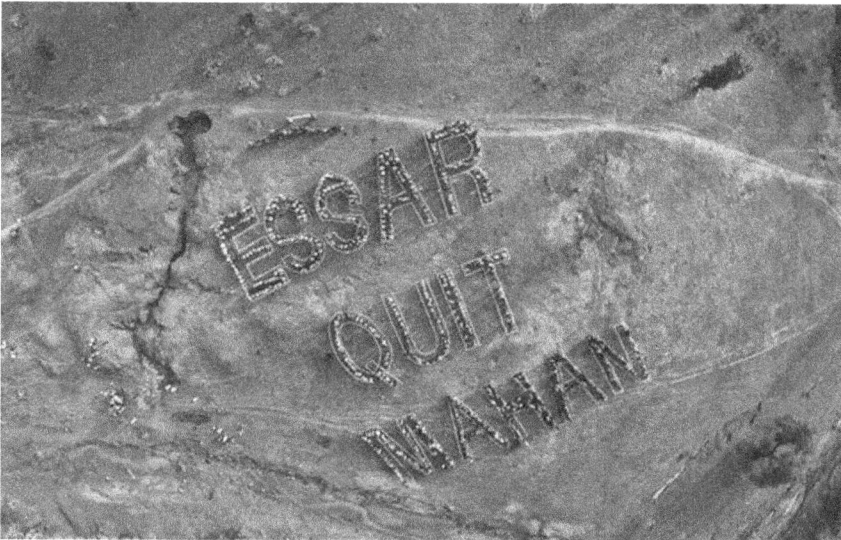

Figure 5.6 Mahan Sangharsh Samiti members form a human sign 'Essar Quit Mahan' in a protest coordinated by Greenpeace in Mahan. Photo by Sajan Ponappa/ Greenpeace India.

All the MSS participants in the banner-drop action were arrested and detained overnight, along with 30 Greenpeace activists and volunteers. The trial continued till 2018, needing 17 out of the 27 MSS members to undertake inter-state travels up to four times a year for court hearings in Mumbai. Narayan treated the challenge of travelling for the hearings as par for the course; in fact, he wished he could have participated in more of such Greenpeace 'stunts'. He saw windmills for the first time on one of the many multi-day journeys in a crowded public bus from Amelia to Mumbai. When he asked co-passengers what they were and learnt that they generated electricity from wind, he wondered why the government could not install windmills in Mahan instead of pushing coal mining?

Greenpeace and the MSS undertook considerable risks and were jointly implicated in multiple legal cases. There were five ongoing cases as of June 2018. One of the Greenpeace campaigners working in Mahan was arrested twice and charged thrice on the baseless allegations of obstructing government officials from doing their duty, even looting and robbery. Two MSS members were also detained along with him. The arrests came within three months of each other in 2014. They were devised to thwart the movement when people started protecting the trees in the Mahan forests from being felled (Greenpeace India 2014c).

Dayanath's diary entries from this period mention that the community persisted with the forest blockades for up to five months. Sunita described the curious incidents that led to the arrests. One day, the company workers who were marking the trees for felling in the forest promptly departed at lunchtime, leaving behind all their equipment. Seeing that they had not returned, MSS members deposited the equipment at the police station. But at midnight, the police arrested two MSS members from Amelia, and two Greenpeace staff from Waidhan, without a warrant, and charged them with looting the company's equipment.

While the others were released the next day, the senior citizen Ramlal was detained for 28 days. MSS and Greenpeace protested this clampdown by organising a demonstration; people gathered in a human sign that said 'Essar Quit Mahan' in a large field. Eight hundred people in Mahan also held a candlelight vigil and a march. These incidents highlighted the community's daily struggles to preserve their ancestral forests from coal mining, and the State's corporate bias (Greenpeace India 2014g).

Beyond the obvious aim to stop the destruction of people's forests and livelihoods, the local movement's actions also reflected an element of dissent. The second Intelligence Bureau report (Intelligence Bureau 2014) that singled out Greenpeace's anti-coal activism also implicated the MSS leaders. MSS members understood that their fight to stop the coal mine signified a struggle to democratise development. Through a series of assertions on Radio Sangharsh, MSS challenged Greenpeace's unfair treatment at the hands of the Indian government and asserted the need to preserve the space for dissent.

After the cancellation of the coal block, the Mahan resistance broadened its horizons by joining with the National Association of People's Movements (NAPM), an umbrella group of people's movements from around the country. In

December 2012, MSS members travelled to the state capital of Bhopal for the 40th anniversary of the 1984 Union Carbide gas tragedy. Speaking on Radio Sangharsh on his return, Dayanath pleaded to his community not to allow a company without accountability to enter Mahan.

In 2014, Greenpeace organised mahua-collection camps in the Mahan forests for its urban youth volunteers, between the months of April and May, which is the peak season for gathering these flowers (Greenpeace India 2014d). The Greenpeace lead campaigner told me that the initiative was meant both as an act of urban–rural solidarity and a fundraising initiative to help establish the MSS as an independent grassroots organisation. The people of Mahan have contributed a cluster of mahua trees in the forest as an economic resource to the MSS. The movement can collect, dry and sell mahua flowers from their forest cluster to support their activities.

The Greenpeace–MSS alliance challenged the developmental paradigm by joining a chorus of rural and urban voices against coal mining, at a time when the public costs of India's rapid growth had begun to be debated. The MSS's voice strengthened the environmental argument for protecting Asia's oldest Sal forests with people's concerns for livelihoods and the rights to decide about their forests. The MSS's actions transformed, even if in small measures, the everyday experience of communities living in a place marked for destructive development by powerful private interests. The resourcefulness of individuals within the MSS and the institutional influence that came with the involvement of a global NGO all contributed towards achieving this objective.

5.5 A celebration of people's forest rights

I first visited Amelia in March 2017. On 31 March, the people of Mahan commemorated the second anniversary of their victory over coal mining in the local forests. The second-anniversary celebration had been linked with international climate actions organised under the global initiative to 'break free from fossil fuels'. Inside the big tent that had been set up for the event, a large yellow banner framing the stage read *Vishal Van Adhikar Sammelan* (massive forest rights gathering), indicating the centrality of forest rights in the lives and politics of the MSS movement.

Late springtime is the peak of mahua gathering season for forest-dependent communities in central India. Mahua trees are native to central Indian forests and serve as the economic and cultural lifeline for forest-dependent communities. Families are known to set up camp in the forests for weeks to collect the yellow mahua blossoms. The harvest is dried and sold for herbal and medicinal needs. Although a large crowd of over 1000 locals had gathered for the anniversary, the presence of far fewer women, at one-fourth that of men, meant that they had gone to the forest to harvest mahua.

The bright green MSS logo was conspicuous on badges worn by locals, and on little triangular flags held by the children. Greenpeace signage was conspicuously absent. Solidarity for forest and democratic rights was demonstrated through the large turnout and speeches over four hours. The MSS leaders spoke about the need to stay united. One of the movement's most frequently used banners, that said *Loktantra Zindabad* (long live democracy) painted in black on a yellow background, was displayed on the tent wall.

A blown-up image of MSS members standing holding the banner under a mahua tree used to hang in Greenpeace's New Delhi office. The photograph was taken to celebrate Mahan's victory over coal mining after the Modi government removed the Mahan coal block from the auction list on 30 March 2015. On that day, the people of Mahan had joined with other local people's movements to mark the victory as 'democracy day'.

The cancellation of the Mahan coal block following the 2014 Supreme Court order had brought immense relief to the Mahan community. Celebratory singing, dancing and feasting had continued through the night, after the government handed down its decision on 30 March 2015. The next morning, they made an offering at the hilltop shrine of *Dih Baba,* their forest god. The victory brought a sense of empowerment and confidence. *Loktantra Zindabad* became the MSS's most definitive slogan. It reflected the coming together of the environmental demands of an international ENGO with people's rights over forests and livelihoods. For the time being, Mahan's old forests and its livelihood-dependent people had repulsed the long arm of destructive industrial development.

Back inside the tent of the second-anniversary celebration, the hot hours of the day and litany of speeches were frequently interrupted with slogans asserting people's forest rights under the Forest Rights Act. *Jangal hamara apka hai, nahi*

Figure 5.7 Mahan Sangharsh Samiti members under a Mahua tree in Amelia village holding a banner saying, 'Long Live Democracy'. Photo by Greenpeace.

kisi ke baap ka hai (the forests belong to you and me, not to the government or the company) was one of the most popular chants. People of Mahan have been intimate with the forests for generations, and have many tales by which to remember these connections. Rajji's face creased up with countless lines as he smiled in recollection of childhood memories. He used to keep an orphaned jaguar cub that his father found in the forest as a pet.

The assertion *Purkho ka naata nahi todenge, jangal zameen nahi chodenge* (we will respect our ancestral land, we will not give up our forests) had grown to become an anthem for the movement. Women led the crowd in chanting the slogan in between the speeches. Based on an understanding that their historic connection with the forest is now recognised by law, the word *adhikaar* (right) had entered the movement lexicon and become the MSS's chief instrument against coal mining.

Various other banners displayed in the tent also expressed the centrality of forest rights for the movement: *'Jan Jan ka naara hai, van adhikar hamara hai'* (there is a people's chorus for forest rights), and *'gaon gaon ki yahii pukaar, le ke rahenge van adhikar'* (village after village will claim their forest rights). Dayanath had put his musical skills to the movement's cause by creating compositions about the struggle. At the second anniversary, he and his band sang about the highs and the lows of the struggle. The crowd cheered enthusiastically when they sang about the forgery of signatures on the mining resolution in 2013.

I visited Mahan three more times after the spring of 2017. Each visit raised further uncertainty about the future viability of the people's movement due to a combination of factors: Greenpeace's inability to keep fully functioning campaigns running after the crackdown, the atmosphere of attack on the protest movement under the Modi government, and its attempts at diluting the FRA, as well as the FRA's violation by the local administration.[10] Could MSS members stay united when faced with constant interference and intimidation by company officials?

The Modi Government cancelled the Mahan coal block on account of it being 'inviolate' for coal mining because of the high quality of its forests. But in 2016 allocated the adjoining Amelia North coal block that also falls in the Mahan forests to a public sector hydro corporation, to supply coal to the Khurja Super Thermal Power Project (KSTPP) (Tribune 2016). The proposal risked the lands and forests surrounding Pidarwah village, located approximately 15 kilometres from Amelia, across dense forest (Sanghera 2020). It demonstrated that yet again the Indian government failed to apply a common rationale in making decisions that are critical for the environment and communities (Pillai 2017).

Further, in 2023, the Modi-led Indian government upturned its own coal block cancellation when it reinstated Mahan and another adjoining Mara II Mahan coal block, which also contains high-quality forests and where forest cover extends to 90% of the coal block, for auction for commercial coal mining by private corporations (Jalihal 2023a). It has been reported that it is likely that the Indian government will allocate the Mara II Mahan coal block to Mahan Energen Limited, a subsidiary of Adani Power Limited that acquired the nearby Mahan Thermal Power Plant from Essar Power in 2022, for underground coal mining (Jalihal 2023b). No likely allocation has yet been reported for the Mahan coal block.[11]

For Greenpeace, stopping the mine had proven a symbolic victory over coal in India. But the Mahan campaign had proven risky for the organisation. Greenpeace had become a primary target for the Indian government's attack on NGOs. Successive crackdowns and the freezing of funds since 2014 forced Greenpeace to reduce its activities and staffing (Talukdar 2019 It reoriented its anti-coal climate activism by not attempting further community mobilisations, such as against the next proposed coal mine in the Mahan forests. It took the less risky approach of exposing the economic implausibility (and ecological impacts) of new coal-fired investments such as the Khurja thermal power plant, through high-level analysis reports.[12]

Greenpeace closed down the campaign office in Waidhan in 2018. Another attack on Greenpeace's funding by the Modi government in November 2018 forced the organisation to close down several of its long-running campaigns and let go of many staff. It has been operating at a significantly reduced capacity in India since then (Talukdar 2019). Members of the Greenpeace Mahan team continued supporting the local movement in their individual capacities as MSS members. Whether Greenpeace could have continued collaborating with MSS after the risk of coal mining had been eliminated, towards long-term community development and sustainable livelihoods measures, remains indeterminate given Greenpeace's forced exit after government crackdowns.

The state of play has left many questions unanswered about the future of Mahan and the MSS. The forests have been saved for the time being, but those already displaced by the Essar Thermal Plant are continuing to struggle for basic rights and employment opportunities (Sanghera 2020). The state of play also raises questions about the nature of new relations forged between grassroots communities in the global South and international environmental actors, whether such relations can endure beyond achieving instrumentalist aims of End(ing) Coal, both due to a fundamental mismatch of worldviews and risks of anti-coal activism.

Did MSS see that there were likely to be more battles ahead? Back at the second-anniversary celebration, enthused by the festive spirit, Narayan told me that he was prepared to fight again. As the event came to an end, the crowd gravitated towards the centre of the tent chanting *ladenge, jitenge* (we will fight, we will win) in a rising chorus. A hot breeze had started blowing across the open plains. It carried the thick fragrance of mahua flowers that covered the trees and carpeted the red earth all around the tent. At that moment, the spirit of the people of Mahan, who had learnt about their rights and fought a coal mine to protect their forests, felt strongly reassuring.

Analysis: significance of forest rights in India's energy capital

The visions that governments have promised through the large-scale industrialisation of Singrauli have borne no resemblance to the ecological and social impacts that these ambitions have wrought on the ground. While announcing the carving out of Singrauli as a separate district in 2008, ostensibly to facilitate private coal mining, the Chief Minister of Madhya Pradesh Shivraj Chauhan had promised to turn the region into India's Singapore. Narayan reflected that *Nangapur* was a more fitting description of what Singrauli had been reduced to. The word *nanga* means naked in Hindi. Narayan's observation is a close approximation to *Bhikharipore* – the Hindi word *bhikhari* meaning a beggar – that project-affected people have referred to Chauhan's false promise as, in Lina Dokuzovic's account of the cost of development in India's energy capital (Dokuzovic 2012).

Singrauli bore one of the worst brunt of India's developmental aspirations through three waves of mass displacements over six decades that reflected the changing paradigm of large-scale energy development. The aspirations of the Indian government transformed the region by resource extraction and coal mining-focused policies over decades. These aspirations were matched by the eagerness of the state governments in Madhya Pradesh to attract foreign investments and mining revenues in the neoliberal era. The State attempted to disrupt the possibility of an alternative outcome for communities to emerge, by fabricating the referendum process for mining under the Forest Rights Act.

The genesis of Mahan's conflict goes back to the year 2006 when the Indian government allocated the Mahan coal block through a corrupt auctioning process. In the same year, the government also passed the landmark Forest Rights Act. When Essar's Thermal Plant was set up, it triggered a first wave of land-related disputes in the immediate vicinity of Mahan. With the power plant came the fly ash and mosquitoes. Its constant hum drowned out the sounds of the forests around which people had woven tales for generations.

The lack of awareness of the Forest Rights Act amongst communities in coal-bearing regions is one of the primary reasons for its very low implementation. But communities in coal-bearing regions had begun contesting the loss of communal and grazing lands to coal mining even without prior knowledge of the FRA (Kohli et al. 2012). Greenpeace's strategic decision to mobilise people in Mahan through building awareness of the FRA had taken these realities into account. The Greenpeace Mahan team was able to engage and successfully mobilise the community, and use reflexive tactics such as mobilising women separately, addressing domestic violence and encouraging a conscious rejection of caste discrimination, to make the movement stronger and inclusive.

For the people of Mahan, a sense of ownership also brought pride in the forest's wealth and richness. I came across an account of the relevance of the Mahan forests for the locals in Dayanath's diary entry in Hindi from 3 May 2014, titled *What do I get from the forests?* The time of the diary entry coincided with one of the peak moments in the movement, when the company was preparing to begin logging the forest, and MSS members, particularly women, were guarding the trees. He told me that he felt a sense of loss and regret at that time, that future generations would not have a chance to value the forests, and share the knowledge that had been passed down over generations. The reflection prompted him to write down the seasonal fruits, flowers and medicines from Mahan. His recording of the wealth of the Mahan forests is also captured in one of his protest songs against the mining company.

India's national rhetoric about double-digit growth and political arguments for large-scale energy generation remain alarmingly disconnected from the disaffection that such projects generate on the ground. The fate of the Mahan forests still hangs in the balance, with attempts to develop the adjoining Amelia coal block, and the future prospect of underground mining if the other adjoining Mara II Mahan block is allocated. This is the case despite coal's global decline, and even though not all the coal being mined in India is reaching the power stations, indicating a surplus extraction of coal (Gross 2019). The ability of the people of Mahan to hold on to their forest rights is the only possible means for an alternative narrative to mainstream development to emerge from Singrauli's coal-ravaged landscape.

In the evening after the second-anniversary celebrations, I travelled with the Greenpeace team and the six MSS leaders to a spot by the Mahan River. To my remark that the bend of the shallow river, flanked by Sal trees on both sides, was beautiful, Deendayal said slowly and conclusively, '*Jo bhi hai, hamara hai*', meaning it does not matter that it is beautiful, what is important is that it is ours! Even if the victory ultimately proves to be a mere pause to hectic coal extraction that affects the ecology and livelihoods in Singrauli, it will most likely remain with the locals as a new experience of *adhikaar* and a language of justice that was a long time coming.

Conclusion

The social landscape of the region known as India's energy capital points to the paradox of coal-led development in a global South context. Even though the Singrauli region supplies electricity to sixteen Indian states, several of its village-based communities still live without electricity. The long arc of large industrial development, from dams to State-led coal and thermal power production, to private coal mining and thermal power generation, tells a repetitive story of community dispossession and loss of livelihoods.

As the Mahan case study shows, although subsistence communities now have democratic bargaining power in the process of industrialisation through legal rights in the land acquisition process and legal forest custodianship, their rights are constantly at risk of being eroded or violated by the State–corporate apparatus. Historically, Singrauli's ecologically dependent communities could not determine their own future and were often forced to eke out an existence on the margins of the coal economy. As in other forested parts of India affected by coal mining, this has often meant illegally collecting the very substance 'in lieu' of earlier forest products, as Lahiri-Dutt (2016, p. 204) poignantly puts it in her ethnographic account of the subsistence economy of coal in India. Against this stark context, the grassroots resistance at Mahan is important on many levels.

As a successful rights-based movement in a region decimated by 50 years of coal extraction, the actions of the MSS symbolised empowerment and historic land justice at the regional level. At the national level, Mahan's livelihood movement challenged the dominant coal-led development paradigm. Through the resources, visibility and support of an international NGO, they were able to draw attention to the true cost of coal to subsistence communities in India. Finally, even though climate change was not a mobilising factor at Mahan, the actions of the movement were directly relevant to the global environmental effort to 'keep coal in the ground'.

The various material realities, socio-political contexts, and claims of frontline resistances in the global South and their critical differences from their counterparts in the global North point to a diversity in anti-coal climate justice politics. This diversity needs to be heeded in order for an inclusive and just transition away from coal and towards renewable energy sources. To be inclusive of the challenges of people's movements in the global South will require reflexivity on the part of environmentalism's new approach, given the persistent malaise of land and Adivasi rights violations by mainstream industrial development and what that signifies about the state of relations between the State and Indigenous peoples in a developing context as seen in the Indian case. In Chapter 9, I discuss a reflexive approach to global climate activism that builds solidarity for Indigenous grassroots struggles by keeping in mind the state of relations between Indigenous peoples and the State and being cognisant of the variations in State–Indigenous relations in global North and South contexts.

Notes

1 People from the first five villages came together in 2012 under the banner of the Mahan Sangharsh Samiti to assert their forest rights and oppose the Mahan coal mine. The remaining six villages joined the movement in August 2013 (Greenpeace India 2014f).

2 It offered market rates for the acquired lands, spacious plots for every adult in each affected family, and unemployment allowance of Rs 4000 (A$74) per month to all displaced persons between 18 and 50 years of age who cannot be employed at the Power Plant, till such time that the project succeeded in offering them a regular job.

3 Only 1050 of the 11,500 displaced persons received regular jobs after the 1988 protests (Sharma and Singh 2009).

4 As per World Bank guidelines, project affected families should have received free electricity, education and medical treatment, should have been guaranteed employment (however limited according to restrictive clauses of rehabilitation schemes), plus resettlement and rehabilitation packages. Further examples of human rights violations by NTPC Ltd. in Singrauli range from the company's bulldozers running over people's crops and homesteads to its dump trucks running over a protestor (Clark 2003), and high levels of corruption through officials charging bribes from displaced people for 'fake interviews' that did not result in employment (Kohli 1997).

5 Misleadingly, when World Bank officials visited Singrauli in 1992, they referred to NTPC's land-based compensation policy as exemplary, although this was far from the reality on the ground (Kohli 1997).

6 The NMP, while streamlining the granting of mineral concessions, also aimed to develop a sustainable framework for the optimum utilisation of mineral resources and equally importantly, sought to improve the lives of people in mining areas.

7 The report states that, according to the District Collector of Singrauli, although 4000 individual rights have been issued under the FRA between 2008 and 2010, with 7000–8000

other applications in the pipeline, only 64 community rights applications had been made, and hardly any granted, during the same period.

8 While inaugurating the Rihand Dam that began environmental destruction and social upheaval in the Singrauli region, Prime Minister Nehru had described Singrauli as India's Switzerland (Singh 2015). A reference and explanation for Singrauli as Singapore is contained in the analysis section.

9 The government had misjudged the catchment area of the Gobind Pant Reservoir. People who had settled close to it after the 1960 displacement were faced with a second dam-related displacement in 1962, when their settlement areas were flooded by the reservoir (Pillai et al. 2011, p. 11).

10 The Narendra Modi government is diluting legal rights and consent-requirements of Adivasi and other forest-dwelling communities before forests can be destroyed for industrial projects (including coal mines) through changes to India's Forest Conservation Rules in June 2022 that contradict the FRA by allowing the government to issue forest clearance for projects on Adivasi lands even before securing consent from *gram sabhas* (Joshi and Sethi 2022).

11 These events – reinstating Mahan on the coal block auction list and the possible allocation of the adjoining Mara II Mahan for underground mining – fall far outside the timeframe of the research undertaken for this book; therefore, I have not discussed their implication for communities' futures here in this chapter.

12 A Greenpeace analysis argued that the KSCTPP and its captive coal mine project was financially risky, unnecessary, and posed a great health risk through worsening air pollution in North India (Greenpeace India 2018). An analysis by the Institute of Energy Economics and Financial Analysis (IEEFA) showed that India's existing supply glut of electricity and the rapidly declining cost of renewable energy made the current project economically illogical (Buckley et al. 2018).

References

Buckley, T., Shah, K. & Garg, V. 2018, The Khurja Power Project: A Recipe for an Indian Stranded Asset, Institute of Energy Economics and Financial Analysis, October 2018, viewed 20 September 2020, <http://ieefa.org/wp-content/uploads/2018/10/Khurja-Thermal-Power-Project_10.2018.pdf>.

Chakravartty, A. 2011, 'Mahan at all costs', *Down to Earth*, New Delhi, 31 October, viewed 14 August 2019, <www.downtoearth.org.in/news/mahan-at-all-costs-34230>.

Chowdhury, C. & Aga. A. 2020, 'Manufacturing Consent: Mining, Bureaucratic Sabotage and the Forest Rights Act in India', Capitalism, Nature, Socialism, vol. 31, 2020, no. 2, pp. 70-90.

Clark, D. 2003, 'Singrauli: an unfulfilled struggle for justice', in D. Clark, J. Fox & K. Treakle (eds), *Demanding Accountability: Civil Society Claims and the World Bank Inspection*, Rowman and Littlefield, pp. 167–190.

Dokuzovic, L. 2012, *Bhikharipore Singrauli: A Case for Just Development*, 30 September, viewed 27 September 2019, <http://sanhati.com/excerpted/5621/>.

Dutta, A. 2020, 'Fly ash slurry in Singrauli contaminates water reservoir', *Mongabay Environment and Health Series*, 29 April, viewed 20 June 2021, <https://india.mongabay.com/2020/04/fly-ash-slurry-in-singrauli-contaminates-water-reservoir-after-taking-lives-and-homes/>.

Ghatwai, M. 2015, 'Essar, Hindalco's Mahan coal block: lobbying for 3 yrs, Greenpeace claims clearance given under pressure', *Financial Express*, 13 January, viewed 27 September 2019, <www.financialexpress.com/industry/companies/essar-hindalcos-mahan-coal-block-lobbying-for-3-yrs-greenpeace-claims-clearance-given-under-press ure/29279/>.

Greenpeace India. 2013a, 'Tribal affairs minister K C Deo says FRA violations in Mahan will be investigated', *Press Release*, July 19, viewed 27 September 2019, <www.greenpe ace.org/india/en/press/3106/tribal-affairs-minister-kc-deo-says-fra-violations-in-mahan-will-be-investigated/>.

Greenpeace India. 2013b, "We will not allow mining in Mahan': Mahan Sangharsh Samiti', *Press Release*, 4 August, viewed 20 September 2019, <www.greenpeace.org/india/en/ press/3100/we-will-not-allow-mining-in-mahan-mahan-sangharsh-samiti/>.

Greenpeace India. 2014a, 'Greenpeace forest activists scale Essar HQ', *Press Release*, 22 January, viewed 20 September 2019, <www.greenpeace.org/india/en/press/2804/greenpe ace-forest-activists-scale-essar-hq/>.

Greenpeace India. 2014b, 'Greenpeace activists granted bail after being detained by Mumbai police', *Press Release*, 23 January, viewed 20 September 2019, <www.greenpeace.org/ india/en/press/2800/greenpeace-activists-granted-bail-after-being-detained-by-mumbai-police/>.

Greenpeace India. 2014c, 'Greenpeace will prove defamation charges wrong. Billionaire corporation sues environment watchdog for Rs. 500 crore and demands a gag order on contents against the company', *Press Release*, 29 January, viewed 20 September 2019, <www.greenpeace.org/india/en/press/2789/greenpeace-will-prove-defamation-charges-wrong-billionaire-corporation-sues-environment-watchdog-for-rs-500-crore-and-dema nds-a-gag-order-on-contents-against-the-company/>.

Greenpeace India. 2014d, 'Pledge to save Mahan forests on earth day: young urban volunteers come together in a show of support to save Mahan forests in Singrauli', *Press Release*, 22 April, viewed 27 September 2019, <www.greenpeace.org/india/en/press/ 2633/pledge-to-save-mahan-on-earth-day-young-urban-volunteers-come-together-in-a-show-of-support-to-save-mahan-forests-in-singrauli/>.

Greenpeace India. 2014e, 'The people of Mahan have prevailed', *Web Story*, 23 July, viewed 20 September 2010, <www.greenpeace.org/india/en/story/2571/the-people-of-mahan-have-prevailed/>.

Greenpeace India. 2014f, 'Civil Society urges the centre not to trample over the rights of forest dwellers in Mahan. Implement FRA in 54 villages of Mahan, demand Mahan Sangharsh Samiti', *Press Release*, 1 August, viewed 20 September 2019, <www.gre enpeace.org/india/en/press/2537/civil-society-urges-the-centre-not-to-trample-over-the-rights-of-forest-dwellers-in-mahan-implement-fra-in-54-villages-of-mahan-demand-mahan-sangharsh-samiti/>.

Greenpeace India. 2014g, "Systematic clampdown on People's Movements Unacceptable': amidst police excesses, MSS demands that the CM should revoke mining licence given to Essar and protect the rights of forest dwellers in Mahan', *Press Release*, 5 August, viewed 20 September 2019, <www.greenpeace.org/india/en/press/2531/sys tematic-clampdown-on-peoples-movements-unacceptable-mss-amidst-police-excesses-mss-demands-that-the-cm-should-revoke-mining-license-given-to-essar-and-protect-the-right/>.

Greenpeace India. 2018, *Khurja Coal Plant Unviable: Solar Project Would Benefit Region More*, October 2018, viewed 20 March 2019, <www.greenpeace.org/india/en/story/3255/analysis-on-khurja-supercritical-thermal-power-plant-proves-solar-will-be-a-better-investment/>.

Gross, S. 2019, 'Coal is king in India – and will likely remain so', *Brookings*, 8 March, viewed 20 September 2019, <www.brookings.edu/blog/planetpolicy/2019/03/08/coal-is-king-in-india-and-will-likely-remain-so/>.

Intelligence Bureau. 2014, Subject: FCRA Registered Greenpeace Spearheading a Concerted Campaign against India's Energy Expansion Plans, viewed 14 August 2019, <www.documentcloud.org/documents/1201267-greenpeace-ib-report.html>.

Jalihal, S. 2023a, 'Advantage Adani: power industry lobbies, coal ministry unlocks dense forests for mining', *The Reporters Collective*, 9 October, viewed 20 November 2022, <www.reporters-collective.in/trc/coal-forests-part-1>.

Jalihal, S. 2023b, *Coal Reform Overturned: After Failed Auctions, Centre Hands Out Coal Blocks At Discretion*, 11 October, viewed 20 December 2023, <www.reporters-collective.in/trc/coal-forest-part-2>.

Joshi, M. & Sethi, N. 2022, 'Government to approve cutting down forests without consent of tribals and forest dwellers', *News Laundry*, 7 July, viewed 20 November 2022, <www.newslaundry.com/2022/07/07/government-to-approve-cutting-down-of-forests-without-consent-from-tribals-and-forest-dwellers>.

Kohli, M. 1997, *Request for Inspection NTC Power Generation Project Cr.* 3632.

Kohli, K., Kothari, A. & Pillai, P. 2012, *Countering Coal?*, Discussion paper by Kalpavriksh and Greenpeace India, New Delhi/Pune and Bangalore, viewed 15 September 2019, <www.greenpeace.org/india/en/publication/989/countering-coal-community-forest-rights-and-coal-mining-regions-of-india/>.

Lahiri-Dutt, K. 2016, 'The diverse worlds of coal in India: energising the nation, energising livelihoods', *Energy Policy*, vol. 99, pp. 203–213.

Ministry of Environment and Forests (2009). 'Diversion of forest land for non-forest purposes under the Forest (Conservation) Act, 1980 – ensuring compliance of the Scheduled Tribes and Other Traditional Forest Dwellers (Recognition of Forest Rights) Act 2006', *Circular*, 7 July, Available from: https://forestsclearance.nic.in/writereaddata/public_display/schemes/981969732$3rdAugust2009.pdf [Accessed 22 November 2022]

Pillai, P. 2017, 'An ongoing battle', *Frontline,* 15 September, viewed 27 September 2019, <https://frontline.thehindu.com/social-issues/an-ongoing-battle/article9831547.ece>.

Pillai, P. 2019, 'Coal mining and ecological fragility', in P. Ray (ed.), *Women Speak Nation: Gender, Culture, Politics*, Routledge.

Pillai, P., Gopal, V. & Kohli, K. 2011, 'Singrauli: The Coal Curse – a fact finding report on the impact of coalmining on the people and environment of Singrauli', *Greenpeace India*, September 2011, viewed 27 September 2019, <www.greenpeace.org/india/en/publication/1006/singrauli-the-coal-curse/>.

Pioneer. 2014, *HC Asks SP to Probe Gram Sabha Forgery*, 16 July, viewed 27 September 2019 <www.dailypioneer.com/2014/state-editions/hc-asks-sp-to-probe-gram-sabha-forgery.html>.

Sanghera, T. 2020, 'This coal plant endangers forests, wildlife, people. And India doesn't even need it, experts say', *IndiaSpend,* 25 May, viewed 20 October 2020, <www.indiaspend.com/this-coal-plant-endangers-forests-wildlife-people-and-india-doesnt-even-need-it-experts-say/>.

Sharma, R. N. & Singh, S. R. 2009, 'Displacement in Singrauli region: entitlements and rehabilitation', *Economic and Political Weekly*, vol. 44, no. 51, pp. 62–69.

Singh, S. R. 2015, 'Neither Switzerland nor Singapore but a site of daily struggle', *The Wire*, 24 July, viewed 30 September 2019, <thewire.in/economy/neither-switzerland-nor-singapore-but-a-site-of-daily-struggle>.

Talukdar, R. 2018, 'Democracy Zindabad! A day in the life of an anti-coal resistance in India's energy capital', *New Matilda*, February 10, viewed 20 March 2020, <https://new matilda.com/2018/02/10/democracy-zindabad-day-life-anti-coal-resistance-indias-ene rgy-capital/>.

Talukdar, R. 2019, Talukdar, R. 2019, 'Profit Before People: Why India Has Silenced Greenpeace', *New Matilda*, 29 March, viewed 20 June 2020, <https://newmatilda.com/2019/03/29/profit-before-people-why-india-has-silenced-greenpeace/?fbclid=IwAR2N wczXun5q1Ug-g3kSxvaOJLInTfwNlnS6Up3Xxts7A7XZF_wbvOsHKYc>.

Talukdar, R. & Pillai, P. 2022, 'Countering the coal curse through community rights: stopping coal-extraction through forest rights in Singrauli, Central India', *Journal of Australian Political Economy*, no. 89, pp. 90–113.

Trivedi, S. 2014, 'Mahan Coal Mines: activists claim harassment by police, authorities', *Business Standard*, Bhopal, 5 August, viewed 30 September 2019, <www.business-stand ard.com/article/companies/mahan-coal-mines-activists-claim-harassment-by-police-auth orities-114080501315_1.html>.

Legislations

Forest Rights *Act 2006*
National Mineral Policy (NMP) 2008
National Rehabilitation *and* Resettlement Policy *1997*
Panchayat Extension to Schedule Areas Act 1996
Right to Fair Compensation and Transparency in Land Acquisition, Rehabilitation and Resettlement Act 2013

6 Environmentalism in the era of Australia's minerals boom

> Minjerriba was a giant in the sun
> His green back coated with cyprus and gum
> Belly brimming with waters so cool,
> He stretched for miles in the sun...
> But minjerriba's back is now broken:
> Men came and tore out his guts;
> Stole his rich grains of sand,
> Stripped his cloak of cyprus and gum...
> Oh man! With your machinery and science,
> Your greed and callous disregard.
> When your savage lying and looting is done,
> Will the Gods in the future
> If future there is
> Spare you your place in the Sun?
>   Noonuccal/Walker (1977)

The resistance to the Carmichael coal mine in the Galilee Basin consisted of sev-eral collaborations between a national environmental mobilisation, a local farmers' resistance, and opposition from the traditional owners of the land. In Chapter 2, I discussed the characteristics of an earlier Australian environmentalism. Although it contained various perspectives and strands, wilderness concerns tended to dominate over other manifestations and discourses of the Australian environ-ment movement (Eckersley 1992). Australia's minerals export boom at a time of exacerbating climate change required the dominant Australian environmentalism to recast its visions and politics to transform Australia's energy economy through stopping coal.

The State did not loom large in the lives of the primarily urban, tertiary educated citizens who were part of the environmental movement, as compared to Indigenous communities with whom environmentalists forged relations. However, the State still played a central role in how the earlier movement framed conflicts and advocated for environmental solutions. Understanding how the actions of the State shape movement objectives and politics is even more necessary in the current era

DOI: 10.4324/9781003410416-6

when Australia's economic pathway of mineral extraction and export including that of coal, and the global need to keep coal in the ground to avoid climate change, are on a collision path. This chapter lays out the political and economic contexts for environmentalism, and Indigenous and farmers' resistances that environmentalists have built alliances with, in the present resource boom era.

This chapter answers the first of the three main questions about the respective case studies in the book: how have the discourses and politics of environmentalism transformed from the previous era? In responding to my question of how environmentalism in Australia has been transformed from its previous version during the resource boom, my respondents traced environmentalism's journey towards becoming dominantly an anti-coal activism to counteract Australia's booming coal export trade, over a 10-year period from 2006 to 2016, even as the world was moving towards a Climate Agreement in Paris in 2015. The thematic structure of this chapter reflects this understanding of Australian environmentalism's progress. It reflects a parallel progress in the movement for Indigenous land rights in Australia, and highlights cases of Green–Black collaboration in environmental resistances both in the past and present era of Australian environmentalism. And it shows the emergence of a third constituency – Australian farmers – in land and environmental resistances during the resource boom.

The first section traces the environmental and economic contradictions inherent in Australian federal governments during the mid-1970s until the mid-1990s, and how these shaped environmental governance and Indigenous land reforms. The second section traces the economic and political contradictions in the Australian State during the resource boom from the mid-1990s, which was characterised by unprecedented extraction of coal and gas. It traces how these contradictions generated new risks and exacerbated old ones, for the environment, for Indigenous land rights, and for Australia's farming sector. The third section traces how these transformations shaped Australian environmentalism, how they created a new disaffected constituency amongst farmers, and how this period of extensive native title deal making between Indigenous groups and corporations transformed Indigenous–green relations. The Analysis discusses the critical transformations to Australian environmentalism produced by the imperatives of Australia's resource boom and advancing climate change.

6.1 Contradictions and unevenness of the Australia State

Even though the Australian State retained its centrality in either mitigating or exacerbating both environmental and Indigenous injustice, its actions had a different bearing on the two strikingly different demographics – environmental activists and Indigenous groups – who attempted to build tactical relations to respond to the State–corporate complex. The two parts of this section trace the significant environmental legislations and governance measures, and land rights reforms introduced from the 1970s to the mid–1990s before Australia commenced extensive deregulation under Prime Minister John Howard. It spans the Prime Ministerial terms of the Australian Labor Party's (ALP's) Gough Whitlam (1972–1975), Liberal-National

Coalition's Malcom Fraser (1975–1983) and once again ALP's Bob Hawke (1983–1991) and Paul Keating (1985–1996).

Environmental governance

Environmental issues gained greater prominence in Australia's social and political arena from the 1970s. Environmental conflicts erupted between state and federal governments during this period owing to the Australian Constitution's near silence on the matter of whether the Commonwealth or states are responsible for environmental issues (Christoff 2015). The federal government's approach towards addressing environmental concerns was marked by two contradictory movements. On the one hand, calls by environmentalists for the federal government's intervention in industrial projects undertaken by state governments were combined with the federal government's own attempts at extending various powers over the states. On the other, the federal government sought institutional reforms that recognised the concurrent nature of federalism (Kellow 1996).

The mineral-rich peripheral states in particular – Queensland, Tasmania and Western Australia – demonstrated a tendency towards what Aynsley Kellow has characterised as 'colonial socialism' even after federation in 1901, pursuing regional development at the expense of the national interest (1996, p. 138). It has been argued that colonial socialism was reinforced through the need for capital formation in these thinly populated regions. And the State's dominant role in penal colonies that were forerunners of many Australia states (Butlin et al. 1982; Eggleston 1932).

Labor Prime Minister Gough Whitlam (1972–1975) sought to resolve the federal–state conflict over Lake Pedder in Tasmania through inducements for preservation, but eventually failed to intervene to prevent it being flooded. The Queensland government under Premier Joh Bjelke-Petersen had zoned 80% of the Great Barrier Reef (GBR) for oil drilling and limestone mining. A prolonged 'Save the Reef' campaign led to the creation of the Great Barrier Reef Marine Park under Whitlam in 1975. Next, Prime Minister Malcolm Fraser from the Liberal–National Coalition (1975–1983), even though reluctant to expand commonwealth jurisdiction over environmental matters, ironically used direct coercion to stop sand mining on Queensland's Fraser Island (Kellow 1996).[1,2] Under Malcolm Fraser the Great Barrier Reef became Queensland's first World Heritage area in recognition of its outstanding natural values in 1981. Labor's Bob Hawke (1983–1991) further developed the Commonwealth's environmental activism by using its external affairs power to protect Australia's international commitments to stop the dam on the World Heritage listed Franklin River (Kellow 1989).[3,4] Passing the *World Heritage Properties Conservation Act (Cth) 1983* immediately on assuming power gave Hawke the legal means to stop the dam on the Franklin (Christoff 2015).

This process of intergovernmental escalation of conflict contributed to causing a weakness in institutional and policy reform at the state level in Australia.

Consequently, Australia emerged both as a site of environmental conflicts and a laggard in international best practice in institutional reform (Kellow 1990).[5] The Hawke government took the dual approach of conflict resolution as well as regaining the national environmental agenda through the process of the National Ecologically Sustainable Development Strategy to introduce sustainable measures in key industries (Christoff 2015). During the combined period of Bob Hawke and Paul Keating's Labor governments (1985–1996) Australia signed several international environmental treaties beyond World Heritage and passed subsequent national legislation to give them effect (Christoff 2015). These included the *Protection of the Seas (Prevention of Pollution from Ships) Act 1983, Antarctic Mining Prohibition Act 1989,* the *Ozone Protection* Act 1989, *Hazardous Waste (Regulation of Exports and Imports)* Act 1989 and the *Endangered Species Protection Act 1992.*

The two decades from the Whitlam government's term to the Hawke–Keating era witnessed a wave of national environmental legislation and policy. Whitlam adopted a number of statutes – the *Environment Protection (Impact of Proposals) Act 1974,* the *Australian National Parks and Wildlife Conservation Act 1975,* the *Great Barrier Reef Marine Park Act 1975,* and the *Australian Heritage Commission Act 1975* – that served as the foundation of Commonwealth environmental authority (Kellow 1996). Although the Fraser Government also signed several international treaties, and maintained environmental progress, it attempted a nationally integrated strategic approach to environment policy through announcement of plans for a National Conservation Strategy of Australia (NCSA) in 1980 (Christoff 2015). Fraser also legislated the *Environment Protection (Nuclear Codes) Act 1978* and *Environment Protection (Sea Dumping) Act 1981.*

Alongside measures for a national takeover of the environmental agenda, the Labor government of Bob Hawke also moved towards a 'new federalism' through which 'the environment must increasingly become an area in which common ground and common purpose come to replace controversy and confrontation' (Hawke, quoted in Galligan and Fletcher 1993, p. 14). Environmental initiatives towards federalism included the development of national standards for air and water quality, the Landcare programme and the management of areas within the jurisdiction of the Commonwealth and one state government, such as the Great Barrier Reef Marine Park and Tasmanian World Heritage Area (Kellow 1996).

Institutionalisation under Hawke and Keating through processes such as the cross-sectoral Ecologically Sustainable Working Groups effectively brought parts of the environment movement under the ambit of the State (Hutton and Connors 1999). The inclusion of the environmental agenda by business also created a mainstreaming effect on environmentalism that resulted in both opportunities and weaknesses. At the same time, the environment movement's outsider strategies, whether that of indirect influence or challenging policies and actions of the State, attested to the State's centrality in implementing better legislation and in environmental management (Doyle 2010).

Indigenous land reforms

What would redefine the politics of environmentalism during the subsequent resource boom – the power and influence of the mining sector over Australian governments – was already shaping the politics of Indigenous land reforms from the mid-1970s. The politics of land reforms from the 1970s to the 1990s serves as a crucial background to understand the contexts of present Indigenous resistance to mining, and the very different places that environmental activists and Indigenous groups come from to form alliances.

Since the 1967 referendum that led to the amendment of the Australian Constitution to include Aboriginal Australians in the census, the ability of the Commonwealth to legislate on Indigenous affairs has borne critical relevance for addressing colonial dispossession (Altman 2009b). A historic strike by Aboriginal stockmen in 1966, which began with families walking off the Wave Hill pastoral station in the Northern Territory, ultimately triggered a movement for Indigenous land rights. From 1972 self-determination replaced assimilation as the central policy approach of the federal government towards Indigenous people (Altman 2012). The Whitlam Labor government's attempt at land reforms subsequently led to the passing of the *Aboriginal Land Rights Act 1983* (ALRA) in the Northern Territory under the Liberal-National Coalition's Fraser Government (Mercer 1987). The right to veto was made integral to ALRA based on the understanding that 'to deny aborigines (sic) the right to prevent mining on their land is to deny the reality of land rights' (Woodward 1974, p. 108). This made ALRA Australia's most extensive land-rights regime.[6]

The mining industry responded to land reforms through a 'relentless campaign to oppose land rights legislation of any kind and to dismantle the Northern Territory Act' (Mercer 1987, p. 174). The political backlash from the industry's campaign in the biggest mining states Western Australia and Queensland (see Gurr 1983) forced the Hawke Labor government to withdraw its election commitment towards national land reforms (Gurr 1983; Mercer 1993).[7] Land rights activists regarded the Mabo decision (Mabo and Ors v Queensland 1992) that followed in 1992 as 'more acceptable solution to the quandary of aboriginal land rights' that came at no real cost to the Labor government (Foley 2013, para 4).

In the Mabo decision the High Court overturned Australia's founding legal fiction of *terra nullius* and ruled that denying Indigenous land rights contravened international human rights, particularly racial equality, guaranteed under the *Australian Racial Discrimination Act 1975* (RDA).[8] Even though the Mabo decision recognised an extremely limited form of native title that granted limited occupation not even akin to a standard lease, another misleading mining industry campaign warned that native title could create a 'national crisis' (Short 2007).[9]

Paul Keating's government passed the *Native Title Act (Cth) 1993* following the Mabo decision. Property rights provided under the NTA's future acts regime were weaker than the ALRA as native title withheld Indigenous groups the right to refuse consent to mining and did not guarantee statutory royalty equivalents from mining on their lands (Altman 2012).[10] The NTA recognised that native title, essentially a form of Indigenous title to land, may continue to exist in areas where Indigenous people still occupied and could display a continuing association with their pre-colonial traditional land, with the caveat that the rights of native title holders would have to yield in case of conflicts with non-Indigenous interests.

A limited national land rights regime contradicted the efforts of successive Labor governments for Indigenous reconciliation. Federal Labor had instigated the process of official Reconciliation that accepted the equality and equity of Aboriginal and Torres Strait Islanders through recognition and commitment to uphold their unique rights (Short 2003). As compared to racial and ethnic minority immigrants in settler-colonial nations, Indigenous people never willingly ceded their lands or political autonomy (Short 2007). This made both returning land and granting rights towards sovereignty necessary reparations for historic dispossessions and freedom from present colonial realities (Gilbert 1994).

Institutionalisation of land rights through native title can be understood as what Turner and Rojeck describe as 'the frequent tension between national systems of rights and international human rights' (2001, p. 127). The United Nations Declaration on the Rights of Indigenous Peoples (UNDRIP) emphasises free, prior and informed consent; it makes land central to Indigenous culture through articles 3 and 31 (see Daes 1999); and defines self-determination as a remedial political right of distinct peoples and nations with a history of dispossession through article 26.[11] Owing to the NTA's limitations, many Indigenous groups turn to the UNDRIP as an accurate articulation of self-determination and lands rights (Short 2007).

Literature on land use agreement-making by mining companies with traditional owner groups suggests that the mining sector's anti-NTA campaign ran simultaneous to its negotiations with First Nations communities (O'Faircheallaigh 1995).[12] Although not opposed to native title in principle, the mining industry regarded the Native Title Act as a risk. Its political posturing rather than ideological opposition over native title was consistent with its responses to other government policies regarded as detrimental to its interests and within its sphere of influence (Lavelle 2010).

Liberal Coalition Prime Minister John Howard's 'Ten Point Plan' outlined amendments to the NTA to rectify a seeming imbalance created by the privileging of 'minority' interests by earlier governments that risked obstructing the free functioning of market forces (Howard-Wagner 2008). In 1996, the Australian High Court's landmark 'Wik decision' on native title ruled that the granting of pastoral leases did not extinguish Indigenous people's native title rights over land and did not confer exclusive possession on pastoralists.[13] A hostile media campaign to the Wik decision by mining companies and pastoralists, particularly the National Farmers Federation, claimed that the Aboriginal population could now lay claim to anyone's backyard (Kuhn 1998). John Howard commented that 'the pendulum had swung too far in favour of Aboriginal people' through the 'potential right to veto over 78% of the landmass of Australia' (Howard 1997a, 1987b). The *Native Title Amendment Act 1998* (NTAA) was introduced; it tightened the registration process for native title claimants contained in the original legislation (Lavelle 2010).

The State's neoconservative and neoliberal approach deepened the conflict between Australia's original intention of Indigenous reconciliation and the pre-eminence of mining in the economy. While the State loomed large in the lives of Aboriginal people (Altman 2009a), the influence of mining loomed large for the Australian State. This power dynamic determined the extent of rights enabled under the combined state and federal land rights regimes instituted in Australia between the mid-1970s and the mid-1990s. The neoliberalisation process set in motion in the last part of this period paved the way for extensive deregulation under Prime Minister Howard. The State significantly withdrew from environmental governance and demonstrated what anthropologist and economist Jon Altman calls a conflicted approach towards a growing Indigenous estate: on the one hand, the State ascribed the values of tradition and conservation and, on the other, it encouraged mining within the geography of the Indigenous estate (Altman 2012a).

6.2 Minerals boom and contradictions of the Australian State

Australia's economic prosperity is historically linked with export-oriented mining. Fostering the competitiveness of its mineral resources has been a longstanding economic policy of the State (Rosewarne 2016). It is the world's largest exporter of black coal (along with Indonesia), iron ore, lead, zinc and aluminium, and the second largest exporter of uranium (Minerals Council of Australia (MCA) 2010). Although mining is undertaken in all six states and the Northern Territory, Western Australia and Queensland dominate the sector and generate three-quarters of the country's mineral resources (Topp et al. 2008).

Australia's mineral resources relationship began prior to federation, with the gold rush that began in the 1850s (Hajkowicz et al. 2011). The 1930's economic recession and World Wars slowed Australia's mineral production, with mining and energy representing only 1% of the GDP and 5% of exports by 1960 (Maxwell 2006). Subsequent expansions were enabled from the mid-1960s owing to increased global economic activity and the rise of Asian markets, with Australia's political stability proving an advantage for overseas mining corporations (McKay et al. 2000). By the 1990s a broadly defined category of mining products contributed 10% of Australia's national economic output (Maxwell 2006).

From the 1990s, China's rise as the global manufacturing workshop drove a demand for minerals on a scale previously inconceivable. This phenomenon led to an unprecedented increase in Australia's mineral extraction and export since the mid-1990s, with far-reaching environmental, social and economic impacts on other sectors.

Economic contradictions and socio-ecological effects

The minerals boom affected the balance between mining and the other sectors of the Australian economy.[14] Being tied to the mineral resources super-cycle connected to the industrialisation of the BRIC economies (Brazil, Russia, India and China), the price of Australian minerals more than tripled between 2002 and

2012 (Tulip 2014). In fact, mining was regarded as key to Australia having avoided the recession in 2008 (Perlich 2013). However, in a major socio-economic fallout from mining's unprecedented growth (as a result of high mineral prices) Australian manufacturing exports consequently declined (Mitchell and Bill 2006).[15]

Massive capital influx into mining made possible new technologies including hydraulic fracturing and horizontal drilling used for coal seam gas (CSG) extraction, increased mechanisation and increased sizes of mining projects, drawing rural and remote regions into the global commodity chain (Bridge 2008). Supersized projects integrated multiple mining operations and associated port, rail and pipeline infrastructures. Richardson and Dennis (2011), amongst others, have shown that, although the scale of projects transformed landscapes, they did not lead to a proportionate increase in employment.[16] Bell and York (2010), amongst others, have highlighted the risk of cumulative impacts and ill-defined risks from new innovations that massive mining projects exposed societies and environments in rural and remote regions to. A rapid expansion in mining meant that new technologies were deployed before their environmental impacts could be fully understood (Bridge 2008).

By 2012, all of Australia's 373 active mine complexes were being expanded in size alongside new projects and entire greenfield sites being developed for extraction. Encroachment of mining on agricultural regions generated a clash of values, land uses, water needs and skilled labour (Everingham 2016). In *Minefield* journalist Paul Cleary describes the scale of the geographic transformation:

> The Australian investment pipeline is claimed to be the world's largest... Massive mining and energy projects are being rolled out with military precision and concentrated in three states. With 41 such projects, Western Australia is regarded as the powerhouse resource state, but Queensland and New South Wales are not far behind, with 29 and 18 projects respectively. In Western Australia, most of the projects are being built in remote areas like the Pilbara; in Queensland and New South Wales, this investment intersects with prime farmland.
>
> Cleary (2012, p. 6)

Minerals and energy constituted half the value of exports by 2010–2011, with the mining sector becoming the fourth largest contributor to Australia's GDP at 8% as the Australian Bureau of Statistics (ABS) figures show (ABS 2012).[17] Australia became one of only three industrialised economies with significantly high proportions of mining exports (Steven and Dietsche 2008). Although mining's share of the Australian GDP had recorded a peak of 15% during the 1861 gold rush, it had only grown to 6% between 1966 and 1975 (Cook and Porter 1984). These statistics indicate the historical scale of the minerals boom. The majority of increased investments were in coal, oil and gas, which were energy sources responsible for major greenhouse emissions (Goodman and Worth 2008).

Mining effectively created what Mitchell and Bill (2006), amongst many others, have called a two-speed economy. The scale and nature of the resources boom reopened an old debate on mining's costs and benefits, particularly around their uneven distribution. It ingrained inequalities and social divisions across states and regions owing to spatially concentrated employment (Goodman and Worth 2008). Mining benefits were largely accrued by cities where the majority of jobs and profits flowed, while rural communities and regional areas incurred social and ecological costs from massive mining projects including through direct displacement for mining operations, creating winners and losers from the minerals boom (Everingham 2016).

The mineral boom's impacts on Indigenous communities remain highly contested. On the one hand, intra-Indigenous divisions over securing mining rights with repercussions for community capacity have emerged as a key area of social contention during the boom (Doyle 2002). On the other, economic participation through employment and royalty payments have provided some Indigenous communities an opportunity to escape the 'drudgery of the welfare economy' and achieve a measure of prosperity (Langton 2012).

The role of the Australian State during the resource boom can be seen in the context of what Marxist Economic Geographer David Harvey (amongst others) qualifies as a neoliberalising political economy that increasingly facilitates internationalised production (Harvey 2011). From the mid-1980s under Labor's Hawke and Keating governments, and under Liberal-National Coalition Prime Minister John Howard from the mid-1990s, the State was consistently weakened by the culture of free markets and globalisation (Doyle 2010).

Mining's high foreign ownership, export focus, large inflow and outflow of capital, and reliance on imported equipment rendered Australia as a client State whose 'main function was to shape the future development of the economy such that profits of foreign corporations had first priority, and needs of people the last priority' (Crough et al. 1983, p. 35).[18] For instance, a high foreign ownership structure meant that a high GDP growth from mining exports did not translate into a proportional wealth distribution across households owing to over 80% of 'windfall gains' from high commodity prices going offshore (Richardson and Dennis 2011).

An ownership structure in Australian mining that is concentrated amongst a small number of transnationalised corporations creates the effect of an oligarchy and affects the free and fair functioning of democracy in Australia through the ability of a few transnational corporations to influence Australian governments (Goodman and Worth 2008).[19] One of the most obvious examples of this situation has been Australia's inability to adequately act on climate change, owing to deeply entrenched mining interests in politics (Baer 2016). The economic, social, political and ecological changes ushered in by the mining boom have been said to demonstrate effects of the resource curse that I now discuss (Goodman and Worth 2008).

Political contradictions: how the State facilitated the minerals boom

The resource curse thesis emerged in the aftermath of the 1960s global economic boom and has primarily focussed on the experiences of resource-rich developing countries where resource extraction and exports produced underdevelopment in the long term (Oskarsson and Ottosen 2010). The phenomenon is also acknowledged to afflict industrialised countries where it can be understood through the 'Dutch disease', a term coined by *The Economist* to describe the experience in Netherlands in the 1960s when the country prioritised the export of newly discovered natural gas. The Dutch disease led to a rise in exchange rates that ultimately rendered the country's manufacturing exports economically unviable (Auty and Warhurst 1993).

The theory has been supported by a variety of empirical evidence of socio-economic, political, ecological and livelihood impacts of nature dependent communities in mineral-rich countries (Sachs and Warner 1997, 2001). Disproportionate costs of resource extraction on certain communities are often cited as a central problem of the resource curse (Martinez-Alier 2002). In regard to settler colonial states like Australia, apart from transforming the ecology and society, mining is understood to extend 'the long arc of violent extractivist frontiers and resource colonialism that has dispossessed Indigenous people' (Parson and Ray 2016, p. 5).

Goodman and Worth (2008) identify three curses of Australia's minerals boom: de-industrialisation and social division from the decline of other industries that cause macro and micro economic effects, capture of Australian Federal Policy by mining demonstrated through diplomatic and military interventions, and ecological degradation and exhaustion through impacts on environments and people. Resource curse literature has largely assumed mineral wealth to be State owned (Luong and Weinthal 2006). As Australia did not nationalise its mining sector in the 1960s and 1970s like various resource rich States, Goodman and Worth (2008) argue that a key determinant of whether it avoided or was afflicted by the resource curse was how the government taxed mining and used the revenues for public good.

ABS data for the periods 2001–2002 and 2004–2005 show that taxation rates for minerals, oil and gas in fact fell as commodity prices and mining profits grew, resulting in windfall tax gains for the mining sector (ABS 2007a). Goodman and Worth (2008) explain that a lack of savings mechanisms for mining rents meant governments spent surplus funds generated from mining rents (during record high commodity prices) for political gains through tax cuts. This pattern was clearly observable in John Howard's government.

Scholars and activists contend that the subsidisation of coal, natural gas and mineral resources mining since the mid-1980s demonstrates the power of the mining sector over both major Australian parties. Despite Australia's significant renewables potential and even as their uptake increased globally, fossil fuel subsidies obstructed the growth in demand for renewable energy during the first two decades of the resource boom (Christoff 2009).[20] This political situation brought into focus the transformation of the State under the combined effect of increasing globalisation and extractivism and underscored the process of the neoliberalising political economy.

A neoliberalising political economy characterises the complex and variable process through which global pressures reconfigure the national, regional and local (Harvey 2011). As political economy involves material realities, neoliberalising political economies effectively consist of 'patterns of…power' and governance regimes influenced by 'agents of transnational capital' (Gill 1995, p. 4, p. 10). By being inextricably bound up in and actively promoting the process of globalisation, the State's political autonomy and steering capacity stand undermined, and consequently the neoliberalising process becomes a driver of ecological destruction (Eckersley 2004). Instead of a simple withdrawal through deregulation, under the process of neoliberalising political economy, the State becomes directly responsible for socio-environmental impacts.

Through a combination of neoliberal and neoconservative values, Prime Minister John Howard attacked social and environmental agendas with ideological purity (Doyle 2010). Howard removed community participation in environmental regulation, streamlined approvals through the *Environment Conservation and Biodiversity Protection Act 1999* (EPBC) that handed areas of Commonwealth determination to states, de-funded ENGOs, and removed global issues like climate change from the national agenda. Compromising the project approval process, especially by fast-tracking super-sized mining projects with high socio-environmental impacts, underscored the State's contradiction under neoliberalisation. Fast tracking was justified by deeming such projects as critical for the national interest. Governments attempted to balance the socio-environmental impacts from fast-tracking major projects by issuing approvals with a long list of conditions attached, without any certainty that companies could meet these conditions, or capacity on its own part to monitor and regulate environmental violations by corporations (Cleary 2012).

Capture of federal policy: minerals boom versus climate action

Starting from 2007 Australia entered a turbulent decade of politics that demonstrated the effect of the resource curse on federal policy and held national climate action to ransom. The coal and thermal power industries formed the 'Green House Mafia' that dominated climate policy (Pearse 2009, p. 41). The nexus between fossil fuel and politics was strengthened through lobbying, political donations and the Rupert Murdoch-owned Newscorp news media that enjoys a broad subscription in Australia (Baer 2016).

In 2007, Labor's Kevin Rudd replaced John Howard as Prime Minister in what was considered Australia's first climate election. The Rudd government ratified the Kyoto Protocol that John Howard had refused to sign and attempted to legislate emissions reduction.[21] A multimillion-dollar advertisement campaign by the MCA spread misinformation about the loss of jobs and competitiveness for companies from Labor's Carbon Pollution Reduction Scheme (CPRS).[22] The Australian Green Party and the Liberal-National Coalition both voted against the CPRS in 2009, albeit for opposite reasons.[23] Another multimillion-dollar advertising attack by the resources sector on Kevin Rudd's proposal for a 40% tax on mining profits is understood to have led to his replacement by Julia Gillard as Prime Minister in 2010. Prime Minister Gillard negotiated a reduced tax of 30% that would only apply to iron ore and coal, Australia's two largest export minerals (Manne 2011).

In 2011, Julia Gillard's minority Labor government passed the 'Price on Pollution' scheme that proposed a fixed carbon price for three to five years before moving to a cap-and-trade system. The Coalition opposition under Tony Abbott ran a misleading attack alleging mass job losses, soaring electricity bills and skyrocketing living costs.[24] Within the first 100 days of winning the 2013 elections, Tony Abbott's Liberal-National Coalition government repealed Labor's carbon scheme. Other attacks on the environment during this time included attempts to de-list World Heritage forests, hand over matters of federal decision-making under the EPBC Act to states, and attacks on the tax-deductibility status of environmental charities to affect their funding and disrupt their climate and environmental advocacy (Wilkinson et al. 2020). Attacks on renewable policy included cutting the Renewable Energy Target (RET) that had enjoyed bipartisan political support. Ironically, the RET was introduced by John Howard, a former Prime Minister from the same political party, demonstrating the extent of ideological opposition to carbon emissions reduction under Tony Abbott.

Parts of the energy industry changed their position on an emissions trading scheme between 2009 and 2015 in response to a global trend away from fossil fuels. However, Liberal Coalition's Malcolm Turnbull who became Prime Minister in 2015, left Tony Abbott's climate policy unchanged owing to the undue influence of climate doubters in the party, and did not introduce an emissions trading scheme despite reassuring businesses (Taylor 2017). Without bi-partisan political support on the climate issue on account of the influence of conservative members within the Coalition, Malcolm Turnbull's signing of the Paris Agreement raised concerns about Australia's ability to comply with its requirements in the future (Stephens 2016). Experts warned that without a long-term policy to phase out coal, Australia would risk failing its emissions-reduction commitments towards keeping global warming within two degrees centigrade (Taylor 2016).

Although outside the time frame of this research, it is worth noting that in 2023, the Labor party which returned to power federally legislated significant amendments to the Liberal Coalition's existing 'safeguard mechanism' climate policy to reduce Australia's emissions by 43% from 2005 levels by 2030. Although it was the most significant Australian climate policy in a decade and sets a hard cap on industrial carbon emissions, it does not ban new coal mines and gas projects, and relies on controversial carbon offsets to reduce emissions. Finally, it does not offer a clear pathway for phasing out Australia's coal exports.

Conflicts, constituents and geographies of the minerals boom

The challenges of the minerals boom and dialectics with State planning and regulatory processes around mining projects followed different trajectories for environmentalists, Indigenous groups and farmers; the three constituent groups that contested the destruction of land, water and environments during the minerals boom. Their campaigns converged in critical geographies that were threatened and transformed by the scale of resource extraction. I discuss some place-based contestations below. They also formed critical alliances that I discuss in Section 6.3.

Political conundrum of the Great Barrier Reef

Australia's conundrum between driving the resource boom and acting on the climate became evident through the State's failure to take science-based actions for the GBR. It is the world's largest natural reef system and an extensive World Heritage listed area stretching 2000–3000 kilometres along the Queensland coastline from the Torres Straits in the North to Fraser Island in the South. By 2012, the Reef had lost more than half its coral cover since the mid-1980s, largely on account of long exposure to chemical runoff from agriculture from the mainland, and the bleaching effects from warming oceans (De'ath et al. 2012). Scientists estimate that business as usual global emissions would possibly shrink the Reef to one-tenth its size by 2050, and completely kill it by 2100 (Del Monaco et al. 2017).

Some of Australia's largest coal ports lie adjacent to the GBR along Queensland's central coast. The resource boom raised the prospect of a sixfold increase in port capacity along the Reef; a Greenpeace report estimated that a full-capacity port-expansion in Queensland would see more than 10,000 coal-laden ships a year cross the GBR by 2020, a 480% increase from the 1722 ships that entered the Reef's World Heritage area in 2011 (Greenpeace Australia 2012). A UNESCO report warned that the Reef was in danger of losing its heritage status unless Australia acted to protect it, especially from 'threatening new port-developments along the coast' (UNESCO 2012). The United Nations (UN) was specifically responding to the development of the Wiggins Island Coal Export Terminal (WICET) in Gladstone Harbour on Queensland's Central Coast that had involved significant dredging and caused major environmental problems through the dumping of dredge-spoils on the GBR.

The Queensland and Federal governments published a joint report 'Reef 2050 Long-Term Sustainability Plan' in response to the UNESCO World Heritage Centre (Commonwealth of Australia 2015). The report was criticised by the Australian Academy of Science for representing business as usual and only addressing short term issues like agricultural run-offs while ignoring long-term climate impacts on corals, and of mining and coastal infrastructure developments on the Reef (Philips 2014). Referring to proposed mega-mines in the Galilee Basin, scientists asked the Government to choose between 'coal mines operating in 60 years' time or a 'healthy reef' (Norman et al. 2015). ENGOs similarly argued that it is far too risky to develop huge new coal mines, build the world's largest coal ports, dump unprecedented volumes of dredge spoil at sea, and still have aspirations to protect the GBR, maintain its World Heritage status, and secure reef-dependent tourism and fishing' (Hughes 2014).

Successive mass bleaching events in 2016 and 2017 killed half the coral in the GBR (Slezak 2016; Knaus and Evershed 2017). Experts warned that with such events were likely to increase, the Plan's central aim would be unachievable and the Reef risked being listed as a World Heritage site in danger (Slezak 2017). The State's politics on the sustainable management of the Reef came under further criticism on account of significant federal funding being provided to an unknown entity, the Great Barrier Reef Foundation, ostensibly for 'conservation' even as governments cracked down on ENGOs and activists (Slezak 2019). The Reef has shown a measure of recovery since the mass bleaching events, but scientists warn that increasing water temperatures owing to global warming are likely to cause its bleaching every year (Australian Institute of Marine Science 2023).

Changing relations between State, mining and Indigenous groups

The actions of Aboriginal peoples towards land and economic justice emerged as key players in Australia's mining economy during the resource boom and transformed the nature of land debates (Norman 2016). Despite legislative and judicial limitations of the concept of native title, the *Native Title Act 1993* placed obligations on mining companies to negotiate with Indigenous claimants and created what Langton (2012) calls an 'era of agreements making'.[25] It shifted mining companies away from 'bareknuckle' racism towards remote Aboriginal communities during previous booms when mining occurred on traditional lands without negotiations or benefits to Aboriginal people.[26] Over 30 years of multiple land rights regimes contributed to an Indigenous estate covering 20% of Australia and containing some of the most ecologically intact landscapes in the North of the continent (Altman et al. 2007).

In contrast to mining's rhetorical transformation, the Australian State under Prime Minister Howard assumed a paternalistic view towards Indigenous development by de-emphasising land rights and singularly emphasising full-scale economic participation through mining for closing the socio-economic gap between Indigenous and non-Indigenous Australia (Altman 2009b). The 1998 NTA amendments indicated that the State approached Indigenous affairs through economic liberalism, emphasising economic freedom rather than civil liberties (Quiggin 2005). The 2006 amendments to the Northern Territory land rights, ALRA, noted that the 'the principal objectives [of amendments] are to improve access to Aboriginal land for development, especially mining' (ALRA Amendment Bill 2006, p. 30). The State also appeared conflicted in its approach to the development of the growing Indigenous estate with the onset of the resource boom (Altman 2012b).

Jon Altman has contested the assumption that benefits from mining will eventually trickle through to remote Indigenous communities with considerable development challenges on two significant grounds. Altman emphasises that the socio-economic benefits from mining can only be assessed at the local and case-by-case basis, making it inaccurate to generalise on this basis (Altman 2009a, 2009b). Also, the presence of a customary market in remote locations demonstrates the failure of free market alone to deliver favourable outcomes for remote communities (Altman et al. 2006; Altman 2007). Altman argues that these factors indicate the need for a 'hybrid economy model' where the State plays a role alongside the market and communities to ensure economic diversity in remote regions (Altman 2005).

Altman further argues that the government's emphasis on mining for remote Indigenous communities in the resource boom era was in conflict with its own approach on land rights, which is based on a discourse of tradition, continuity and a connection to country that augments the conservation values of the Indigenous

estate (Altman 2012b). The Australian government supported the Indigenous conservation initiative through the formation of Indigenous Protected Areas (IPA) and
instituting the Working on Country Program to employ Indigenous rangers, demonstrating a hybrid economic model.[27] However, as Altman's argument suggests,
increasing pressure for Indigenous communities to participate in mining enabled
through government policies, threatens to destroy these environmental and cultural
values and extinguish native title and land rights (Altman 2012b).

Coal seam gas versus agriculture and politics of water

Although mining has historically coexisted with agriculture in Australia, its unprecedented increase during the resource boom led to an encroachment on prime
agricultural lands, sparking land and water concerns amongst farmers and directly threatening Australia's second largest export industry. In particular, the drive
for CSG exposed prime agricultural land in Queensland, New South Wales and
Victoria to drilling operations.[28] Resistances in the more populated farming regions
of New South Wales and Victoria were able to check the explosion of the CSG
industry in the early stages of exploration. Queensland however became the epicentre of CSG production, with vast distances between rural towns and smaller
populations making the grassroots resistance less effective.

By 2011, the Surat Basin in Southern Queensland, a region containing some of
the state's most productive farmlands, and the Bowen Basin, Queensland's largest
coal mining region, collectively faced the prospect of 40,000 CSG wells being
sunk into the landscape. Pipelines 500 kilometres long linked the wells to processing plants on Curtis Island on the Central Coast, creating a 'spiderweb effect'
across the landscape (Cleary 2012). New export terminals at Gladstone Harbour to
transport liquefied gas affected water quality with consequences for the adjoining
Great Barrier Reef (Hunt 2011).

Hydraulic fracturing or fracking is a process of CSG extraction that involves
injecting millions of litres of chemically treated water deep underground to
release the gas trapped in the coal seams under pressure. Apart from the destruction and fragmentation of farmland through the digging of wells and associated
infrastructure, fracking risks dewatering and contaminating underground aquifers,
threatening the water supplies of farmers in the interiors of the dry Australian continent. Disposal of saline CSG water posed an additional problem, with no impact-
free method having been identified.

Concern of farmers, scientists and conservation groups about the impacts of
CSG and large coal projects on water sources, and significant resistances in New
South Wales and Victoria, forced the federal government to add a 'water trigger'
to the EPBC Act in 2012. Under the water trigger, CSG and large coal mining
projects with significant water impacts required federal approval, which in turn
needed to be informed by recommendations from the Independent Expert Scientific
Committee (IESC) on CSG and Large Coal Mining. The role of the State in the
CSG conflict indicates a double movement on its part. On the one hand, States
approved CSG projects that deployed new technologies with unknown risks in

violation of the precautionary principle through fast tracking and often without its own capacity for monitoring (Cleary 2012). On the other, the federal government was forced to capitulate under widespread protests from the agriculture sector and pass ameliorative legislation.

6.3 Narratives, politics and alliances of environmentalism during the resource boom

This section looks at three areas pertinent to the transformation of Australian environmentalism during the resource boom: grassroots anti-fossil fuel resistances and their linkages with global anti-fossil-fuel activism, the alliance between environmentalists and farmers against coal and gas, and the alliances of environmentalists with Indigenous communities fighting extraction.

The first four subsections chronologically outline how the environment movement's politics transformed from that of wilderness protection towards that of 'End(ing) Coal'. The shift was generated through mass movement formations at sites of massive coal extraction, and became evident through new narratives, activisms and organisational structures of ENGOs, particularly in the period between the Copenhagen Climate Summit in 2009 and the Paris Climate summit in 2015. The shift was also marked by the emergence of new forms of activisms and models of mobilisations that were interlinked with global anti-fossil-fuel networks.

The fifth subsection looks at politics, narratives and activisms emerging from its engagement with the newly disaffected community of farmers impacted by coal and CSG mining. Sixth, seventh and eighth subsections look at three Indigenous land-rights conflicts during the resource boom. These indicate a transformation of the dynamics between the State, corporations and Indigenous groups during the minerals boom. They also add new political dimensions and narratives to Indigenous–green alliances through deeper relations and articulations of a shared future founded on cultural and historical obligations of Indigenous communities towards Country.

Grassroots anti-coal activism before Copenhagen

From the 1980s, the integration of the Hunter Valley in New South Wales into the global coal economy has been bringing up environmental and health impacts from mining and thermal power generation, but transformations during the minerals boom proved unprecedented. During the first decade of the minerals boom, the Hunter Valley became Australia's largest coal producing region, losing its previous economic balance between coal mining and wine, tourism, defence and thoroughbreds (Cleary 2012). Grassroots and local resistances that emerged during this time challenged the legitimacy of coal mining in Australia's oldest coal-exporting region and represented the 'shifting grounds of environmental knowledge and oppositional practices' by coal-affected residents and environmentalists (Connor et al. 2009, p. 496).

While local groups Minewatch and Singleton Healthy Air raised local environmental concerns, Rising Tide's actions demonstrated a militant participatory democracy in response to climate change where citizens directly stopped economic activity for both local and global sustainability. The global anti-capitalist Rising Tide network first formed in 2000 in the United Kingdom with the stated purpose of organising for social justice and sustainability (Evans 2010). Rising Tide Newcastle that was established in 2004 reflected a new politics in Australian environmental activism at a time when professionalised ENGOs had not yet begun intervening against Australia's coal exports. A member of The Sunrise Project told me during an interview in Sydney that:

> Fifteen years ago big NGOs working on climate change focused on Howard not ratifying Kyoto. The emphasis was on renewable energy because of small successes that were possible in that area. Targeting 'big polluters' was not yet a strategy. The energy export industry was seen as too powerful to touch. Under Kevin Rudd, there was optimism with the policy approach and the movement lacked a political edge, getting involved in technical emissions arguments instead. After Inconvenient Truth increased public awareness and led to a blossoming of local groups, we (big NGOs) trained new local networks. But overall, there were very few political actions; grassroots actions of Rising Tide were the only early politically focussed actions.
>
> Interview (20/10/2017)

Rising Tide's 2006 campaign against New South Wales's largest proposed coal mine at Anvil Hill was Australia's first direct opposition to coal mining on the basis of climate change; it made Anvil Hill an 'icon of the climate issue' in public awareness (Woods 2007). The network organised annual 'People's Blockades of the World's Biggest Coal Port' consisting of flotillas of hundreds of canoes that halted shipping traffic at the port of Newcastle, and annual 'Climate Camps' of radical activists that culminated in peaceful protests involving rail and coal train blockades (Evans 2010).[29] The opposition to the construction of a third coal-export terminal (T3) at Newcastle Harbour in 2008 turned out to be the largest direct action in the climate movement (Rosewarne et al. 2014). Protestors blockaded the train line to the port for a whole day and disrupted economic activity to the tune of A$1.2 million dollars (Talanoa 2008).

However, the failure to stop the development of the mine or port projects revealed the mining sector's power and the legal system's ineffectiveness to protect the public interest. The Sunrise Project member said:

> Anvill Hill basically taught us that environmental laws are ineffective, that projects are never stopped and that the companies are so powerful they can get laws changed anyway. Community campaigns like Rising Tide's opposition to the T3 coal terminal expansion at Newcastle harbour hardly put a dent in the projects.
>
> Interview (20/10/2017)

A landmark decision by the New South Wales Land and Environment Court in the *Anvil Hill Case* (Gray v Minister for Planning 2006) ruled that the government should have included 'indirect emissions' associated with the burning of coal from the mine either in Australia or overseas in the overall environmental assessment, but it could not stop the mine (Connor et al. 2009). Subsequent legal challenges against the mine also proved unsuccessful (Strachan 2007).

Towards a national anti-coal environmentalism

The fertile lands of the Liverpool Plains to the north west of Hunter Valley became the next big region for the encroachment of coal mining and CSG drilling, directly threatening farmlands and the agricultural exports market.

With the size of mining projects increasing between the first and second decades of the resource boom, the Whitehaven Coal owned Maules Creek coal mine in the Gunnedah Basin that commenced operations by 2014 emerged as New South Wales's largest coal mine. State and federal governments approved the coal mine despite the project's financial unviability owing to a structural decline in coal, creating a risk of the project becoming a stranded asset (Greenpeace Australia 2015). A lonely battle by local farmers over several years to save their farmlands was followed by the forging of alliances with environmental groups and the Gomeroi traditional owners against mining at Maules Creek (Evans 2014).

The idea to blockade at Maules Creek grew over years to become a protest gathering of climate and forest campaigners from around Australia who had a 'big picture view of the world' (Hunter Valley Campaigner interview 23/10/2017). What began in 2012 as a small protest camp in the Leard Forests, the last remaining ancient woodlands in the region, grew to become the Frontline Action on Coal (FLAC), a community-driven frontline anti-coal protest group (Evans 2014). Big NGOs eventually 'got involved when the movement gathered steam' (Hunter Valley Campaigner interview 23/10/2017). The multi-pronged resistance to the coal mine set a precedent for future anti-coal campaigns. A Stop Adani Campaigner told me during an interview in Mackay, central Queensland, that:

> We did not know if we would win, but we wanted to make it a watershed moment for the climate movement, when many people stepped up and held off coal. So the anti-Adani base got built there. We had multiple strategies, such as financial disruption, and diverse public participation, such as with faith groups getting involved. The climate movement got its act together and delayed the mine.
>
> Interview (27/10/2017)

Several big ENGOs who represented pragmatist tendencies within Australia's climate and environmental movements had grown disillusioned by the medium range of Kevin Rudd's climate policies (Rosewarne et al. 2014). A split amongst large ENGOs over the Rudd government's second CPRS legislation in 2009,[30] followed by the failure of the 2009 Copenhagen Climate Summit to deliver global commitments on tackling emissions, turned several ENGOs towards direct disruption of coal mining. A campaigner with experience both in grassroots and international ENGOs who works in the Hunter Valley told me during an interview in Sydney that:

> After a split on the CPRS, the politics got ugly and optimism turned to pessimism. There was a real explosion of actions by coal-affected communities between 2010 and 2012. In new areas where coal was expanding, they had not seen the dynamics of governments before. First they were shocked they (governments) can act against all scientific evidence. In a way it was necessary to break the faith of people in government and planning, in order to build movement.
>
> Interview (23/10/2017)

By 2012, a multi-pronged strategy of anti-coal activism consisting of on-ground protests, economic disruptions through financial targeting of investors and shareholders, diverse civil society participation and the involvement of large groups began to emerge out of nearly a decade of localised anti-coal and gas protests. By this time, Australia was poised for an even further unprecedented scale of fossil fuel expansion on account of record high mineral prices for a decade and a half, forcing a complete shift in the movement's outlook. The Sunrise Project member told me:

> Oil terminals on Curtis Island and Gladstone harbour got approved and there was no opposition. Everyone said the industry was so big, so powerful, how can we even begin to challenge it? Our imagination was colonised! But Gladstone was a turning point...I got a call from Drew Hutton, 'There is a massive gas hub coming up can you help? That was the beginning of LTG (Lock the Gate)...And in 2011 a philanthropist asked 'what is happening with export coal, Australia's biggest problem'? Targeting coal finally became a national strategy in 2012.
>
> Interview (20/10/2017)

'Save the Reef': success and concerns of proxy anti-coal activism

A 2012 Greenpeace report *Boom Goes the Reef* estimated a sixfold increase in coal traffic across the Great Barrier Reef World Heritage Area from six proposed coal port expansions along Queensland's central coast (Greenpeace 2012). The visual scale of the expansion of Queensland's coal exports has been described as 'a sea of ships waiting to get out and not enough ports' (Australian Marine Conservation Society (AMCS) GBR Campaign Director interview 15/10/2017). The sea of ships started collecting after the expansion of the Bowen Basin in Central Queensland, making it Australia's largest black coal-producing region by 2010 (Maddison 2011). The proposals included ambitious expansions at Abbott Point near Bowen – comprising four new terminals with an additional handling capacity of $120 million tonnes – to service proposed coal mines from the greenfield Galilee Basin in Central Queensland.[31] It would make Abbott Point the world's biggest coal port and would come at an environmental cost of five million tonnes of dredged sediment from the seabed adjoining the GBR (Greenpeace 2012).

Proposals for this sixfold increase in Queensland's coal ports came around the same time that the UN expressed concerns about the Reef's deterioration and over Queensland allowing a Liquefied Natural Gas Plant to be built within the Reef's designated World Heritage Area (Hoegh-Guldberg 2012). In Queensland, a very high dependency on the international resources market and the historic power of the mining industry over the State resulted in weak environmental governance that continued and was further compromised during the minerals boom. The Campaigns Director at the Wilderness Society told me during an interview in Sydney that:

> The narrative was that Queensland has to catch up with the rest of Australia. A whole political class was created on the exploitation of nature, and foreign capital and cheap tools went together to serve the purpose. Bowen had open-cut coalmines, Cape York had Bauxite mines. That was the Queensland model of development; a quarry! When the GFC shrunk the economy to half after 2008, the government got desperate and largely suspended all environmental planning and regulation. It became about development at any cost. Gas processing at Curtis Island went ahead even before verifying if there was adequate CSG resource.
>
> Interview (15/10/2017)

Extending anti-coal activisms that were shaped over a decade of environmental conflicts in the Hunter Valley and Liverpool Plains in New South Wales to Queensland required two approaches. The national environment movement worked to improve the capacity of local, regional and state-based environmental groups and activist networks in Queensland through campaign resources and national and international funding (ex-Coordinator Mackay Conservation Group (MCG) interview 28/10/2017). And, instead of directly tackling the problem of coal mining and exports expansions from Queensland, the movement strategically chose an indirect political approach to challenge the state's coal-expansion issue through a national campaign to save the GBR. The movement considered this approach as necessary on account of Queensland's weaker democratic structures in comparison to New South Wales, raising the need to exert public pressure from the outside. The Sunrise Project member said:

> We could see around 100 mines, 9 ports and rail expansions across both states. In New South Wales the key issue was the expansion on agriculture. We launched the Land, Water, Futures campaign there. New South Wales has a diverse economy, and two houses of Parliament. The Greens are an institutional power in Upper House; there would be a much better chance of outcomes from this campaign in New South Wales than Queensland. In Queensland we focussed on the Reef as a strategic focus for coal exports from Queensland, that has a wider, a national and international appeal.
>
> Interview (20/10/2017)

During the peak of Queensland's coal boom in 2012, national and regional ENGOs organised themselves into the 'Fight for our Reef' movement aimed at stopping Queensland's coal expansions. This campaign targeted new port developments along the GBR. The GBR Campaign Director at the Australian Marine Conservation Society (AMCS) told me during an interview in Sydney that:

> We were talking about possibly millions of cubic metres of dredge spoils being dumped on the Reef! There were fish kills from water pollution from the LNG terminal on Curtis Island; no one could fish. A major donor who was worried about the dredging funded a significant joint campaign. Our campaign aligned with the World Heritage Committee notice in 2012. The WHC came over in 2013; IUCN also came to examine the Reef. AMCS ran the public campaign and WWF did the inside lobbying-work. Our logic was that if we stopped port-expansions we stopped coal-exports. We pushed for the Reef to be considered for the In Danger List (of World Heritage sites).
>
> Interview (15/10/2017)

As the only national ENGO with a prominent presence in Queensland, with staff and volunteers at multiple locations including Mackay, Cairns, Whitsundays and Brisbane, the Australian Marine Conservation Society was key to the running of the anti-coal campaign on the ground (The Sunrise Project member interview 20/10/2017). Government proposals for dredging and dredge spoils dumping operations within the GBR's World Heritage Area were one of the 'Fight for the Reef's' strongest contentions and the subject of several legal challenges.[32] Constituency based mobilisations against the ideological environmental attacks by the state and federal Liberal governments proved effective in increasing the reach of the movement. The AMCS coordinator for the Airlie Beach area in central Queensland told me during an interview in Mackay:

> It was easier to generate outrage against the Abbott and Newman governments. People got motivated to agitate at the thought of dredge spoils being dumped in the World Heritage Area! Along the coast, community groups like the Whitsundays Residents against dumping (WRAD) cropped up. Thousands joined in from around Australia, I think it became Australia's biggest conservation campaign at one point! Reef Tourism operators were worried. Some are wary to talk about Reef damage; they don't want to drive tourists away, specially in Townsville and Cairns where they are more commercial. But outrage at what happened at Gladstone helped to galvanise the sector. They did their lobbying and threatened to take Greg Hunt to court, after which the Minister pivoted away from dumping on the Reef. We succeeded in making investors pull out from port investments.
>
> Interview (30/10/2017)

'Fight for the Reef' combined the multi-pronged approach of earlier anti-coal activisms in New South Wales, in which legal challenges and divestment activism played crucial roles alongside community mobilisations. GetUp! ran an international online campaign that could 'mobilise even in Europe', targeting Deutsche Bank's investment in the massive, proposed expansions of the Abbott Point coal port on the Central Coast (GetUp! Queensland Coordinator interview 14/06/2018). The German investor eventually pulled out of the Abbott Point expansion project in 2014 (Jewel 2014).

Regional and local groups brought four legal cases against coal port expansions and the state and federal government proposals to dump dredge spoils in the World Heritage Area of the GBR (Environmental Law Australia 2016). The local group Whitsunday Residents against Dumping went into voluntary liquidation in 2016 after losing its legal challenge against the expansion of the Abbott Point coal port, but was reformed as 'Reef Action Whitsundays', demonstrating a continuance of grassroots resistances. Activist groups organising blockade camps in New South Wales started mobilising non-violent disruptive actions at strategic sites along Queensland's Central Coast. The Stop Adani Campaigner said:

> In 2015 the Reef Defenders who were basically people from Front Line Action on Coal organised a 'Listen Up' with a First Nations elder from the Birri tribe and his family on their Country near Bowen, near Abbott Point. Between 200 to 300 people attended from all over. It was three days of training and planning. We did a symbolic blockade. We handed a pledge at Abbot Point saying we will continue fighting peacefully.
>
> Interview (27/10/2017)

'Fight for the Reef' reached a significant milestone in 2015 when Queensland abandoned proposals to dump dredge spoils in the GBR World Heritage Area. Significant investment withdrawals from coal-port expansions owing to both activist pressure and turn in the global outlook for coal drastically shrunk the government's ambition for coal-port expansions. The GBR Campaign Director at the AMCS told me:

> By 2014 most port-developments had stopped, only extensions could go ahead. They changed the dredge spoil dumping site to the Caley Valley wetlands, but pressure continued, and even that idea was abandoned. The WHC met in July 2015, the Reef was on the brink of being put on the In Danger List. The Federal and Queensland governments saved the day with the Reef 2050 plan. It had 150 actions; one was to restrict port-developments, basically, no new Greenfield ports. The Queensland State Port Authority (QSPA) restricted port development; it included criteria for what kind of port-development can occur. Adani's coal mine scaled down, so Abbott Point didn't need a new terminal. There was no immediate threat of dredging, so we had won the dredging and dumping campaign! We did not have a message anymore.
>
> Interview (15/10/2017)

This movement milestone was, however, confronted with the reality of back-to-back and extensive bleaching events of the GBR in 2016 and 2017, forcing a need to move towards a more direct narrative of saying 'no' to coal in the next wave of the anti-coal movement's focus in Queensland. The GBR Campaign Director at AMCS said:

We saw intense coral bleaching in 2016 and 2017. 30% of all shallow-water corals died, within a depth of 5 to 10 meters. 2016 was worse in Port Douglas, and 2017 between Cairns and Townsville. 1500 kilometres severely bleached across two events. We had continuous engagement with the World Heritage Committee during the bleaching events. There was despair about the Reef, but no direct action after the threat of dredging was put aside. What would be our new message? We have reached a point where coal is seen as toxic…So it used to be 'Fight for the Reef' and then it became 'Stop Adani'.

Interview (15/10/2017)

Activists within the movement also expressed concerns about the ineffectiveness of the 'Fight for the Reef' narrative to tackle Queensland's coal-export problem. One of the two primary concerns involves the movement's centralised model and message. The former Coordinator of the central Queensland-based Mackay Conservation Group told me:

We are exporting more coal now than we ever have. We are not saying that upfront. National groups need national issues, so they picked the Reef issue, while the little community groups are not being heard; there's a form of colonisation going on. The Sunrise Project tries to break down the centralised model and have people and funding for very local mining issues, but still, they are finally governed by funding interests.

Interview (28/10/2017)

Another concern with the national campaign was around the movement failing to create change from the ground-up in Queensland. The Campaigns Director at The Wilderness Society said:

The Reef crying for a helpline is a mickey-mouse frame. There is a tension between the local messages and national messages and the issue is that the solutions for coal are not being driven out of Queensland. Right now the movement seems to carry the DNA of the US movements! It is hyper partisan.

Interview (15/10/2017)

A third inter-related tension within the environment movement arose between outlooks of activists that prioritised a region's just transition from coal versus those who responded to the urgency to stop Australia's coal exports. Particularly between those groups and activists strongly invested in local communities and regions versus those driving the movement's national objectives of stopping coal (Hunter Valley Campaigner interview 23/10/2017).

A new environmentalism of 'End(ing) Coal!'

Anti-coal resistances in New South Wales had honed global environmental activism's new dual approach: that of targeting and getting financial institutions to divest from fossil fuel projects and build grassroots resistances at the local level (Strategist in the climate movement interview 18/06/2018). Divestment, community mobilisations and mass civil disobedience actions have been used in resistances against massive fossil fuel projects in North America, such as the Keystone XL Pipeline and Tar sands, and Dakota Access Pipeline. The CEO of the Australian arm of the international ENGO 350.org told me during an interview in Sydney:

> The Tar sands protests set a big example for us. Bill Mckibben called out to people to come and get arrested and they all came. 1500 people got arrested.
> Interview (10/10/2017)

350 Australia ran a 'Divest from Fossil Fuels' campaign between 2012 and 2015 in the lead up to the Paris Climate Summit. The CEO of 350.org Australia said:

> It started as a moral issue, 'will you stop supporting big polluters' and divest from the fossil fuel industry?' but it grew into one of the most effective forms of campaigning by making it about the money. Even someone like the Financial Review acknowledged it be the most effective green campaign. We targeted Super Funds, Universities, the 'big four' Australian banks – Commonwealth, National Australia Bank, Australia New Zealand Bank and Westpac – and got support from several communities, faith-based and religious groups, and churches. Thirty-five of Australian's biggest councils pledged to divest.
> Interview (10/10/2017)

Conservation focussed Australian ENGOs were most affected by this transformation; they were 'revitalised' through a big shift in their staffing and organisational structures to include a newer generation of activists focussed on building community power, and collaborating with the wider movement, rather than their older approach of policy expertise and lobbying in Parliament (The Sunrise Project member interview 20/10/2017). This transformation led to the emergence of new organisations with a different organisational scale, scope and structure compared to older ENGOs; they approached community organising differently, often performing niche and specialised functions within Australia's new anti-coal activism focussed environmental and climate movement.

The small divestment-activism-focussed organisation Market Forces targeted shareholders and investors of fossil fuel projects (www.marketforces.org.au). The

Sunrise Project was formed after the Australian environment movement formed its national anti-coal strategy and focussed on coordinating grassroots resistances to coal across multiple locations in Australia through providing funding and strategy-based support, helping groups to 'basically find their niche' (The Sunrise Project member interview 20/10/2017).

The political activism group GetUp! started in 2005 with a model of online mobilisation driven through email lists and databases for online activism. The Coordinator for Queensland for GetUp! told me during a phone interview that:

> We worked on only one thing at one time for maximum impact. We have fundraised for legal actions and to support small groups. We aimed to build electoral power; so marginal electorates were a focus and we worked to build up a member-base there. We also know what the high-density GetUp! member areas are, from where we can make the most impact. Members use our calling technology to call MPs. It can be easily done from home. Our strength is a large membership and good audio and video technology; that's how members take action. And we share this with other groups.
>
> Interview (14/06/2018)

Several activists I interviewed had moved across multiple new organisations within a short span of time, indicating a strong current of dynamism in the environment movement in the period after the Copenhagen Summit. It also indicated the availability of considerable seed funding and major donor grants for the environment movement's new objective of transitioning Australia away from coal exports. It indicated a rapid evolution in anti-coal activist approaches, models of community organising and mobilisation, and consequently a dynamic process of formation of multiple organisations with a niche focus and an agility to respond to the dynamic campaigning needs of anti-coal activism.

These transformations of the environment movement occurred within the atmosphere of consistent attacks on the environment movement by state and federal governments. Government funding was withdrawn from the Environment Defenders Offices (EDO), the Australian environmental legal network that helps environmental, climate and community groups take legal action. The EDO network demonstrated resilience and flexibility in being able to transition to new business models including relying on public donations to provide legal services in various legal challenges brought by communities and local groups (Solicitor, Queensland Environment Defenders Office interview 25/09/2017).

The federal government's attack on the charitable status of ENGOs under the Australian Charities Act that allows financial supporters of registered charities to make tax-deductible donations affected all the large environmental groups at the state and national level. ENGOs have been susceptible to attacks by Liberal-National Coalition governments that have tried to undermine their activism through draconian measures such as the requirement to spend a quarter of their budgets on environmental remediation (Walker 2017).

Smaller and newer activist groups with alternative funding options were able to operate with greater freedom, bringing more interdependence and complementarity within the environment movement, effecting a shift both in its culture and practice (Strategist in the climate movement interview 18/06/2018). The online-activist group GetUp! is a case in point: 'Since we do not have a DGR status we can tell people how to vote during elections' (GetUp! Queensland Coordinator interview 14/06/2018). The acronym DGR stands for deductible gift recipient and the term DGR status refers to the charitable status of not-for-profit organisations under the Australian Charities Act discussed in the previous paragraph.

Over 10 years, while Australia's political pendulum on climate action swung from the left to right, the environment movement moved the debate on coal mining, and the condition of the GBR, to a point where these two came to be intricately linked in people's minds as the cause and effect of climate change. These skirmishes between political and coal interests on one side, and a rapidly organising environment movement against coal on the other, constituted the backdrop within which the Carmichael project travelled through the various hoops of government approvals from 2012. The build up of a national anti-coal environmental movement also set the ground for the resistance to Adani's Carmichael mine that I discuss in the next chapter.

Lock the Gate! Farmer–environmentalist alliance against mining

Encroachment of mining on productive farmlands during the resource boom reshaped regional economies and brought cumulative impacts from massive mining projects and risky technologies on rural communities (Everingham 2016). Fast-paced approvals for mega coal and CSG mining projects on fertile farmlands imposed a sudden sense of shock on farming communities (McAdam 2017). The scale and spread of extractive projects with impacts on farmland, property and water catalysed mobilisations of farming communities against coal and gas, in Queensland, New South Wales and Victoria.

The conflict led to the formation of an unusual alliance between farmers and environmentalists coming from opposite ends of the political spectrum, sparking the creation of the farmers-driven grassroots network 'Lock the Gate' (LTG). Environmentalist Drew Hutton emphasises that farmers collaborated with environmental activists 'out of desperation' at the government disregarding their social, ecological and economic concerns while promoting mining interests. LTG first formed as an organisation to help farming communities to mobilise against hydraulic fracking and coal mining during the peak of Queensland's coal-seam gas boom (Hutton 2013). While it could not arrest Queensland's CSG boom, between 2008 and 2010 LTG succeeded in building an effective organising network in parts of regional New South Wales and Victoria that were anticipating coal and gas projects. The Campaigns Coordinator at Friends of the Earth Australia told me during an interview in the ENGO's Melbourne office that:

We recognised that the horse has bolted in Queensland. There the environmental governance is weak when it comes to fossil fuel extraction. But there was a good opportunity in New South Wales and Victoria. There is already a history of 40 to 50 years of movements in northern New South Wales where the gas fields are being developed. Land tenure is different across Queensland and New South Wales. Inland Queensland has a lot of open country, large properties and few towns. New South Wales has fertile land and more towns and farmlands dotted across a smaller landscape. In 2009 we started working earnestly with regional communities in Victoria, it was very early and very opportune here.

Interview (20/11/2017)

From the perspective of environmentalists, the collaborations were born out of environmentalism's strategic shift after Copenhagen. The collaboration evolved to generate successful new models for grassroots resistance. The Friends of the Earth Campaigns Coordinator told me:

Since 2009 we focussed our energies on 'new constituencies'. We met people where they were and took it from there. The rest is history. The approach was not NIMBYism, but bioregionalism which connected to state-level political activism. It asked the question: what do people want for their region? The 'Gas-field free Organising Model' emerged out of this exercise. First, we define the boundary of the community, then we door-knock to understand community sentiment, and collect data, then develop the narrative, then we make a gas-field free declaration. Bit by bit we block off the land from the miners, watershed by watershed. This model originally came from Northern New South Wales and flowed on to Victoria, South Australia, Tasmania and Western Australia, and even in Queensland.

Interview (20/11/2017)

LTG did not reflect the values of environmentalists and unlike ENGOs did not assume a directly anti-corporate stance (Hutton 2012, p. 15). LTG coordinators saw no value in taking environmentalism's cultural values upfront into farming communities, focussing instead on finding common purpose in resistance and making models of solidarity scalable across places while bearing in mind critical differences. The Friends of the Earth Campaigns Coordinator said:

Lock the Gate's Gasfield Free Communities model worked in rural and smaller regional towns with less than 2000 people. It needed close to 100% engagement. But in bigger communities with more than 6000 we came up with a different model – streets declared themselves gasfield free – so that became another stone in the pond causing ripples across the state. Building power in the grassroots is our aim, as opposed to directed network campaigning that some activists talk about. Ours is a political philosophy predicated on solidarity – what happens if we lock up industry here?

Interview (20/11/2017)

The alliance refrained from putting climate change at the centre of their actions as a reflection of its diverse support base and a compromise between the need to stop mining on farming land and historically different values and orientations of farmers and environmentalists. The Hunter Valley Campaigner said:

Climate change might not be the main driver of concern for many in the grassroots. Although they care about climate change, they might focus on water and land as the main concerns, being closest, as compared to environmentalists in cities, to impacts of coal and gas extraction on water and land.

Interview (23/10/2017)

Like other anti-fossil-fuel resistances during the minerals boom, the tactics of LTG combined grassroots mobilising along with political advocacy and divestment activism. Describing the Victorian league of LTG, the Friends of the Earth Campaigns Coordinator said:

We also did 'inside track' work with lobbying of both federal and state governments. For grassroots networks it gives a sense of agency to people. They cannot think only about the region – we always urged them to 'look up' to the region, the state and also national levels. So groups like Market Forces were critical to our research and economic activism campaigns. Rhizomatic organising is how we see the VIC campaign having worked, and grassroots fibrous networks that collect all this information.

Interview (20/11/2017)

Farmer–environmentalist collaborations enabled through LTG and its new models of anti-mining resistance built a large grassroots movement that was distinct from urban mobilisations involving ENGOs. The Hunter Valley Campaigner said:

In fact there were three separate movements – Beyond Coal and Gas Movement, the Climate Action Movement and the Climate Justice Movement. In a way they separated the grassroots from the big NGOs. There was not too much 'movement' between the Beyond Coal and Gas movement and Climate movement. Before the Beyond Coal and Gas movement there was Lock the Gate. It is Lock the Gate's work that built Beyond Coal and Gas up.

Interview (23/10/2017)

Although the resistances raised possibilities of recasting the political field through society's reoccupation of and a consequent democratisation of politics, Zuleika Arashiro warns against idealising these struggles due to the striking presence of neoliberal logics in community discourses in Australia (Arashiro 2017). Unlike in a developing world context where structural inequality and social justice are fundamental to environmental debates and mining conflicts, concerns with public accountability and protection of private goods might not translate into a resistance against capitalism in the Australian case.

Green–Black anti-coal alliances on the Liverpool Plains

I now discuss three important cases of Indigenous–environmental alliances formed in opposition to mining projects during the resource boom that have added new dimensions to historic Green–Black relations in Australia. The first, discussed below, continues the discussion on anti-coal activism on the Liverpool Plains. The other two, an anti-uranium mining campaign and an anti-gas campaign, discussed in the next two subsections, although unrelated to coal extraction, require consideration within the Australian case study as primary examples of Green–Black alliances during the resource boom.

The Liverpool Plains in north-western New South Wales also witnessed new coal mines and CSG projects as well as expansion of existing projects during the minerals boom. The region contains fertile farmlands and falls under the native title claim of the Gomeroi people.[33] The resource boom provided an opportunity for local Aboriginal people to make economic deals based on their land ownership and also effected a renegotiation of farmer's historical relations with local Indigenous communities. Aboriginal Political Historian Heidi Norman notes that the agreement making process between mining corporations and Gomeroi traditional owners was often characterised by lack of transparency on part of companies and confusion, criticism and dissent on the part of native title claimants. As a result, the protection of sacred sites emerged as a dominant imperative for the Gomeroi people's mobilisations against mining and their alliance building with farmers and environmentalists (Norman 2016).

Whitehaven Coal's Maules Creek mine in the Leard State Forest threatened to clear-fell across 4000 acres of culturally significant and biodiverse woodlands. Whitehaven's 'incomplete and disrespectful cultural heritage process' did not allow traditional owners to properly assess cultural values at the project site (Talbot 2013, para 3). The coal mine was challenged by a diverse resistance formed of anti-coal networks, ENGOS, farmers, traditional owners and local alliances. A permanent Leard Forest Alliance campsite was set up for two years. Mass blockades and picket lines delayed mine construction by two years (Greenpeace Australia 2015). Non-Aboriginals showed respect for Gomeroi actions for cultural protection, and signed a 'Protection Treaty' to respect the Leard Forest's cultural significance (Norman 2016).

In the adjoining Pilliga State Forest, Santo's Narrabri Gas Project covering 98,000 hectares, began drilling operations for over ten wells in 2011, aiming to supply 50% of New South Wales's gas needs (Santos 2014). Although majority of Gomeroi elders opposed Santos, their representative body the Narrabri Local Aboriginal Land Council who worked closely with the company supported it, causing internal conflicts in negotiations. In 2013, a meeting of 400 Gomeroi Traditional Owners in Tamworth took a strong stand against CSG exploration by resolving to stop all mining and development on their ancestral lands (Norman 2016).

Aboriginal political historian Heidi Norman notes that even though many rural communities in the region continue to be demarcated along class and race, the encroachment of mining prompted Aboriginal groups to play host to farmers with a renewed sense of alliance to country, thereby bridging a bleak historical divide (Norman 2016). In fact farmers were being called on to protect Aboriginal sites and map cultural heritage on their farms, when in the past they were known to have routinely destroyed such sites (Bryant 2016). Norman argues that while non-Indigenous landholders can be bought out by mining companies or their lands compulsorily acquired by governments, land use change needs to be negotiated with Aboriginal landholders, making their views and cultural continuities critical for farmers (Norman 2016). And observes that alliances and discussions that emerged from the contested situation in north-western New South Wales held one of the best opportunities to recast environmentalism's narratives (Peter Thompson, quoted in Norman 2016).

The three cases of alliance building between Indigenous and non-Indigenous constituents during the minerals boom were characterised by a shared vision for an anti-mining future that embedded Indigenous cultural values and needs for looking after Country at their centre.

Anti-Jabiluka mine campaign

The movement against the Jabiluka uranium mine adjoining the World Heritage listed Kakadu wetlands in the Northern Territory is considered a crowing example of an Indigenous land rights movement that pointed to new forms of empowerment, in part through mining derived income, and new potential recourse to global campaigning through strong alliances (Altman 2012a). The Ranger and Jabiluka mine lease areas are the traditional lands of the Mirrar Gundjeihmi people. In 1996, the newly elected Howard government withdrew Labor's 'Three Mines' Uranium policy that had previously prevented Jabiluka's development. The 'Three Mines' policy had been introduced by the Labor government of Prime Minister Bob Hawke in 1984 to restrict Australia's uranium mining to the three existing mines at that time, at Ranger, Nabarlek and Olympic Dam.

The traditional owner's concerns over mining were shaped through adverse impacts and consequent erosion of cultural life from the Ranger uranium mine established in 1977.[34] They invoked cultural, moral and environmental imperatives through Indigenous heritage protection, national parks and conservation, and anti-nuclear arguments to oppose the Jabiluka mine (Trebeck 2007). The campaign incorporated Federal and High Court actions, blockades at the mine site, mass protests in major cities, engagement from international activist groups, pressuring institutional investors,[35] actions against the mine owners North Ltd and Rio Tinto, national and international speaking tours by Mirrar Gundjeihmi, and appeals to the UNESCO World Heritage Committee (Trebeck 2005).[36,37,38,39,40]

Although Indigenous–green alliances have a fraught history, the presence of a clear and common interest in this case helped the traditional owners to build successful and intricate relations with environmental NGOs and anti-nuclear activists (Altman 2012a). Majority of the Australian Senate and many trade unions supported the movement (Trebeck 2007).[41] The European Parliament passed a resolution in 1998 condemning Australia's decision to mine Kakadu (Gundjeihmi Aboriginal Council (GAC) 2001, p. 73).

The prospects of mining Jabiluka came to an end after Rio Tinto became the major owner partly on account of the company's stated objective to work with Indigenous people. A formalised agreement in 2005 effectively gave traditional owners a veto over the mine (Altman 2012a).

Anti-gas campaign at James Price Point

In 2009, the Western Australian government chose James Price Point on the remote Kimberley coast as the site for one of Australia's largest industrial proposals, a $30 billion processing hub for gas from the Browse Basin (Botsman 2012). The Liberal government threatened compulsory land acquisition for the industrial precinct even while native title claimant groups were negotiating with the proponent Woodside over the proposed development.[42] Both the Goolarabooloo and Jabirr Jabirr people claimed native title over the James Price Point area. In 2011, a majority of native title claimants voted for the project as a pragmatic step given the risk of compulsory acquisition (Altman 2012b).

The Goolarabooloo family group objected that the project location at James Price Point would disrupt their Songlines, burial sites, law and culture (Joseph Roe quoted in Weber 2011). In 1987, Goolarabooloo elder Paddy Roe had established the Lurujarri Heritage Trail along a section of the Song Cycle and containing sacred sites, to share cultural knowledge of the Kimberley coast (Conroy 2017). The cultural value of the diverse dinosaur footprints on the Kimberley was recognised through National Heritage Listing in 2011 (Mills 2011). Residents from Broome, the coastal economic hub in Kimberley, mobilised the 'No Gas' campaign and joined the Goolarabooloo in resistance camps along the Lurujarri Trail. The movement included local, national and international ENGOs, and citizens from all over Australia. Apart from non-violent direct action to delay development at James Price Point, the movement relied on political and corporate lobbying, targeting investors and legal challenges (Counteract 2013).

The movement is considered Australia's most significant and successful Indigenous–green alliance on account of the number of green groups involved, and decades long collaborations between the Goolarabooloo and environmentalists through the Lurujarri Trail (Muecke 2016). By being located along a living heritage trail through which the central institution of the Goolarabooloo, the *Bugarrigarra,* which is understood as the original creative force that gave shape and meaning to the landscape – prevailed, the No Gas movement held Aboriginal modes of belonging as the central purpose of their resistance. Citizen Science projects tracked whales, turtles and endangered bilbies, incorporating ecological science and Indigenous knowledge (Muecke 2016). Although the project received state approval in 2012, Woodside withdrew in April 2013 citing commercial reasons; in August 2013, the Supreme Court of Western Australia blocked further development of the LNG processing plant at James Price Point (Wilderness Society 2013).

Analysis: environmentalism's transformation to End(ing) Coal!

Parallel and mutually conflicting movements of the Australian government on climate change and mineral and fossil fuel extraction from the mid-1990s have forced a critical transformation of Australia's environment movement. Australian environmentalism's dominant concern shifted to stopping climate change through ending coal extraction. The Howard government's argument against GHG reductions on account of Australia's small net emissions had become a hardwired political logic during the following 10 years; this decade can be characterised as the swinging pendulum of Australia's climate politics.

As Australia got drawn into an unprecedented scale of globalised minerals trade from the mid-1990s, its economy lost the previous balance between mining and other exports that was considered as a key factor in its prosperity. During the minerals boom, a high influx of international capital and mega projects by transnational corporations made the State's role in privileging the extractive private sectors while passing on social and environmental risks to citizens increasingly evident (for example, see Bebbington et al. 2008).

Landscape level transformations wrought by a massive increase in fossil-fuel extractions led to widespread local conflicts and resistances. Such local resistances became incorporated into agendas of national and transnational environmental organisations, effectively generating grounds for the democratisation of environmentalism's values through a new relational, alliance-based politics and approach to tackling the environmental challenge. The process of the transformation of environmentalism was extensively shaped by the political economy of Australia's minerals exports, and in turn the movement's new politics had implications for the political economy of coal in Australia (Connor et al. 2009).

Environmentalism's approach now involved a political philosophy predicated on a solidarity that emphasised 'meeting communities where they were and building up a movement from there' (FoEA Campaigns Coordinator interview 20/11/2017). The emergence of a new disaffected constituency of farmers is one of the two aspects of the transformation of the Australian State during the minerals boom that this chapter highlights. The other is the experience of a reconfiguration of the settler-colonial state instead of its disappearance from the point of view of Indigenous communities (Lyons 2019). Both these aspects have been key factors in the environment movement's transformation towards the formation of solidarities during the minerals boom.

The pragmatic approach of Australian environmentalism during the minerals boom attempted to find common ground with other constituents against coal and CSG extraction. Multiple contestations converged around sites of coal and

CSG extractions, making them critical geographies of resistance and giving coal-bearing regions in New South Wales and Queensland a 'scaled' meaning of place during the minerals boom. These sites of resistance served as building grounds for environmentalism's multi-pronged and alliance-centric new approach.

Although mining on Indigenous lands increased significantly during the resource boom, the possibility to negotiate outcomes with mining corporations through the native title regime gave native title groups a bargaining power they did not have before. This has implications for today's Green–Black relations. Articulations of Indigenous climate justice that have emerged during this era linked to notions of sovereignty and Indigenous land rights in the same manner as past Indigenous opposition to mining (Esposito and Neale 2016). As seen in the case of anti-coal, anti-gas and anti-nuclear Green–Black alliances discussed in the previous section, Indigenous–green alliances in the mining-boom era have allowed for a vision of Indigenous futures at their centre, attempting a decolonisation of environmentalism's approach and narratives that was needed.

Environmentalism went back to a grassroots actions-based approach in response to the scale of the challenge of fossil-fuel extraction and the inability of environmentalism's institutionalised structures to influence policy and politics. The emergence of farmers as a disaffected constituency during the resource boom was a key factor in transforming environmentalism's politics towards one of pragmatic solidarity. The extent of disenfranchisement of a politically conservative rural constituent was indicated through their participation in radical and direct actions against coal and CSG mining, using environmentalism's tactics:

> I see Lock the Gate as important because, for the first time, serious environmental issues are being taken up in a really strong way by people in the country, to the point where you've got farmers locking on to machinery and getting arrested in rural parts of New South Wales.
>
> Hutton, quoted in Robertson (2017)

The transformation in environmentalism's politics can be seen through how the environment movement's narratives have changed on the GBR. The Reef has historically been at the centre of State-environmentalist conflicts. The Great Barrier Reef Marine Park was declared in 1975, after a prolonged campaign by conservation groups against its destruction by mining and other extractive activities. Australia's resource boom that unfolded even as climate impacts became more pronounced through coral bleaching and ocean acidification, posed a double risk to the GBR's outstanding natural values.

Old campaigns to Save the Reef as a place worth keeping (Bonyhady 1993) became recast as a national movement to stop climate change, with the Reef being recognised as a barometer for the wellbeing of the whole planet. But in a new era of democratisation of activism following a global strategic shift, with the new anti-coal environmental movement attempting to meet local resistances where they are at rather than taking a top down approach, the grand environmental narrative of Save the Reef met criticisms from local Queensland groups struggling against massive coal mines. The latter regarded 'Fight For the Reef's' national calls to action a form of 'colonisation', which did not do justice to the scale and extent of the State's proposed coal extraction and the potential ecological and social crises. The dilemma of the new anti-coal environment movement's narrative on the GBR reflects the dynamic, contested and deliberative process of its transformation through two decades of Australia's minerals boom.

To conclude this chapter, Australian environmentalism was transformed simultaneously through its focus on climate change and the scale and extent of Australia's minerals boom that was characterised by a massive increase in the extraction of coal and CSG. While climate change per se had an effect on democratising environmentalism's values towards an understanding of environmental crisis as a human rights issue rather than saving nature from humans, the transformation of its politics towards pragmatism and solidarity and a return to grassroots activism was shaped by the scale of fossil fuel mining during the minerals boom and the necessity of confronting massive extractive projects on prime agricultural lands, and where Indigenous worlds are present. Through embracing the struggles of disenfranchised farmers, and through attempts at putting Indigenous visions at the centre its narratives, environmentalism was recast in the era of Australia's resource boom through a deliberative and relational politics emerging from such sites of extraction.

Notes

1 In fact, Malcolm Fraser handed certain Commonwealth powers such as control of the first three miles of coastal water to the states (Cullen 1990).
2 The Fraser government also secured World Heritage Listing for Fraser Island, and stopped oil drilling on the Great Barrier Reef by the Queensland government by establishing the Great Barrier Reef Marine Park and through World Heritage Listing.
3 Although the Fraser Government nominated the southwest Tasmanian Wilderness that included the Franklin River for UNESCO World Heritage Listing in 1981, being strongly in favour of the rights of states it was reluctant to use its external affairs power to stop the hydroelectric project even though it had become a national issue (Christoff 2015).
4 In 1982, the High Court upheld the ability of the Commonwealth to act under the external affairs power to honour international treaty obligations (in *Koowarta vs Bjelke-Petersen* 1982). Next, the decision in the Tasman Dam case (The C/wealth v. Tasmania 1983), upheld the Commonwealth's ability to use the races power, the corporations' power, and external affairs power. In addition to previously confirmed power to deny export licences, this reading considerably empowered the Commonwealth to act for the environment in state jurisdictions.
5 Christoff (2015) argues that in Tasmania, where the success of Green electoral politics and the presence of the Greens in Parliament have been most notable, a diversion from the need to reform institutions and, consequently, a lack of success in institutionalising environmental values has been most marked.
6 It was followed by other significant legislations for the return of Indigenous lands at the state level, in South Australia and New South Wales in particular (Mercer 1993).
7 The backlash was particularly concerted in WA where the mining and pastoralist industries were a strong political force (Foley 2013).

8 Australia enacted the Racial Discrimination Act (RDA) during the Prime Ministership of Gough Whitlam in June 1975 to accord equal treatment under the law to all Australians. It also reflected the *International Convention on the Elimination of All Forms of Racial Discrimination,* adopted by the United Nations in 1969 and ratified by Australia in September 1975.

9 A scare campaign by the mining industry spread the myth of the 'backyard threat' – that people could lose their private backyards to native title – causing the Mabo decision to become a large electoral liability for any political party supporting it. The opposition leader, Liberal-National Coalition's John Hewson, reiterated this argument in the run-up to the 1993 federal election. The Liberal Coalition and industry lobby's awareness of the effectiveness of dubious claims in influencing public opinion is well regarded (Edelman 2001).

10 In case of exclusive possession of land, native title groups had the right to negotiate with resource developers within six months of notification of a proposed mining-development project, after which the matter required arbitration (Altman 2012a). The process of arbitration has largely proved unsympathetic to the wishes of native title groups (Corbett and O'Faircheallaigh 2006).

11 UNDRIP was adopted by the UN General Assembly in 2007. Along with the other Anglophone settler colonial nations – the United States, Canada and New Zealand – Australia first voted against the UNDRIP, arguing that matters of Indigenous self-determination and rights over national resources fell within domestic jurisdiction. Australia finally signed the UNDRIP in 2010 (Ford 2012).

12 Although empirical evidence pointed to a marginal impact, miners made extravagant and persistent claims about significant impacts (Lavelle 2010).

13 The High Court's Wik 'native title' decision in December 1996 related to the question of whether granting pastoral leases extinguished native title rights (ATSIC 1997, 1–6). After unsuccessful claims in the Federal Court on account that pastoral leasehold extinguished native title rights on lands under Queensland laws, the Wik and Thayorre People from Cape York Peninsula received a favourable ruling from the Australian High Court that decided inter alia that pastoral leases did not confer exclusive possession on the pastoralist. The ruling, however, noted that in the event of a conflict between pastoral and native title-holders, the former's rights were upheld under the NTA (Howard Wagner 2008).

14 The complementarity between mining and prosperity in the Australian economy used to be a function of primary commodities from the mining and agriculture sectors running a surplus, that in turn funded deficits in the manufacturing trade (Goodman and Worth 2008).

15 Mining's unprecedented growth from the mid-1990s delivered a shock to the economy by appreciating the Australian dollar and consequently destabilised other trade-exposed industries including manufacturing. RBA modelling indicated that manufacturing output in 2013 was 5% below what would have been achieved without a minerals boom (Tulip 2014).

16 The mining sector has one of the lowest employment rates of 1.9% compared to manufacturing which employs five times more (Richardson and Dennis 2011),. Employment in mining grew rapidly but from a very small original base, with the effect that its net contribution to job increases in the Australian economy stood at a mere 7% between 2005 and 2011 (Richardson and Dennis 2011).

17 While minerals and fuels accounted for 27% of Australian exports during the 1968 resource boom, that figure rose to 39% by 2002 and to 43% by 2007 (Department of Foreign Affairs and Trade 2007; Australian Bureau of Statistics 2007).

18 A 2016 Treasury paper estimated that 86% of investments in major projects are foreign-owned and that only 10% are solely Australian-owned. This includes a 26% ownership from the US and 27% from the UK (Australian Treasury 2016).

19 Australia's mining sector which is 86% foreign-owned has spent $541 million between 2007 and 2017 in lobbying Australian governments (Aulby 2017).

20 The federal government gave an estimated A$4 billion in subsidies and state governments spent A$17.6 billion over a 6-year period to support mining (Dennis 2015).

21 The government's Carbon Pollution Reduction Scheme (CPRS) proposed a cap-and-trade mechanism to reduce Australia's emissions by 5% (of 2000 levels) by 2020.

22 Independent analysis showed that mining sector would prosper even under drastic emissions reductions and that employment would in fact grow by 22,800 jobs by 2025.

23 The Australian Greens criticised that the 5% target aimed to stabilise atmospheric CO_2 levels at 450 parts per million (ppm) as opposed to the scientifically necessary 350 ppm and risked locking in an average temperature rise of 4 degrees centigrade or more compared to the maximum permissible rise of 2 degrees centigrade. The Liberal and National Parties were opposed to a binding commitment to reduce emissions without other big polluters, particularly major developing economies such as China and India coming on board.

24 The Coalition's cost-of-living scare campaign against a carbon price has been criticised as entirely political, one of the 'crudest and most distorted debates' in Australian politics, a complete hoax (Edis 2012).

25 The peak mining industry body Minerals Council of Australia (MCA) estimated over 300 'benefit sharing agreements' between mining companies and Indigenous communities in the first decade of the resource boom (Altman 2009a).

26 The shift in the MCA's public narrative from attacking native title legislation to building sustainable Indigenous communities is also palpable during this period (Altman 2009b).

27 In 2011 the Australian Government listed 50 IPAs covering 24% of the Australian Conservation Estate (Australian Government 2011).

28 Soaring demands from the Asian markets of Japan, China and Korea created opportunities for major Australian oil and gas companies like Santos and Origin to develop coal seam gas wells accompanied by liquefied natural gas (LNG) processing plants and massive export terminals.

29 The Australian Climate Camps started in 2008 with the Newcastle T3 coal terminal action, with the aim of directly disrupting coal production and export. The 2009 Climate Camps in Helensburgh in New South Wales and La Trobe Valley in Victoria targeted the extension of an underground coal mine and the Hazelwood Power Station that the State Government had decided to extend instead of phasing out. They struggled to gain popularity within the broader climate movement in Australia, being considered too radical (Rosewarne et al. 2014).

30 The Australian Conservation Foundation (ACF), World Wildlife Fund (WWF) and the Climate Institute backed the CPRS in return for Labor adopting a conditional 25% emissions-reduction target for 2020. The Greens, Greenpeace, Friends of the Earth (FOE), the Wilderness Society, Australian Youth Climate Coalition (AYCC) and GetUp! opposed the bill on account of the conditions lending uncertainty to whether a 25% target could be achieved (Pearse 2011).

31 In 2011 Abbott Point had two berths and a coal handling capacity of 50 million tonnes per annum.

32 The initial proposal for disposal of dredge spoils from port expansion proposed dumping in the Great Barrier Reef Marine National Park. It was opposed by environmental groups and marine scientists. The second proposal to dump dredge spoils in the coastal Caley Valley wetlands that provide sanctuary to populations of over 40,000 waterbirds in the wet season and serve as a turtle nesting sites was also opposed. Concerted mobilisations coupled with litigation against dumping dredge spoils in the World Heritage area by civil society groups, concerns of marine scientists, and the Great Barrier Reef Marine Park Authority (GBRMPA)'s advice to the Federal Environment Minister not to approve dredging at the coal terminal, thwarted initial attempts at planned expansion of the Abbott Point Coal Terminal. The final proposal, approved by the Federal Environment Minister in December 2015, was a marked improvement, proposing dumping dredge spoils on land immediately adjacent to the existing port. It proposed only one new coal terminal and an increase in the port's coal handling capacity to 120 million tonnes per annum (mtpa), that would require the dredging of 1.1 million cubic metres of seafloor material. As against the previous taxpayer funded expansion proposals, mine proponents would pay for the expansion as per the final proposal. A community group also challenged the federal approval of the third proposal in Court.

33 The Gomeroi people's native title claim to a large part of northwestern New South Wales extending from the Upper Hunter to the Queensland border, to Coonabarabran and up to the Western slopes of New England was registered by the Native Title Tribunal in 2012 (Clifford 2013).

34 Factors contributing to the erosion of cultural life included lack of access to sites of significance within mining leases, desecration of sacred sites and 'exclusion from effective decision-making over the interpretation of what is significant and integral to their living tradition' (GAC 2001, p. 32). The anti-mining stance is characterised by considerable ambivalence to a range of institutions associated with the regulatory regime and mine infrastructure in the region (Trebeck 2005).

35 Pressure was applied further up the supply chain through shareholder activism. Activists targeted institutions that held North Ltd shares with anti-Jabiluka communications, achieving success in shareholding establishments selling their company shares, that eventually caused the company share price to drop by more than 65% in 1999 (Trebeck 2007).

36 In 1998, the mine proponent Energy Resources Australia's attempts at beginning construction with government support was met with a significant blockade of 5000 peaceful protestors at the mine site for eight months. Mobilisation was driven by the Gundjeihmi Aboriginal Council in alliance with environmental groups, political parties including the Greens and Australian Democrats (Trebeck 2005).

37 An Indigenous–environmental coalition brought national attention to the issue through multi-city protests, public meetings and anti-Jabiluka film screenings. Friends of the Earth, Wilderness Society and the Australian Conservation Foundation were the most prominent environmental groups. Many community groups also supported the movement.

38 'Globalisation' of the issue was achieved through international networks such as Friends of the Earth, the Africa–Australia Exchange, and the Global Sisterhood network, that helped to make mining without Indigenous consent a reputational risk for Rio-Tinto, the multi-national ownership–partner at Jabiluka (Trebeck 2007).

39 A significant 4-day blockade was held outside the Melbourne headquarters of North Limited who owned ERA (Trebeck 2007).

40 Engagement with the United Nations World Heritage process included traditional owner Yvonne Margarula presenting the case of cultural destruction from the Ranger mine at the UNESCO meeting in Paris in 1998, followed by the UNESCO World Heritage Mission reporting that there were significant potential threats to Kakadu's World Heritage values from mining, based on a visit (Trebeck 2007).

41 The Senate's Jabiluka Enquiry advised against mining (Trebeck 2007).

42 The 1998 amendments to the Native Title Act under Prime Minister John Howard allowed state governments to extinguish native title within their jurisdictions and to compulsorily acquire native title land for private infrastructure (Botsman 2012). The State's attempt at compulsory acquisition was subsequently invalidated through a Supreme Court ruling in 2011 based on legal action brought by Goolarabooloo and Jabirr Jabirr members (Pickerill 2018).

References

Aboriginal Land Rights (Northern Territory) (ALRA) Amendment Bill. 2006.

Altman, J. 2005, 'Development options on Aboriginal land: sustainable Indigenous hybrid economies in the twenty-first century', in *Centre for Aboriginal Economic Policy Research*, Australian National University, Canberra.

Altman, J. 2007, 'The Howard government's Northern Territory Intervention: are neo-paternalism and Indigenous development compatible?' in *Centre for Aboriginal Economic Policy Research*, Topical Issue 16/2007, Australian National University, Canberra.

Altman, J. 2009a, 'Benefit sharing is no solution to development: experiences from mining on aboriginal land in Australia', in R. Wynberg, D. Schroeder & R. Chennells (eds.), *Indigenous Peoples, Consent and Benefit Sharing: Lessons from the San-Hoodia Case*, Springer, Dordrecht, pp. 285–302.

Altman, J. 2009b, 'Indigenous communities, miners and the state in Australia', in J. Altman & D. Martin (eds), *Power, Culture and Economy*, Australian National University Press, Canberra, pp. 17–50.

Altman, J. 2012a, 'Indigenous rights, mining corporations, and the Australian state in the politics of resource extraction 2012', in S. Sawyer & E. Gomez (eds), *The Politics of Resource Extraction; Indigenous Peoples, Multinational Corporations, and the State*, Palgrave McMillan, pp. 46–74.

Altman, J. 2012b, 'Land rights and development in Australia: caring for, benefitting from, governing the indigenous estate', in L. Ford & T. Rowse (eds), *Between Indigenous and Settler Governance*, Routledge, pp. 121–134.

Altman, J., Buchanan, G., & Larsen, L. 2007, 'The environmental significance of the Indigenous estate: natural resources management as economic development in remote Australia', *Centre for Aboriginal Economic Policy Research*, Discussion Paper 286, Australian National University, Canberra.

Altman, J., Buchanan, G. & Nicholas, B. 2006, 'The 'real' economy in remote Australia', in B. H. Hunter (eds), *Assessing the Evidence on Indigenous Socioeconomic Outcomes: A focus on the 2002 NATSIS*, Australian National University Press, Canberra, pp. 139–152.

Aulby, H. 2017, 'Undermining our democracy', *The Australia Institute*, August 2017, viewed 14 June 2020, <www.tai.org.au/sites/default/files/P307%20Foreign%20influe nce%20on%20Australian%20mining_0.pdf>.

Arashiro, Z. 2017, 'Mining, social contestation and the reclaiming of voice in Australia's democracy', *Social Identities*, vol. 23, no. 1, pp. 1-13.

Australian Bureau of Statistics (ABS). 2007a, *Mining Indicators Australia*, June 2007, ABS, Canberra.

Australian Bureau of Statistics (ABS). 2007b, *Australian Economic Indicators*, December 2007, ABS, Canberra.

Australian Bureau of Statistics (ABS). 2012, *Mining Industry: Economic Contribution*, May 2012, in 1301.0 – Year Book Australia, ABS, Canberra.

Australian Government. 2011, *Indigenous Protected Areas: Indigenous Australians Caring for Country*, online, viewed 20 June 2020, <www.environment.gov.au/indigenous/ipa/pubs/indigenous-protected-area.pdf>.

Australian Institute of Marine Science. 2023, *Annual Summary Report of the Great Barrier Reef Coral Reef Condition 2022/2023: A Pause in Recent Coral Recovery Across Most of the Great Barrier Reef*, 9 August, viewed 20 January 2024, <www.aims.gov.au/sites/default/files/2023-08/AIMS_LTMP_Report_GBR_coral_status_2022_2023_9August2023.pdf>.

Australian Treasury. 2016, *Foreign Investment into Australia: Working Paper*, January 2016, Treasury, Canberra, viewed 20 June 2020, <https://treasury.gov.au/sites/default/files/2019-03/TWP_201601_Foreign_Investment.pdf >.

Auty, R. & Warhurst, A. 1993, 'Sustainable development in mineral exporting economies', *Resources Policy*, vol. 19, no. 1, pp. 14–29.

Baer, H. 2016, 'The nexus of the coal industry and state in Australia: historical dimensions and contemporary challenges', *Energy Policy*, vol. 99, pp. 194–202.

Bebbington, A., Hinojosa, L., Bebbington, D. H., et al. 2008, 'Contention and ambiguity: mining and possibilities of development', *Development and Change*, vol. 39, no. 6, pp. 887–914.

Bell, S. E. & York, R. 2010, 'Community economic identity: the coal industry and ideology construction in West Virginia', *Rural Sociology*, vol. 75, no. 1, pp. 111–143.

Bonyhady, T. 1993, *Places Worth Keeping: Conservationists, Politics and Law*, Allen and Unwin, Sydney.

Botsman, P. 2012, *'Law Below the Top Soil': Walmadany (James Price Point) and the Question of the Browse Basin Gas Resources of North West Australia*, Save the Kimberley, 10 November, viewed 20 June 2020, <www.savethekimberley.com/2012/10/11/law-below-the-top-soil-just-released/>.

Bridge, G. 2008, 'Global production networks and the extractive sector: governing resource-based development', *Journal of Economic Geography*, vol. 8, pp. 389–419.

Bryant, S. 2016, 'Farmers urged to map Aboriginal cultural heritage on their land', *The Country Hour, ABC Radio*, 18 March, viewed 20 June 2020, <www.abc.net.au/news/rural/2016-03-18/farmers-urged-to-map-aboriginal-cultural-heritage/7258592>.

Butlin, N. G., Barnard, A. & Pincus, J. J. 1982, *Public and Private Choice in Twentieth Century Australia: Government and Capitalism*, George Allen and Unwin, Sydney.

Christoff, P. 2009, 'If not now, then when?', in H. Skyes (ed.), *Climate Change for Young and Old*, Future Leaders, Sydney, pp. 29–45.

Christoff, P. 2015, 'Fraser paved the way for a national environment policy', *The Conversation*, 24 March, viewed 15 May 2020, <https://theconversation.com/fraser-paved-the-way-for-a-national-environment-policy-39182>.

Cleary, P. 2012, *Mine-Field: The Dark Side of Australia's Resource Rush*, Black Inc., Collingwood, Melbourne.

Clifford, C. 2013, 'Gomeroi native title claim flushes out 140 respondents', *Australian Broadcasting Corporation*, 23 February, viewed 23 June 2020, <www.abc.net.au/news/2013-02-07/gomeroi-native-title-claim-flushes-out-140-respondents/4507130>.

Commonwealth v. Tasmania 1983, viewed 20 September 2021, <https://www.ato.gov.au/law/view/print?DocID=JUD%2F158CLR1%2F00007&PiT=99991231235958>.

Commonwealth of Australia. 2015, *Reef 2050 Long-Term Sustainability Plan 2015*, viewed 21 June 2020, <www.environment.gov.au/system/files/resources/d98b3e53-146b-4b9c-a84a-2a22454b9a83/files/reef-2050-long-term-sustainability-plan.pdf>.

Connor, L. G., Freeman, S. & Higginbotham, N. 2009, 'Not just a coalmine: shifting grounds of community opposition to coal mining in Southeastern Australia', *Ethnos*, vol. 74, no. 4, pp. 490–513.

Conroy, G. 2017, 'Western Australia is home to 'Australia's Jurassic Park' ', *Australian Geographic*, 8 November, viewed 20 June 2020, <www.australiangeographic.com.au/topics/wildlife/2017/11/western-australia-is-home-to-australias-jurassic-park/>.

Cook, L. H. & Porter, M. G. 1984, *The Minerals Sector and the Australian Economy*, The Centre for Policy Studies, Monash University, George Allen and Unwin, Melbourne.

Corbett, T., & O'Faircheallaigh, C. 2006, 'Unmasking the politics of native title: the national native title tribunal's application of the NTA's arbitration provisions', *University of Western Australia Law Review*, vol. 33, no. 1, pp. 153–172, viewed 14 June 2020, <www.austlii.edu.au/au/journals/UWALawRw/>.

Counteract. 2013, *Case Study: James Price Point/Walmadan – A Huge Win*, November 2013, viewed 6 January 2020, <https://counteract.org.au/wp-content/uploads/2013/11/creating-change-case-study-walmadan.pdf>.

Crough, G. J. & Wheelwright, E. L. 1983, 'Australia: Client State of International Capital: a case study of the mineral industry', in E. L. Wheelwright & K. Buckley (eds), *Essays in the Political Economy of Australian Capitalism*, Australia and New Zealand Book Company, Sydney, pp. 15–42.

Cullen, R. 1990, *Federalism in Action: The Australian and Canadian Offshore Disputes*, Freedom Press, Sydney.

Daes, E. 1999, *Indigenous Peoples and Their Relationship to Land*, Second Progress Report on the Working Paper, 3 June, UN Sub-Commission on Prevention of Discrimination and Protection of Minorities, E/CN.4/Sub.2/1999/18.

De'ath, G., Fabricius, K. E., Sweatman, H. & Puotinen, M. 2012, 'The 27-year decline of coral cover on the Great Barrier Reef and its causes', in P. G. Falkowski (ed.), *Proceedings of the National Academy of Sciences*, Rutgers, The State University of New Jersey, New Brunswick, NJ, viewed 20 June 2020, <10.1073/pnas.1208909109>.

Del Monaco, C., Hay, M., Gartrell, P., et al. 2017, 'Effects of ocean acidification on the potency of macroalgal allelopathy to a common coral', *Scientific Reports*, vol. 7, viewed 20 June 2020, <https://doi.org/10.1038/srep41053>.

Dennis, R. 2015, 'When you're in a hole – stop digging! The economic case for a moratorium on new coal mines', *The Australia Institute*, Discussion Paper, 18 October, viewed 20 June 2020, <www.tai.org.au/content/when-you-are-hole-stop-digging>.

Department of Foreign Affairs and Trade. 2007, *Trade Topics: A Quarterly Review of Australia's International Trade*, September 2007, DFAT, Canberra.

Doyle, T. 2002, 'Environmental campaigns against mining in Australia and the Philippines', *Mobilization*, vol. 7, no. 1, pp. 29–48.

Doyle, T. 2010, 'Surviving the gang bang theory of nature: the environment movement during the Howard years', *Social Movement Studies*, vol. 9, no. 2, pp. 155–169.

Eckersley, R. 1992, *Environmentalism and Political Theory: Toward an Ecocentric Approach*, UCL Press, London.

Eckersley, R. 2004, *The Green State: Rethinking Democracy and Sovereignty*, MIT Press, Cambridge.

Edelman, M. 2001, *The Politics of Misinformation*, Cambridge University Press, Cambridge.

Edis, T. 2012, 'A year in the carbon tax scare campaign', *The Australian Business Review*, 17 December, viewed 20 June 2020, <www.theaustralian.com.au/business/business-spectator/a-year-in-the-carbon-tax-scare-campaign-/news-story/48d3eb1d3726428e2ce15c480f84f9f1>.

Eggleston, F. 1932, *State Socialism in Victoria*, King, London.

Environmental Law Australia. 2016, *Carmichael Coal ("Adani") Mine Cases in Queensland Courts*, viewed 15 March 2020, <http://envlaw.com.au/carmichael-coal-mine-case/>.

Esposito, A. & Neale, T. 2016, 'Never squib the rights issue in favour of conservation', in E. Vincent & T. Neale (eds), *Unstable Relations: Indigenous People and Environmentalism in Contemporary Australia*, UWA Publishing, Crawley.

Evans, G. 2010, 'A rising tide: linking local and global climate justice', *Journal of Australian Political Economy*, vol. 66, pp. 199–221.

Evans, G. 2014, 'Maules Creek mine: frontline action on coal', *Chain Reaction*, November 2014, no. 122, pp. 31–33.

Everingham, J. 2016, 'Transformation of rural society and environments by extraction of mineral and energy resources', in M. Shucksmith & D. Browns (eds), *Routledge International Handbook of Rural Studies*, Routledge, London.

Foley, G. 2013, 'How Bob Hawke killed land rights', *Tracker Magazine*, 13 January, viewed 20 June 2020, <www.kooriweb.org/foley/essays/tracker/tracker19.html>.

Ford, L. 2012, 'Locating indigenous self-determination in the margins of settler sovereignty: an introduction', in L. Ford & T. Rowse (eds), *Between Indigenous and Settler Governance*, Routledge, pp. 1–13.

Gilbert, K. 1994, *Because a White Man'll Never Do It*, Angus and Robertson, 4th Edition, Pymble, New South Wales, Australia.

Gill, S. 1995, 'The global panopticon? The neoliberal state, economic life, and democratic surveillance', *Alternatives*, vol. 20, pp. 1–49.

Goodman, J. & Worth, D. 2008, 'The minerals boom and Australia's 'resource curse' ', *Journal of Australian Political Economy*, vol. 61, pp. 201–219.

Gray v Minister for Planning. [2006] NSWLEC 720.

Greenpeace Australia. 2012, *Boom Goes the Reef: Australia's Coal Export Boom and the Industrialisation of the Great Barrier Reef*, March 2015, viewed 20 June 2020, <www.greenpeace.org.au/news/boom-goes-the-reef/>.

Greenpeace Australia. 2015, *Whitehaven Coal: No Future*, viewed 20 June 2020, <www.greenpeace.org.au/wp/wp-content/uploads/2017/10/GRE01238_Whitehaven_report_final.pdf>.

Gundjeihmi Aboriginal Council (GAC). 2001, Submission by the Mirrar Aboriginal People, Kakadu, Australia, to the Office of the High Commissioner of Human Rights on 'Indigenous peoples, private sector natural resource, energy and mining companies and human rights', Unpublished, Jabiru.

Gundjeihmi Aboriginal Council (GAC). 2006, '

Gurr, T. R. 1983, 'Outcomes of public protest among Australia's aborigines', *American Behavioural Scientist*, vol. 26, no. 3, pp. 353–373.

Hajkowicz, S. A., Heyenga, S. & Moffat, K. 2011, 'The relationship between mining and socio-economic well-being in Australia's regions', *Resources Policy*, vol. 36, no. 1, pp. 30–38.

Harvey, D. 2011, *The Enigma of Capital and the Crisis of Capitalism*, Oxford University Press, Oxford.

Hoegh-Guldberg, O. 2012, 'Is the Great Barrier Reef listing? The UN asks if we're still heritage-worthy', *The Conversation*, 7 March, viewed 20 March 2020, <https://thec onversation.com/is-the-great-barrier-reef-listing-the-un-asks-if-were-still-heritage-wor thy-5401>.

Howard, J. 1997a, 'Transcript of the Prime Minister, the Hon. John Howard', *ABC TV*, Television interview with Kerry O'Brien, The 7:30 Report, 4 September, viewed 24 June 2020, <https://pmtranscripts.pmc.gov.au/sites/default/files/original/00010469.pdf >.

Howard, J. 1997b, 'Mining', *House of Representatives Hansard*, 29 March, p. 4432.

Howard-Wagner, D. 2008. 'Legislating away Indigenous rights', *Law, Text, Culture*, vol. 12, no. 1, p. 5.

Hughes, T. 2014, 'Mounting evidence shows dredge spoils threat to the Great Barrier Reef', *The Conversation*, 18 August, viewed 24 June 2020, <https://theconversation.com/mount ing-evidence-shows-dredge-spoil-threat-to-the-great-barrier-reef-29773>.

Hunt, C. 2011, 'What the Gladstone's LNG development really doing to the environment?', *The Conversation*, 20 October, viewed 20 June 2020, <https://theconversation.com/what-is-gladstones-lng-development-really-doing-to-the-environment-3885>.

Hutton, D. 2012, 'Lessons from the lock the gate movement', *Social Alternatives*, vol. 31, no. 1, pp. 15–19.

Hutton, D. 2013, *Mining: The Queensland Way*, At A Glance Pty Ltd, Queensland.

Hutton, D. & Connors, L. 1999, *A History of the Australian Environment Movement*, Cambridge University Press, Cambridge.

Jewel, C. 2014, 'Deutsche Bank rules out Abbott Point financing following campaign', *The Fifth Estate*, 23 May, viewed 15 June, <www.thefifthestate.com.au/business/investment-deals/deutsche-bank-rules-out-abbot-point-financing-following-campaign/>.

Kellow, A., 1989, 'The dispute over the Franklin River and the South West wilderness area in Tasmania, Australia', *Natural Resource Journal*, vol. 29, p. 129.

Kellow, A. 1990, 'Spoiling for a fight or fighting over the spoils? Resource and environ-mental politics and policies in Australia towards 2000', *Australia Towards 2000*, pp. 198–214.

Kellow, A. 1996, 'Thinking globally and acting federally: intergovernmental relations and environmental protection in Australia', *Contributions in Political Science*, vol. 368, pp. 135–156.

Knaus, C. & Evershed, N. 2017, 'Great Barrier Reef at 'terminal stage': scientists despair at latest coral bleaching data', *The Guardian*, 10 April, viewed 20 June 2020, <www.theg uardian.com/environment/2017/apr/10/great-barrier-reef-terminal-stage-australia-scienti sts-despair-latest-coral-bleaching-data>.

Koowarta v Bjelke-Petersen. (1982), 153 CLR 168.

Kuhn, R. 1998, 'Rural reaction and war on the waterfront in Australia', *Monthly Review*, vol. 50, no. 6, pp. 30–44.

Langton, M. 2012, 'The quiet revolution: Indigenous people and the resources boom', *Boyer Lectures 2012*, ABC Books.

Lavelle, A. 2010, 'The mining industry's campaign against native title: some explanations', *Australian Journal of Political Science*, vol. 36, no. 1, pp. 101–122.

Luong, P. & Weinthal. E. 2006, 'Rethinking the resource curse: ownership structure, insti-tutional capacity and domestic constraints', *Annual Review of Political Science*, vol. 9, pp. 241–263.

Lyons, K. 2019, 'Securing territory for mining when Traditional Owners say 'No': the exceptional case of Wangan and Jagalingou in Australia', *The Extractive Industries and Society*, vol. 6, 2019, pp. 756–766.

Mabo and Ors v Queensland 1992, viewed 20 September 2021, <https://aiatsis.gov.au/ntpd-resource/742>.

Maddison, M. 2011, 'Figures reveal Bowen Basin mining boom', *Australian Broadcasting Corporation*, 8 October, viewed 20 March 2020, <www.abc.net.au/news/2011-08-10/figu res-reveal-bowen-basin-mining-boom/2832436>.

Manne, R. 2011, 'Rudd's downfall: written in the Australian', *ABC News*, 5 September, viewed 20 September 2020, <www.abc.net.au/news/2011-09-05/manne-rudds-downfall-written-in-australian/2869942>.

Martinez-Alier, J. 2002, *The Environmentalism of the Poor. A Study of Ecological Conflicts and Valuation*, Edward Elgar, Cheltemhan.

Maxwell, P. 2006, 'Minerals growth and development', in P. Maxwell (ed.), *Australian Mineral Economics: A Survey of Important Issues*, Monograph 24, AusIMM, Melbourne.

McAdam, D. 2017, 'Social movement theory and the prospects for climate change activism in the United States', *Annual Review of Political Science*, vol. 20, no. 20, pp. 189–208.

McKay, B., Lambert, I. & Miyazaki S. 2000, *Australian Mining Industry, 1998–99*, no. 84140, Australian Bureau of Statistics (ABS), Canberra.

Mercer, D. 1987, 'Patters of protest: native title claims and rights in Australia', *Political Geography Quarterly*, vol. 6, pp. 171–194.

Mercer, D. 1993, 'Terra nullius, aboriginal sovereignty and land rights in Australia: the debate continues', *Political Geography,* vol. 12, no. 4, pp. 299–318.

Mills, V. 2011, 'National heritage listing for the Kimberley', *Australian Broadcasting Corporation Local*, 1 September.

Minerals Council of Australia (MCA). 2010, *The Australian Minerals Industry and the Australian Economy*, MCA, March 2010.

Mitchell, W. & Bill, A. 2006, 'The two-speed Australian economy – the decline of Sydney's labour market', *People and Place*, vol. 14, no. 4, pp. 14–24.

Muecke, S. 2016, 'Indigenous-Green knowledge collaborations and the James Price Point dispute', in E. Vincent & T. Neale (eds), *Unstable Relations: Indigenous People and Environmentalism in Contemporary Australia*, UWA Publishing, Crawley.

Noonuccal, O. 1977, '*Minjerriba', Meanjin*, Summer, viewed 20 September 2022, <https://meanjin.com.au/poetry/minjerriba/>.

Norman, B., McCalman, L. & Hughes, T. 2015, 'Governments unveils 2050 Great Barrier Reef plan: experts react', *The Conversation*, 24 March, viewed 20 June 2020, <https://theconversation.com/government-unveils-2050-great-barrier-reef-plan-expe rts-react-39172>.

Norman, H. 2016, 'Coal mining and coal seam gas on Gomeroi country: sacred lands, economic futures and shifting alliances', *Energy Policy,* vol. 99, pp. 242–251.

O'Faircheallaigh, C. 1995, *Mineral Development Agreements Negotiated by Aboriginal Communities in the 1990s*, Discussion paper no. 85, Centre for Aboriginal Economic Policy Research, Australian National University, Canberra.

O'Faircheallaigh, C. 2006, 'Aborigines, mining companies and the state in contemporary Australia: a new political economy or 'business as usual'?' *Australian Journal of Political Science*, vol. 41, no. 1, pp. 1–22.

Oskarsson, S. & Ottosen, E. 2010, 'Does oil still hinder democracy?' *Journal of Development Studies*, vol. 46, no. 6, pp. 1067–1083.

Parson, S. & Ray, E. 2016, 'Sustainable colonization: tar sands as resource colonialism', *Capitalism Nature Socialism*, vol. 29, no. 3, pp. 1–19.

Pearse, G. 2009, 'Quarry vision: coal, climate change and the end of the resources', *Quarterly Essay*, vol. 33, Black Inc., Melbourne.

Pearse, G. 2011, 'The climate movement', *The Monthly*, 3 June, viewed 6 January 2020, <www.themonthly.com.au/issue/2011/september/1316399650/guy-pearse/climate-movement#mtr>.

Perlich, H. 2013, 'Australia's two-speed economy', *Journal of Australian Political Economy*, vol. 72, pp. 106–126.

Philips, N. 2014, 'Scientific academy slams government's Great Barrier Reef plan', *Sydney Morning Herald*, 27 October, viewed 15 June 2020, <www.smh.com.au/technology/sci entific-academy-slams-governments-great-barrier-reef-plan-20141027-11cjwj.html>.

Pickerill, J. 2018, 'Black and green: the future of indigenous-environmentalist relations in Australia', *Environmental Politics*, vol. 27, no. 6, pp. 1122–1145.

Quiggin, J. 2005, 'Economic liberalism: fall, revival and resistance', in P. Walters & J. Walter (eds), *Ideas and Influence: Social Science and Public Policy in Australia*, UNSW Press, Sydney.

Richardson, D. & Dennis, R. 2011, *Mining the Truth: The Rhetoric and Reality of the Commodities Boom*, The Australia Institute, 8 September, viewed 23 June 2020, <www.tai.org.au/node/1777>.

Robertson, J. 2017d, 'Drew Hutton, how he galvanised the Greens and his unlikely alliance with Alan Jones', *The Guardian*, 23 June, viewed 20 June 2020, <www.theguardian.com/environment/2017/jun/23/drew-hutton-how-he-galvanised-the-greens-and-his-unlikely-alliance-with-alan-jones>.

Rosewarne, S. 2016, 'The transnationalisation of the Indian coal economy and the Australian Political economy: the fusion of regimes of accumulation?', *Energy Policy*, viewed 20 June 2020, <https://doi.org/10.1016/j.enpol.2016.05.022>.

Rosewarne, S., Goodman, J. & Pearse, R. 2014, *Climate Action Upsurge: The Ethnography of Climate Movement Politics*, Routledge, Abingdon, Oxon, UK and New York.

Sachs, J. & Warner, A. 1997, *Natural Resource Abundance and Economic Growth: The Curse of Natural Resources*, National Bureau of Economic Research Working Paper No. 5398, Harvard University, Cambridge, MA.

Sachs, J. & Warner, A. 2001, 'Natural resources and economic development: the curse of natural resources', *European Economic Review*, vol. 45, pp. 827–838.

Santos 2014, *Narrabri Gas Project: Our Plans to Develop Natural Gas for New South Wales*, July 2014, viewed 20 June 2020, <https://narrabrigasproject.com.au/uploads/2014/08/Narrabri_Gas_Project_brochure_2014.pdf>.

Short, D. 2003, 'Australian "Aboriginal" reconciliation: the latest phase in the colonial project', *Citizenship Studies*, vol. 7, no. 3, pp. 291–392.

Short, D. 2007, 'The social construction of Indigenous 'Native Title Land Rights' in Australia', *Current Sociology*, vol. 55, no. 1, p. 857.

Slezak, M. 2016, 'Agencies say 22% of Barrier Reef coral is dead, correcting 'misinterpretation'', *The Guardian*, 3 June, viewed 20 June 2020, <www.theguardian.com/environment/2016/jun/03/agencies-say-22-of-barrier-reef-coral-is-dead-correcting-misint erpretation>.

Slezak, M. 2017, 'Great Barrier Reef 2050 plan no longer achievable due to climate change, experts say', *The Guardian*, 25 May, viewed 20 June 2020, <www.theguardian.com/environment/2017/may/25/great-barrier-reef-2050-plan-no-longer-achievable-due-to-clim ate-change-experts-say>.

Slezak, M. 2019, 'Controversial Great Barrier Reef grant did not comply with transparency rules, National Audit Office says', *Australian Broadcasting Corporation*, 16 January, viewed 20 October 2020, <www.abc.net.au/news/2019-01-16/great-barrier-reef-funding-grant-scrutinised-auditor-general/10720928>.

Stephens, T. 2016, 'Signing the Paris Climate Agreement is easy – what comes next for Australia will be hard', *The Conversation*, 21 April, viewed 20 June 2020, <https://thec onversation.com/signing-the-paris-climate-agreement-is-easy-what-comes-next-for-australia-will-be-hard-58072>.

Stevens, P. & Dietsche, E. 2008, 'Resource curse: an analysis of causes, experiences and possible ways forward', *Energy Policy*, vol. 36, no. 1, pp. 56–65.

Strachan, J. 2007, 'Delay to hearing on mine', *Herald*, 4 July.

Talanoa, S. 2008, 'First Climate Camp starts: coal train and world's largest port blockaded', *Climate Action*, 15 July, 20 June 2020, <www.climateaction.org/news/first_climate_camp_starts_coal_train_and_worlds_largest_port_blockaded>.

Talbot, S. 2013, 'Aboriginal workers walk off the job – Whitehaven in protest at Maules Creek coalmine', *Media Release, Maules Creek Community Council*, 8 July, viewed 20 June 2020, <http://nationalunitygovernment.org/content/gomeroi-protest-against-whi tehaven-coal-gunnedah-13-july-2013>.

Taylor, L. 2016, 'On climate policy, neither time nor Trump are on Turnbull's side', *The Guardian*, 19 November, viewed 22 June 2020, <www.theguardian.com/australia-news/2016/nov/19/on-climate-change-policy-neither-time-nor-trump-are-on-turnbulls-side>.

Taylor, L. 2017, 'Australia's conservative government fiddles on climate policy while the country burns', *The Guardian*, 20 January, viewed 15 June 2020, <www.theguardian.com/environment/commentisfree/2017/jan/20/australias-conservative-government-fidd les-on-climate-policy-while-the-country-burns>.

Topp, V., Soames, L., Parham, D. & Bloch, H. 2008, *Productivity in the Mining Industry: Measurement and Interpretation*, Productivity Commission Staff Working Paper, Productivity Commission, Canberra.

Trebeck, K. A. 2005, *Democratisation through Corporate Social Responsibility? The Case of Miners and Indigenous Australians*, PhD thesis, Political Science, Australian National University, Canberra.

Trebeck, K. A. 2007, 'Tools for the disempowered? Indigenous leverage over mining companies', *Australian Journal of Political Science*, vol. 42, no. 4, pp. 541–562.

Tulip, P. 2014, *The Effect of the Mining Boom on the Australian Economy*, Reserve Bank of Australia, December 2014, Bulletin – December Quarter 2014, viewed 20 June 2020, <www.rba.gov.au/publications/bulletin/2014/dec/pdf/bu-1214-3.pdf>.

Turner, B. S. & Rojeck, C. 2001, *Society and Culture: Principles of Scarcity and Solidarity*, Sage, London.

UNESCO. 2012, *Great Barrier Reef Australia: World Heritage Mission Report*, 14 March, viewed 20 June 2020, <www.google.com/search?client=safari&rls=en&q=unesco+2012+great+barrier+reef&ie=UTF-8&oe=UTF-8>.

Walker, C. 2017, 'Coalition tries again to strip eco-charities of tax-deductibility status', *Michael West Media*, 5 August, viewed 20 June 2020, <www.michaelwest.com.au/coalit ion-tries-again-to-strip-eco-charities-of-tax-deductibility-status/>.

Weber, D. 2011, 'Secret men's business could threaten gas hub: Joseph Roe', *PM, Australian Broadcasting Corporation*, 5 December, viewed 20 June 2020, <www.abc.net.au/pm/cont ent/2011/s3384015.htm>.

Wilderness Society. 2013, 'Victory for Australia's nature: WA Supreme Court rules James Price Point approval 'illegal'', *The Wilderness Society*, viewed 24 June 2020, <www.wil derness.org.au/articles/victory-austrlalia's-nature-wa-supreme-court-rules-james-price-point-approval-'illegal>.

Wilkinson, A. & Austen, N. 2020, *Joint Stakeholder Submission to the Universal Periodic Review of Australia*, Environmental Justice Australia, Environment Defenders Office,

Earth Justice, 8 July 2020, viewed 20 August 2020, <https://earthjustice.org/sites/default/files/files/joint_submission_edo_eja_ej_to_australias_upr_8_jul_2020_002.pdf>.

Woods, G. 2007, 'Latest Anvil Hill legal case fails, but campaign continues!', *Greenpeace Blog*, viewed 20 June 2020, <www.greenpeace.org.au/blog/latest-anvil-hill-legal-case-fails-but-campaign-continues/>.

Woodward, E. 1974, *Aboriginal Land Rights Commission: Second Report*, Australian Government Publishing Services, Canberra.

Legislations

Aboriginal Land Rights Act 1983.
Australian Heritage Commission Act 1975.
Antarctic Mining Prohibition Act 1989.
Australian National Parks and Wildlife Conservation Act 1975.
Australian Racial Discrimination Act 1975.
Endangered Species Protection Act 1992.
Environment Conservation and Biodiversity Protection Act 1999.
Environment Protection (Impact of Proposals) Act 1974.
Environment Protection (Nuclear Codes) Act 1978.
Environment Protection (Sea Dumping) Act 1981.
Great Barrier Reef Marine Park Act 1975.
Hazardous Waste (Regulation of Exports and Imports) Act 1989.
Native Title Act (Cth) 1993.
Native Title Amendment Act 1998.
Ozone Protection Act 1989.
Protection of the Seas (Prevention of Pollution from Ships) Act 1983.

Website

www.marketforces.org.au

7 Countering coal in Australia

The politics of the Carmichael coal mine

> It (exporting coal from Adani's Carmichael coalmine in Queensland) will lift hundreds
> of millions of people out of energy poverty, not just in India but right across the world.
>
> Australian Energy Minister Josh Frydenberg (2015)

In 2010, at the peak of Australia's mining boom, the Indian conglomerate Adani Enterprises embarked on an ambitious mission of developing the largest coal mine in the southern hemisphere. That year, the corporation acquired 7.9 billion tonnes of coal assets in the Galilee Basin from Linc Energy (Murphy 2010). There was a possibility of up to nine mega coal mines being developed in the previously untapped Galilee Basin in Queensland's central west, approximately 400 kilometres inland from the Great Barrier Reef. Greenpeace Australia estimated that coal extracted and burnt at full capacity from all proposed coal mines in the Galilee would make the Basin responsible for the seventh highest emissions in the world, with the rest of Australia following way behind at the 14th place (Greenpeace Australia 2012a).

The opening up of the Galilee Basin would deplete groundwater and aquifers of the Great Artesian Basin (GAB), affect the land rights of the Wangan and Jagalingou (W&J) traditional owners, risk native vegetation and threatened wildlife species, and physically harm the Great Barrier Reef and its associated coastal wetlands through port expansion and increased coal traffic (Environment Law Australia 2016).

Cumulative climate and environmental risks from opening up the Galilee Basin to coal mining mobilised Australia's biggest environmental movement: it consisted of over 30 national and state-based organisations and over 400 local groups that came together under the banner 'Stop Adani'.[1] The movement aimed to stop the first coal mine – the Adani Enterprises-owned Carmichael – as a strategic move to stop the opening up of the entire coal region to mining. Impacts on groundwater mobilised farmers in the Galilee region during one of the worst droughts in the continent's interiors. Adani Australia's 'disrespectful' dealings with the W&J and the State's approval and support for the Carmichael project without Indigenous consent generated a sustained Indigenous resistance under the campaign slogan 'Adani, No Means No!'.

DOI: 10.4324/9781003410416-7

The Galilee Basin is a semi-desert region that spreads over 250,000 square kilometres in Queensland's central west, roughly the size of the United Kingdom. It is one of the world's largest coal basins holding an estimated 27 billion tonnes of thermal coal (Huleatt 1991). Plans to develop it did not eventuate until the higher coal demand and prices resulting in the minerals made extraction of coal from this remote location economically viable. Paradoxically, Australia's coal exports started declining in 2015 even as the Galilee mega mines passed through state and federal approvals, raising questions about their long-term viability (Buckley et al. 2018). Underneath the Galilee's coal seams lies the Great Artesian Basin, an ancient body of water stretching across 22% of Australia's interior and supplying freshwater to remote parts of four states. Farmers west of the Great Dividing Range rely on water supplies from aquifers connected to the Great Artesian Basin for their livelihoods.

The area for the proposed Carmichael mine in the Galilee Basin contains the traditional lands of the W&J people, whose custodianship has extended for 'untold thousands of years' (Burragubba 2018, p. viii). The W&J's traditional law, the *Kubbah* or native Bee in their *Wiirdi* language, holds them responsible for protecting the sacred Doongmabulla Springs formed of 60 freshwater springs fed through underground aquifers by the GAB. Their cultural survival is tied to the health of the Springs through the dreaming totem *Mundunjudra* or Rainbow Serpent Water Spirit, which is believed to have emerged from here to give shape to the land, rivers and waterholes of the dry Australian continent (Burragubba 2018).

Carmichael, the largest proposed coal mine, comprising six open-cut and five underground mines, would cover an area of 28,000 hectares, roughly five times the size of the Sydney Harbour. Initially estimated to be operational by 2014 and reach full production capacity by 2022, Carmichael would export an estimated 60 million tonnes of coal each year, operating over a lifespan of 90 years (Rolfe 2014). The ambition of the Adani Group's A16.5 billion Australian venture became evident through its proposed scale of vertical integration; a 400-kilometer-long railway line would connect the mine to the Abbott Point coal port near Bowen (Queensland Government 2016). In 2011, the Adani Group acquired a 99-year lease on the Abbott Point X50 Coal terminal (APCT) (Grant-Taylor 2011). A new coal terminal was proposed at Abbott Point port to handle the increased volumes of coal exports. Coal from Carmichael would be shipped to India (Elliot 2017).[2]

Although the environmental and Indigenous resistance and local farmers' objections could not stop the Carmichael project, they provided crucial challenges through legal cases, political disruption and making major investors abandon the project. The various resistances delayed the Carmichael mine by five years, during which the size and scope of the project were significantly reduced, thereby reflecting coal's structural decline. This chapter traces the build up of one of Australia's largest environmental movements, mobilisations of politically conservative farmers affected by coal mining in Central Queensland, and an extensive Indigenous land-rights resistance, between 2014 and 2018, a critical period in the political economy of Australian coal. During this period, the decline in global coal demand, timed with increasing civil society demands for climate action after the Paris Agreement, turned a long-held political optimism about the longevity of

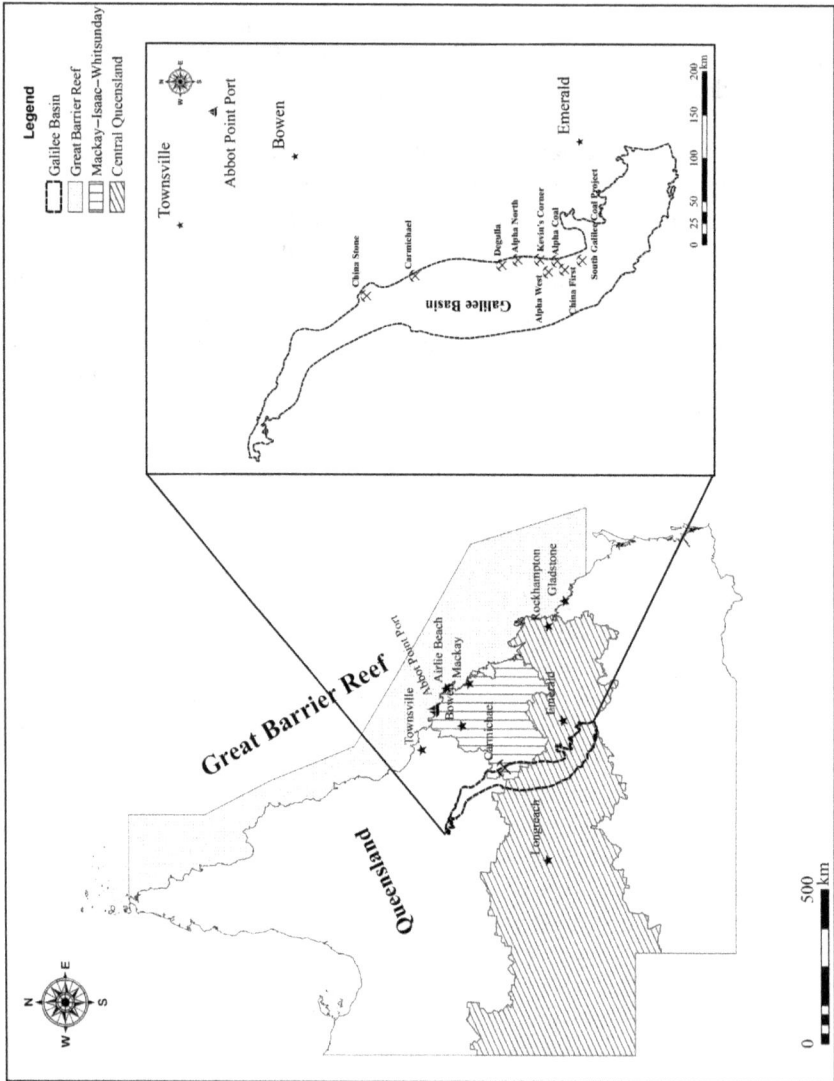

Figure 7.1 Central Queensland with Galilee Basin.

Figure 7.2 Wangan and Jagalingou Country, Galilee Basin.

Source: Photo by Anthony Esposito.

Australia's high-quality coal exports into desperate political measures for rescuing Australia's largest proposed coal mine. It answers the second question of this book with regard to anti-coal activisms in Australia: what is the State and civil society dialectic and how has coal extraction been countered?

The first section outlines the policy and political support for coal mining in Australia, both historically and during the resource boom, leading to the grand ambition of developing the Galilee Basin at the time of declining global demand for coal and worsening effects of climate change. The second section traces the build up of the various forms of activism of the Stop Adani environmental movement – legal, financial, grassroots and local – in response to the championing of the Adani project by successive federal and Queensland governments, between 2014 and 2018. It outlines the emergence of farmers' dissent from the Galilee Basin: local farmers spoke about the impacts of coal mining on groundwater sources that are critical for agriculture, criticised political inaction on climate change, and brought legal actions against coal mines.

The third section traces the build up of the land rights conflict of the Carmichael mine and the W&J traditional owners' land rights resistance through sustained legal campaigns, political disruptions, appeals to the United Nations and international advocacy of financial institutions associated with the Carmichael project. The conflict was shaped through the W&J people's experiences of marginalisation by the

State and the 'inherent racism' of native title institutions that favoured the interest of mining corporations. Summarising from the chapter, the Analysis discusses various ways in which Australian coal was delegitimised by civil society through the conflict over the Carmichael project. The chapter concludes by identifying how the anti-coal discourses of the groups resisting Carmichael are crucial to a global narrative against coal and climate change.

7.1 Political economy of coal in Australia

Coal is Australia's second-largest export commodity after iron-ore, with 75% of coal mined in Australia exported as sea-borne coal primarily to Japan, China, India and South East Asia. A significant proportion of Australia's coal exports are high-quality metallurgical coal used for manufacturing steel. Large-scale exports occur from the coal-bearing regions and adjoining ports on Australia's eastern seaboard; from the Hunter Valley in New South Wales through the world's largest coal port of Newcastle; and from the Bowen and Surat Basins through Gladstone, Hay, and Abbott Point ports on Queensland's central and north coasts. To a smaller extent, brown coal is exported from Victoria's La Trobe Valley. Domestic coal is primarily used for thermal power, with 47% of Australia's electricity coming from coal (Australian Government n.d.).

Australian coal proved valuable for sustaining the British Empire's steam-powered sea routes in the Indian and South Atlantic Oceans. Coal from Hunter Valley in New South Wales was the first commodity to be successfully exported from colonial-era Australia, making this region Australia's first export-bound commercial coal hub (Comerford 1997). The first British settlement in the Hunter Valley was established as a convict camp for coal mining within decades following the arrival of the first fleet at Sydney Cove in 1788 (Evans 2010). The use of convict labour for coal mining in colonial New South Wales and Tasmania indicates the nexus between the coal industry and the Australian State from the earliest times (Martin et al. 1993). By the mid-1800s, the colonial industries of agriculture and coal mining had dispossessed many Indigenous peoples of their lands (Turner and Blyton 1985).

Building 'world-class' coal exports

The Commonwealth Coal Industry Act (1946) was created in response to coal mining union agitations for the nationalisation or heavy regulations of Australia's coal industry to protect against the 'cyclical swing in investment and capital utilisation' (Lee & Draper 1988, p. 45). After World War II, Australia created the Bureau of Mineral Resources, which sponsored geological surveys and caused a surge in mineral discoveries including coal (Coal and Mineral Industries Division 1998–1999). Australian coal exports increased in the 1950s and 1960s, partly to the reindustrialisation of Japan after World War II. Increasing coal exports to Japan were facilitated by cheaper shipping costs and lower overall prices making Australian coal a serious contender to the United States' east coast coal trade (Anderson 1971).

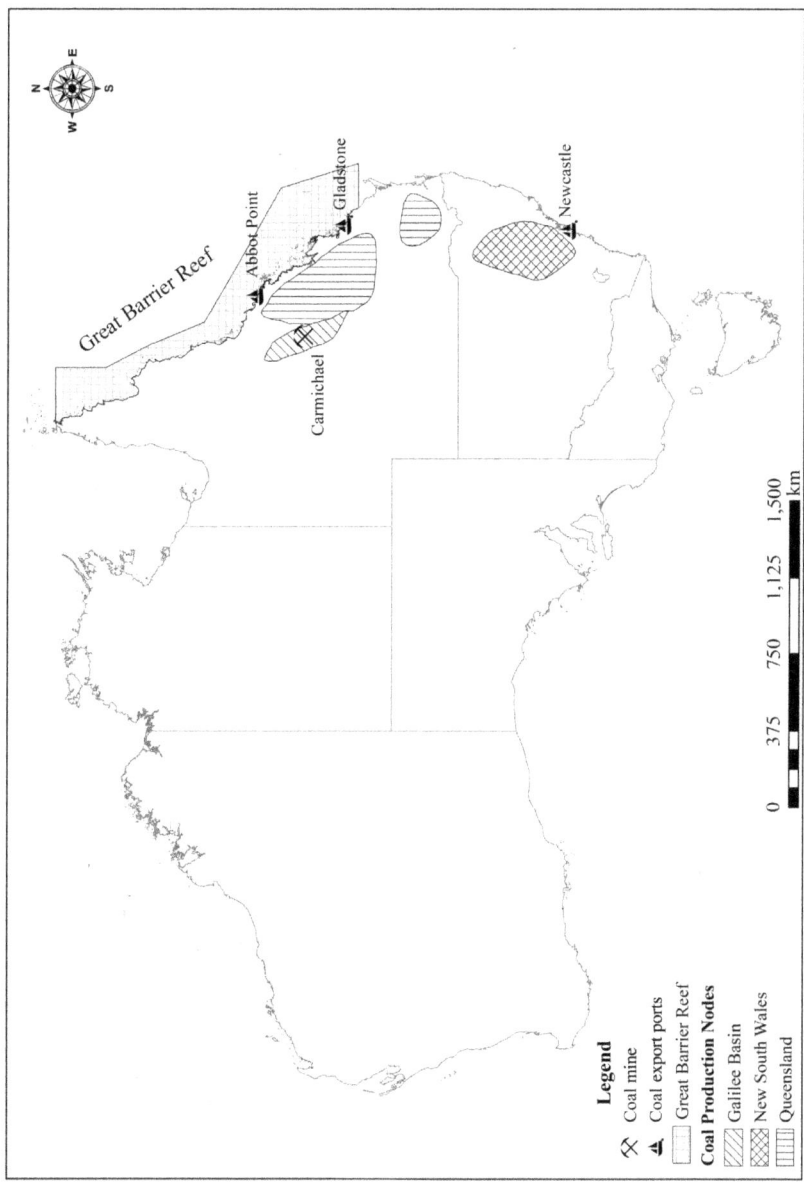

Figure 7.3 Major coal regions and coal ports in New South Wales and Queensland.

Unsuccessful attempts were made during the short-lived Gough Whitlam Labor government (1972–1975) to nationalise energy resources to avoid profits from going overseas (Baer 2016). However, the efforts of 'nationalistic Australian capitalists' failed on account of the Australian business class joining hands with powerful foreign investors (Crough et al. 1983, p. 35). Governments from both major political parties have since incentivised private coal production through the approach of subsidising mining and related infrastructural developments (Baer 2016). Large investments by state governments in port and harbour infrastructures also encouraged mineral exports (Fagan and Bryan 1991). In NSW and Queensland, State-owned railways carried coal from the mines to the ports, with Queensland Rail being created in 1995, and Freight Corp in NSW in 1996 (Energy Minerals Branch 1999).

Governments increased their support for coal and natural gas expansions after the 1979 oil shock (Corrighan 1980). In 1984 Australia became the world's largest coal exporter overtaking the United States (Owen, A.D. 1988). Queensland's Country Party, which later became the Liberal National Party, aligned its programme of State developmentalism with the economic agendas of US coal multinationals Utah Development Company and Thiess Brothers. The Department of National Development is understood to have been instrumental in the prosperity of Utah. Through this model, Queensland built a 'world-class' coal export industry along with new port and rail infrastructure and new towns (Galligan 1989, p. 121).

A close friendship between Queensland's longest-serving Premier the conservative Joh-Bjelke-Petersen (1968–1987) and industrialist Les Thiess is considered a significant factor in the development of Queensland's large-scale coal industry. Large mines were governed by exclusive special agreement acts that stipulated some conditions but overall made environmental regulation a complicated affair, particularly owing to a lack of enforcement and lack of rehabilitation by the mining companies. A strong representation of farming interests and a largely unionised rural workforce in Queensland's political economy was transformed from the 1980s through the prominence of large mining corporations with significant power over the government (Hutton 2013).

The Bob Hawke ALP federal government ignored the demands of the labour movement in 1987 for a National Coal Authority to achieve direct control over mining operations and create centralised planning of investment, mine development and productive capacity. Export controls were implemented in response to price cutting by Utah and in response to actions by mining unions. However, in 1986 the Hawke government worked against unions by curtailing the regulation of coal exports (Lee and Draper 1988).

Drew Hutton notes in *Mining: The Queensland Way* (2013) that traditional approaches to analysing the Queensland government's mining-led development were either romantic accounts of the evolution of pioneering mining companies and towns, or accounts of royalties, infrastructure and benefits to local communities. Later, some critical accounts by historians such as Ross Fitzgerald have focussed on Queensland's corrupt culture of developmentalism that promoted mining projects that caused high social and environmental impacts (Fitzgerald

1984). The forced resettlement of the Aboriginal community of Mapoon in western Cape York Peninsula in Far North Queensland in 1963, and the subsequent closure of the Aboriginal Mission at Mapoon, after the state government granted leases for bauxite mining to Comalco corporation (which later became Rio Tinto Alcan), is considered one of the most significant social impacts of Queensland's corrupt developmental approach (Fitzgerald 1984).

Engineering Australia's coal boom (1996–2011)

Although Australia's unprecedented coal expansions and exports from the mid-1990s were built on existent economic and infrastructural support from governments, increasing global prices of coal and massive foreign investments in Australian coal mining transformed the Australian political economy on a scale not seen before. During the minerals boom, major coal-bearing regions in New South Wales and Queensland were reorganised as global sites of massive coal operations and caused pervasive social and environmental effects.

Within a decade the mineral boom transformed the Hunter Valley into what activists called the 'Carbon Valley', responsible for one of the world's highest per capita emissions of greenhouse gases (Ray 2005).[3] As Australia's most intensive coal mining region, 30 coal mines in the Hunter produced 100 million tonnes of saleable coal per year, which was a quarter of Australia's annual coal production (Cleary 2012). Newcastle became the world's largest black coal exporting port. Despite the significant new wealth from coal mining in the Hunter Valley, those residing close to mine sites and coal-fired power stations struggled against harmful effects on health, rural livelihoods and the environment (Connor et al. 2009).[4]

In Queensland, coal is the largest export commodity, constituting 40% of the state's exports. Up to 71% of these exports are made up of high-quality coking coal that generates three times as much royalty for the Queensland government as the lower-grade thermal coal variety (Buckley and Nicholas 2019). Coal expansions particularly in the second decade of the coal boom developed the Central Queensland region into Australia's largest coal-producing region containing some of the largest coal mines with high foreign ownership. This led to the transformation of coastal cities such as Mackay from a sugarcane farming and processing centre into a coal-export hub. The Bowen Basin region witnessed massive coal developments by foreign corporations such as Anglo Coal, Xstrata and Peabody Energy making it the site of some of the world's largest coal mines (Hutton 2013).

Despite declaring climate change to be 'the greatest moral challenge of our time', the Kevin Rudd Labor government gave federal approval for the multibillion dollar expansion of Queensland's Gladstone Harbour in 2008 to enable the export of 84 million tonnes of additional coal annually (Baer 2016). Queensland spent A$5.4 billion on a Coal Transport Infrastructure Program and the Rudd government also announced taxpayer-funded subsidies of A$580 million to expand coal and port operations in NSW to export more than 100 million tonnes of additional coal per annum through Newcastle (Pearse 2009). These expansions included approvals for a new A$900 billion coal port that would allow for the export of 66 million tonnes of coal per annum, which when burnt in overseas power stations would release 174 million tonnes of greenhouse gases (NSW Department of Planning, quoted in Cubby and Environment Reporter 2009).

The grounded effects of the resource boom and the risk of climate change led to the emergence of other social impacts of coal beyond the long-held narrative of prosperity and economic growth through a booming export industry. Environmentalists argued that there are ethical and moral costs associated with Australia exporting coal, a 'dirty energy' to the world (Anvil Hill Project Watch Association 2006). Coal and coal seam gas expansions have been equated to 'an invasion of our country, a taking over of land, and a clearing out of people' (Munro 2012, p. 1). Such descriptions of the extreme social and environmental effects of Australia's coal boom evoke earlier colonial dispossessions of Indigenous people by the colonial-era industries of agriculture and mining.

Previously latent contradictions of the coal-export industry became pervasive during the coal mining boom. Coal mining in Australia is characterised by high levels of foreign ownership and control and an export focus, but it generates only a relatively small number of jobs (Davidson and de Silva 2013). In the previous era, 'state governments insisted that corporations build housing, street, schools, hospitals and recreation facilities if they wanted a mining licence' but during the minerals boom, coal companies were able to establish a 'fly-in-fly-out' system for mining that helped them to avoid building infrastructures for local communities (Pearse et al. 2013, p. 52). This difference indicates that the power imbalance in the state–industry nexus increased during Australia's minerals boom.

State–industry nexus and Australia's coal optimism

Kenworthy and Gordon (2011) amongst others argue that the nexus of State and industry has facilitated coal expansions and simultaneously undermined efforts to reduce greenhouse emissions. This nexus has also sustained a coal optimism in Australian politics that is not commensurate with the trajectory of economic decline of coal globally since 2012. Greenhouse emissions from coal are justified on the pretext that the sector is generating significant economic benefits, and is on the cusp of resolving the emissions problem through 'clean coal' (Pearse et al 2013).[5] Activists saw the promotion of the unproven and energy-intensive Carbon Capture and Storage (CCS) technology as a demonstration of the State–corporate nexus on coal (Milne 2008).[6] Whether through the pragmatism of Labor governments or evangelism of Liberal governments, Australia's political narratives demonstrate faith in new markets for Australian coal.

Anthropologist Hans Baer writes in the paper 'The nexus of the coal industry and the State in Australia: Historical dimensions and contemporary challenges' (2016) that parts of the government effectively became a branch of the mining industry. Political donations and the revolving door between the coal industry and the State emerged as dominant features of the nexus between the State and the coal industry. The Australian Coal Association that promoted coal interests from NSW and Queensland was part of the 'Greenhouse Mafia' and a major donor to both the

Liberal and Labor parties. Another network, the Australian Greenhouse Network (AIGN) included peak bodies from fossil fuel-dependent industries and large fossil fuel corporations whose lobbyists had previously shaped Australia's coal-export policies within the Hawke and Keating administrations. Journalist Guy Pearse observes that 'when carbon lobby recruits are not moving through the revolving door between government and industry, they're often moving side-ways between industry associations in a game of musical chairs' (Pearse 2009, pp. 40–41).

Liberal Prime Minister Tony Abbott's term proved a period of coal idealism in Australia. Abbott declared that 'coal is good for humanity', 'vital for the future energy needs of the world', and 'should not be demonised' (Massola et al. 2014). Resources Minister Josh Frydenberg declared coal mining a moral imperative for Australia to 'help lift hundreds of millions of people out of energy poverty' (quoted in Kelly 2015). Tony Abbott's coal idealism was influenced by the global relations strategy for America's largest coal corporation Peabody Energy, whose 'Advanced Energy for Life' campaign spread awareness about coal's potential to solve energy poverty in the global South. In a close approximation to Peabody's narrative, a report by the right-wing Australian think tank The Institute of Public Affairs (IPA) titled *The Life Saving Potential of Coal* made a case for how Australian coal could help 82 million Indians access electricity (Hogan 2015). The IPA has been found to shape the policies of the Abbott Government (Ghoukassian and Crook 2015).

Although outside the timelines of this research, it is worth noting here that the propaganda that Australian coal will lift South Asians out of poverty has been proven baseless through an independent economic assessment in 2022, of the cost of power generation from coal shipped from Carmichael in central Queensland, burnt in Adani's thermal power plant in Godda, Jharkhand in eastern India, and finally transmitted to Bangladesh on the basis of an overpriced power purchase agreement between India and Bangladesh. The Institute for Energy Economics and Financial Analysis estimates that this power will be available in Bangladesh at almost three times the price at which the Bangladesh Power Development Board sells power to distributors, which can ultimately increase power tariffs for consumers (Nicholas 2022). The assessment lays bare the social, economic and environmental costs of the long arc of Adani's coal from Australia to South Asia. And how power and a State–corporate nexus dominate the global political economy of coal, often at the expense of the public interest.

Returning to the politics of coal in Australia in 2014, Prime Minister Tony Abbott talked up an altruistic vision for Australian coal and free trade:

Australia is poised to turbo-charge a rapid escalation in living standards in India through the supply of affordable and abundant energy such as natural gas, coal, and uranium…that will power the lives of 100 million Indians. It's one of the minor miracles of our time: that Australian coal could improve the lives of 100 million Indians, and it just goes to show what good that freer trade can do for the whole world'.

Abbott quoted in Kenny (2014)

The fetishisation of coal continued through industry lobby groups and under subsequent Liberal governments. The Minerals Council of Australia released a multi-media advertising campaign labelled 'The Little Black Rock' (https://litt leblackrock.com.au) extolling the benefits of coal to the economy as well as its ability to 'now reduce its emissions by 40%', which was criticised by environment groups as 'desperate' and demonstrating an '18th-century' vision (Milman 2015). In another 'stunt', Treasurer Scott Morrison in Malcolm Turnbull's government brandished a lump of coal in Parliament in response to the backlash against coal by green groups saying 'Don't be afraid, it's just coal!' The lump was then passed around amongst Cabinet Ministers (Hamilton 2017).

The Liberal–National Coalition government under Prime Minister John Howard had emphasised the increasing of combustion efficiency worldwide by using 'coals with high calorific values' available in Australia, as an effective way of reducing carbon emissions (Energy Minerals Branch 1999, p. 14). These approaches were a consequence of the industry's influence on the State. The 'no-regrets' approach towards carbon emissions and a confidence in Australia's high-quality coal has remained a key political stand.[7] During the mining boom, successive governments of both the Liberal National and Labor parties in Queensland assumed a 'no regrets policy' approach towards keeping Queensland in the export-coal business, despite the possibility of severe risks to the Great Barrier Reef. Premier Campbell Newman said in response to the 2012 UN Report on the condition of the Great Barrier Reef:

> We are in the coal business. If we want decent hospitals, schools and police on the beat we all need to understand that.
>
> Newman, quoted in Australian Broadcasting Corporation (2012)

Because of the high quality of most of its coal exports, Newman and other premiers have remained confident that 'we have a competitive advantage' in a world where 'coal is going to be needed for many, many decades to come' (Newman, cited in Remeikis 2014). This contradicted a 2012 government survey of Queenslanders on a 'vision for the state beyond the next ten years' that did not feature coal as a choice for Queensland's future economy (Cole 2014). The gap indicated a widening rift between coal's industry-influenced political rhetoric and a democratic vision for Australia's energy future. The aspirations of NSW's Hunter Valley residents for a post-carbon society are demonstrated through iconic actions such as coming together in a 'Beyond Coal' human sign on a Newcastle beach on International Climate Day in November 2006 (Evans 2010).

Risky politics of planning the Galilee Basin (2012–2018)

In 2011, proposals for 120 new coal mines or extensions and a massive rail and port expansion put Australia on the verge of achieving a near threefold increase in coal exports by 2020 (Hepburn et al. 2011). The largest of such proposed coal developments was in the greenfield site of the Galilee Basin in Central Queensland where nine new mega mines with associated rail and port infrastructures were planned. The Queensland government also proposed the expansion of the Abbott Point coal port near Bowen by private developers at an estimated cost of A\$6.2 million to service the Galilee Basin coal mines (Paton 2011).

Carmichael was the fourth of the nine Galilee mega mines to be approved at the state and federal levels in 2014. Alpha, the first to be approved in 2012, is a joint venture of the Indian conglomerate GVK and Australian Gina Rhinehart's Hancock Prospecting and was expected to reach a full production capacity of 30 million tonnes per annum.[8] The mine would require its own rail line to Abbott Point (Greenpeace Australia 2012b).[9] The Alpha mine site lay adjacent to two underground coal projects, Kevin's Corner and Alpha West, with production capacities of 27 and 24 million tonnes per annum (mtpa), respectively. The first was fully owned by GVK and the second held at a 79–21% ownership structure between GVK and Hancock similar to Alpha (Greenpeace Australian 2012b).

The Alpha North and China First sites that lay adjacent to the Carmichael mine areas were estimated to reach full production capacity of 40 mtpa. They were owned by Australian mining magnate Clive Palmer through his company Waratah Coal. The Waratah projects obtained federal approval in 2018, years after Carmichael, and were reliant on the construction of the crucial 400 kilometres North Galilee rail link connecting the mines to the coast, reaffirming the centrality of the Carmichael project to opening up the Galilee Basin (Slezak 2018b).

The proposed China Stone mine by Macmines Austasia of the Chinese-owned Menjin Energy Group equalled the size of Carmichael, with an estimated production capacity of 60 mtpa. The South Galilee Coal project, a joint venture between Australian mining exploration company Bandanna Energy and private equity firm AMCI capital, constituted the smallest of the nine Galilee mega mines, at an estimated total production capacity of 14 mtpa. The Degulla mega mine proposed by the Brazilian corporation Vale, with an estimated production capacity of 35 mtpa, made up a list of five Galilee mega mines that would be larger than any current Australian coal mine (Greenpeace Australia 2012b).[10]

Although Australian governments have maintained optimism about the future of coal through a rhetoric of altruistic resource internationalism, economic assessments have cautioned that new massive mines risk becoming stranded assets due to decreasing coal prices, increasing affordability of solar technology, and international momentum to reduce greenhouse emissions. (Dennis 2015).[11,12] China's coal demand falling in absolute terms after 2012 was regarded as a significant factor in declining sea-borne coal trade (Parker and Chang 2014).[13] The price of Australian coal rose from 2015 not on account of a growing coal sector, but from withdrawal of investment in thermal coal mining, raising questions about Australian coal's long-term viability (Buckley et al. 2018).

The Galilee mines presented an economic risk to Australia's existent major coal mining regions in the Hunter Valley in New South Wales and the Surat and Bowen Basins in Queensland. Proposed Galilee mines going into production would lead to an estimated reduction of up to 115 mtpa of coal output from these regions and would cost up to 12,500 jobs (Murray et al. 2018).

However, Queensland's economic policies optimistically estimated that opening the Galilee Basin would generate a one-third increase in Australia's sea-borne coal trade. Galilee constituted one of the four pillars for Queensland's growth with an estimated A$60 billion in revenue generation and the creation of 15,000 jobs (Queensland Government 2013). The Galilee Basin Development Strategy (2013) waived mining royalties, streamlined land acquisition and fast-tracked project approvals.[14] The State facilitated railways and port infrastructure developments for the greenfield mining region by the declaration of the Galilee Basin State Development Area (SDA) in 2014. Stretching over 100,000 hectares from Abbott Point to the proposed Carmichael site, it was Queensland's largest SDA, dwarfing previous ones by more than a factor of 10 (Lyons 2017a).

The Newman government proposed spending hundreds of millions of taxpayer funds for the Galilee's rail infrastructure (Cox 2015a). It proposed subsidising the expansion of the Abbott Point coal port by spending public money for seabed dredging and dumping dredge spoils inside the Great Barrier Reef World Heritage Area (Market Forces 2014). More spending on mining by the Newman government meant less spending on social infrastructures including hospitals and schools (Dennis 2015).

Globally, climate activists recognised the Galilee Basin as on the 'frontline of expansion' of the coal industry, stopping which should be made a priority to keep global warming below the required 2 degrees limit to meet the requirements of the Paris Climate Agreement (Muttitt 2016). The risks posed by the Galilee mega mines to the global climate, led to them being referred as 'carbon bombs', with the Carmichael mine imagined as a 'line in the sand'. Allowing it to go ahead would risk the world's climate 'tipping over' (Mckibben 2017).

The Carmichael coal mine stood out as the only significant export-oriented project from a new coal basin anywhere in the world (International Energy Agency 2017). This fact underscored the economic risk of developing the Galilee coal region. Hunter Valley in New South Wales had already started facing the impacts of declining coal demand even as the Carmichael coal mine progressed through state and federal approvals. The drop in global coal prices in 2017 forced Australia's major coal producing regions to undertake production cuts (Cox 2015c). Especially after the 2015 Paris Agreement, resonances for withdrawal from coal started coming from within the coal industry, with Newcastle, the world's largest coal port, preparing to diversify beyond coal (Smyth 2017). And Australia's largest coal mining company Glencore committing to cap its coal output (Khadem 2019). The 2017 climate policy of Australia's second largest bank, Westpac, limited its financing of new thermal coal projects to existing coal-producing basins and did not invest in greenfield coal mining (Westpac 2017).

By the end of 2018, the ambition of Queensland and Australian governments to unleash the continent's largest coal frontier remained unrealised. Several other project developers in the Galilee Basin had either been waiting for years or had sold their stake in Galilee by this time. Vale, a Brazilian mining company that was one of the original nine, put its lease up for sale in 2013. The licence for the China

Stone mega mine adjoining Carmichael was quietly withdrawn, putting thousands of the promised Galilee jobs at doubt (Gartry 2019).

The only other projects gearing up for approvals from the Queensland government when the Adani coal mine finally started in 2019 were those that belonged to Clive Palmers' Waratah Coal Company. Shrinking investments in coal projects also affected prospects of expansions at the Abbott Point coal port, with large infrastructure developers such as Land Lease and large mining corporations such as BHP Billiton and Rio Tinto withdrawing as developers due to uncertainties in the global market (South Asia Times 2014).

The coal industry's influence on Australian politics was revealed through the sweeping success of the Liberal–National Coalition in Queensland in the May 2019 Australian federal election. Following the election outcome, the state Labor government granted the last remaining environmental approvals for Adani, allowing the mine to officially commence. During the federal election campaign in Queensland, mining magnate Clive Palmer with high stakes in coal mining in the Galilee funded a misleading advertising campaign against the Labor Party to polarise the electorate (Crowe 2019). The News Corp media's monopoly throughout the region also proved effective in influencing opinions; its newspapers had been mounting a sustained propaganda campaign for the Carmichael mine (Wilson 2018). Australia's most controversial coal mine started despite its economic unviability. The political manoeuvres and the compromising of democracy that was needed to initiate coal mining in the Galilee Basin demonstrated the grip of the coal industry on Australian politics despite its sinking global future.

7.2 Environmental politics of the Carmichael coal mine (2012–2018)

As the largest proposed coal mine, and the one seen as unlocking other mining operations in the Basin, the Carmichael project triggered all the environmental risks that concerned scientists and ENGOs. To its opponents, the sheer scale of the combined impacts from the Carmichael mine, rail and port project and calculated emissions from burning its coal implied it should never be approved (Waters 2015). Coal from the Carmichael mine alone would create an estimated average annual emission of 79 million tonnes of carbon dioxide, a figure greater than the average emissions of a small South Asian country like Bangladesh or Sri Lanka (Taylor 2015).

The Independent Expert Scientific Committee (IESC) on coal seam gas and large coal mining developments to the Queensland and Federal governments expressed concerns about the impacts of the Carmichael mine's water use on the GAB. The Committee said that scientific uncertainty around the architecture of the GAB and its connected aquifers that hydrologists had not yet fully mapped made it difficult to ascertain the full water impacts of the Galilee mines (IESC 2013). Uncertainty prevailed over assessing the impacts of the Carmichael mine on the source of the nationally important wetlands the Doongmabulla Springs complex that was sacred to the W&J people. An IESC assessment warned that as not enough was known about how coal seams connected to the GAB, the true impacts of mining on such a scale could not be determined. It also stated it had little confidence in Adani's modelling.

The better-known hydrological risks of the Galilee projects pertained to the quantity of water required for coal mining. The report *Draining the Lifeblood* by Lock the Gate estimated that the nine mega mines would need 2000 gigalitres of water cumulatively over their lifetimes, a volume higher than two and a half Sydney Harbours. This risked an estimated 400 bores on surrounding farming properties and water supplies for nearby towns of Alpha and Jericho by drawing down the underground water table (Lock the Gate 2013). A former Queensland Water Bureaucrat explained that:

> The pits of the mines are lower than the actual aquifers, so they will drain out. One of the other concerns we have and we think there is a fairly high potential for some impacts on the recharge of the Great Artesian Basin…So far there has been no study done and the approvals are still going ahead…What we should be looking at is a stop on any further approvals until we get a cumulative study done of the impacts.
>
> Crothers (2013)

The Carmichael coal mine would draw an estimated 12 billion litres of water every year (IESC 2013). And a total of 270 billion litres over its lifetime operating at full capacity (Mckeown 2018). The scale of the project's water needs risked causing a one-metre drop in the water table, and draining the aquifers connected to the GAB along with the springs and wetlands that feed off them, during periods of drought. Adani's own experts admitted that even a temporary drying of the Doongmabulla Springs risked causing irreversible ecological changes (*Adani Mining Pty Ltd v Land Services of Coast and Country Inc. & Ors 2015*). Contaminated water from the mine risked polluting creeks and rivers (Mckeown 2018). Destruction of the Springs would threaten the cultural survival of the W&J people and the possibility of downstream effects on other traditional owners with stories and connections through the Water Spirit ancestor would risk fragmenting the 'seamless web of cultural landscape in Aboriginal law and lore' (Burragubba 2018, p. ix).

The project was also dogged by controversies over actual taxes and local jobs it can generate, with legal cross-examinations revealing that the company inflated claims by 270% (Ludlow 2015). Legal challenges by civil society groups to the Carmichael mine and rail exposed the full extent of the impacts of the mine on groundwater and biodiversity, from the burning of Carmichael's coal on global emissions, risks to the Great Barrier Reef from port and coal expansions, and the economic unviability of the mine that the corporation had not disclosed (Environmental Law Australia 2016).[15]

The Carmichael mine and rail proposals emerged and gathered shape during the federal Labor leadership of Prime Minister Julia Gillard (2010–2013) and the Queensland Labor premiership of Anna Bligh (2007–2012). They received approvals under the Liberal Prime Minister Tony Abbott (2013–2015) and Liberal National

Party (LNP) Premier Campbell Newman (2012–2015). It received a second federal approval under Malcolm Turnbull. The project was helped along against growing civil society opposition and financial withdrawal by major investors by successive Liberal Prime Ministers Malcolm Turnbull (2015–2018) and Scott Morrison (2018 onwards) and successive terms of Labor Premier Annastasia Palaszczuk (first term from 2015 to 2017 and re-election in 2017).

A combination of investor and shareholder activism to divest from the Adani project, grassroots mobilisations exerting political pressure at the electoral level in both Labor and Liberal seats around the country, legal actions, alliances with disaffected farmers in Central Queensland sustained a five-year resistance, caused critical delays, and proved effective in significantly reducing the size of the ambitious Carmichael project. A significantly diminished Carmichael project finally commenced after the decisive Australian federal election in May 2019. In the following weeks, the Queensland Labor government controversially approved last remaining environmental and water plans. The following four subsections chronologically trace the political championing of the Carmichael mine at the state and federal levels and account for how the environmental and political conflict over the Carmichael mine shaped Australia's largest environmental and climate mobilisation and a farmers' campaign in Central Queensland.

State and federal Liberal governments' championing Adani

The project received its first environmental approvals in 2014, with the Queensland Coordinator General – responsible for managing assessments and approvals of infrastructure projects including their social and environmental impacts – recommending that the state government approve the project with an 'extensive and wide-ranging' set of conditions to ensure environmental protections under the *Minerals Resources Act 1989* (QLD MRA) and the *Environment Protection Act 1994* (QLD EPA) (Queensland Government 2014). The Federal Environment Minister similarly approved the project under 'the absolute strictest of (36) conditions', in fact the 'strictest conditions in Australian history' under the *Environment Protection and Biodiversity Conservation Act 1999* (EPBC) (Minister Hunt quoted in Hasham 2015).[16]

Queensland's fast-tracked approval process failed to consider the IESC's water advice and to scrutinise the company's track record of environmental harm in India (Environment Justice Australia 2015).[17] The Federal approval for Carmichael was immediately followed by the first legal challenge against the project, with the Mackay Conservation Group (MCG) seeking a judicial review of the Environment Minister's decision on account of the mine's impacts on two vulnerable species – the Yakka Skink and Ornamental Snake – under the EPBC Act. The Federal Court of Queensland set aside Adani's federal approval, requiring the Australian Environment Minister to follow procedural requirements in reissuing an approval (Hepburn et al. 2015).

Prime Minister Tony Abbott labelled environmental litigations against the Carmichael mine as a 'green sabotage' and attempted to repeal section 487 of the

EPBC Act that contains provision to seek judicial reviews, on grounds of protecting jobs (Hepburn 2015).[18,19] Government narratives also presented Carmichael as a solution to India's historic poverty and a milestone in Australia–India relations. This message was reinforced through the News Corp media in an article evocatively titled 'Gautam Adani's dream to light India's darkened nights', in *The Australian*, a national daily with a right-centre bias:

> Last year, Modi promised in his Independence Day speech to provide electricity to the estimated 18,000 villages still in the dark – with a combined population of 300 million people – by 2019. Playing a significant part in delivering this transformation is Adani, and its plans for Australia. Its founder, self-made billionaire university dropout Gautam Adani, wants to build the A$16 billion Carmichael mine in Central Queensland – producing up to 60 million tonnes a year of coal – to help satisfy the growing electricity demand in India.
>
> McKenna (2016)

Political support was also justified through claims that high-quality Galilee coal when burnt in India will improve air quality and lower emissions. A second litigation against Carmichael challenging Queensland's environmental approval brought by the Coast to Country local group challenged such moral overtures around the project (*Adani Mining Pty Ltd v Land Services of Coast and Country Inc. & Ors* 2016). An Indian environmentalist from the Mumbai-based Conservation Action Trust testified that 'coal from Carmichael, when burnt in Adani's power stations in India, will damage the health of the Indian rural poor and the land and water on which they depend for their livelihoods. And they still won't be able to access the electricity generated' (quoted in Environment Justice Australia 2014).

Ironically, a report submitted on behalf of Adani on the case at Queensland's Land Court refuted the high-quality claim for Carmichael's coal, confirming it was a high ash and low-energy product. The publicly claimed 10,000 jobs figures from the project were also disproved through Adani's own modelling presented in the case which revealed the figures as closer to 1500 (Adani Mining Pty Ltd v Land Services of Coast and Country Inc. & Ors 2015).[20] Such findings challenged the government's rhetoric on the economic significance of Carmichael. Queensland had already declared the Carmichael coal and rail plan as a significant project that could generate over 11,000 jobs in 2010 (SBS News 2016). The evidence presented in court indicated that the government had put public money at risk for opening up the Galilee Basin that could fail to deliver what the State and corporation were promising.

During his Australian visit during the G20 Summit in 2014, Indian Prime Minister Narendra Modi reportedly cleared a funding roadblock for Carmichael through a memorandum of understanding (MOU) for an A$1 billion loan from India's public bank the State Bank of India (ENS 2014).[21] The Newman government matched this by announcing an A$450 million investment for the rail project (Dennis 2015). The Newman government's championing of the Carmichael project and subsidising it was contradicted by the concerns of the Queensland Treasury over Adani's high levels of debt, unclear corporate structure and use of offshore tax havens (Cox 2015b). In February 2015, Premier Newman lost to Labor after a single term in one of the biggest swings in Australian politics, with mining and accountability proving key factors in the electoral defeat (Dennis 2015).[22]

Queensland Labor's first term and special favours for Adani

Queensland Labor committed to not subsidise the Carmichael project, requiring the Galilee coal mines to be commercially viable. Although the ideological rhetoric around coal dimmed down during Labor Premier Annastacia Palaszczuk's term, special favours still flowed to the Carmichael project. The legal challenge in the Land Court concluded in December 2015 with a ruling that an environmental authority and mining lease can be granted to Adani subject to further conditions related to monitoring the impacts on the Black-Throated Finch species endemic to the area (Environmental Law Australia 2016). The court verdicts in the first two cases had favoured Adani, indicating that the 'the system is designed to expedite' over a range of social and economic concerns (Lyons 2017a).

The third legal case against Carmichael was launched in November 2015 following its federal reapproval. In *Australian Conservation Foundation Incorporated v Minister for the Environment and Energy [2017] FCAFC 134*, the national ENGO Australian Conservation Foundation challenged that the mining and burning of coal from Carmichael was inconsistent with Australia's obligation to protect the World Heritage Listed Great Barrier Reef under the EPBC Act (van Vonderen 2015). Heeding the court challenge, Commonwealth, a major Australian bank, withdrew from its financial advisory role to the project (West and Cox 2015). Adani reportedly asked Prime Minister Malcolm Turnbull to introduce legislation to prohibit activists from challenging approvals, alleging that judicial reviews had delayed the project by one and a half years and made investors unwilling to associate with it (Cox 2015c).

Queensland steered the project toward completion regardless of investor withdrawals. In February 2016 following the Land Court verdict, Queensland granted Adani's final environmental authority – the licence needed by the proponent to commence a project with significant environmental impacts – with 140 conditions under Queensland's *Environment Protection Act 1994* (Agius 2016). Carmichael was declared a critical infrastructure, a status that had been granted only four other times and never to a private commercial development (EDO Queensland 2016a).[23] The manoeuvre allowed Queensland to fast-track water assessments and exempt Adani's water use from public scrutiny (EDO Queensland 2016b). Queensland-based ENGOs saw this as a problem of the State's accountability. The former coordinator of the central-Queensland-based Mackay Conservation Group told me during an interview in the ENGO's office that:

> The State Development and Public Works Act (under which Adani was granted critical infrastructure status) came under Bjekle Petersen; it was designed in hard times to build public interest infrastructure, even if local communities were

affected. But now it is hollowed out and private operators like Adani benefit. It comes back to accountability issue, to democracy.

Interview (28/10/2017)

The local group Land Services Coast and Country launched the fourth legal challenge, against the granting of the environmental authority for Carmichael, arguing that the Queensland government had not considered the requirement for sustainable development under the state's Environment Protection Act (Environmental Law Australia 2016). In April 2016 Queensland approved Carmichael's mining licences with 200 'strict conditions', emphasising that the 'benefits (from Carmichael) outweigh those challenges' (Australian Broadcasting Corporation 2016). Although state and federal governments cleared all approvals by the end of 2016 by giving a go-ahead to the rail line and mine site construction camp, Adani's April 2017 deadline for starting the project appeared unlikely given the pending third and fourth legal cases (Mitchell-Whittington 2016).

After facing public criticism, Adani changed its initial plan of locating all project offices in the state capital Brisbane, which is located in southern Queensland. Central Queensland was economically prioritised and Townsville was announced as Adani Australia's regional headquarters and remote operations centre (O'Brien and Mellor 2016). The coordinator of the Mackay Conservation Group provided a ground-level account of how Adani influenced city councils in Central Queensland during an interview in the ENGO's office:

Adani has been spinning their tale on the ground, working very intensely with the local councils over the last two years with all the national level resistance going on. With local mines, refineries closing, local government looked at Adani as the saviour. And Adani has been stringing councils along; 'who will get the headquarters?'. There is a soap opera playing out in the local media.

Interview (28/10/2017)

In March 2017 Premier Palaszczuk led a delegation of Mayors from regional centres to Mundra in Gujarat, India, for a tour of Adani's coal port, thermal power plants and the special economic zone. A protest-delegation of two ENGO representatives, a Reef tourism operator, and a Queensland farmer pursued the government delegation to India. The protest team delivered an open letter signed by 90 prominent Australians and traditional owners that asked Gautam Adani to abandon the Carmichael project to the corporate headquarters in Ahmedabad, Gujarat (Cousins 2018). In the same month, various strands of anti-Adani environmental activism, including grassroots and constituency-based opposition, divestment campaigns, and national political advocacy and legal activism, officially came together under the Stop Adani banner. The movement website summed up its approach to activism as 'We must generate unprecedented, relentless and organised political pressure, through targeted community organising and sustained creative mobilising' (www.stopadani.com).

In April 2017, Queensland granted Adani a licence to extract unlimited groundwater (Hannam 2017). While acknowledging that the mine would impact underground water levels, the government defended the licence on the basis that it had set an extensive set of 270 conditions in the approval to protect groundwater (Davison 2017). However, experts pointed out that conditions had not set volumetric limits on water withdrawals or triggers to halt mining operations when required.[24] Adani's groundwater plan ignored the scientific uncertainty of the Artesian Basin and risked the Doongmabulla wetlands being completely drained by mining (Robertson 2018a).[25]

The water licence was granted in the midst of what has been described as the worst drought in living memory in Central Queensland (Australian Broadcasting Corporation Rural 2018). A petition started by a third-generation grazier from Longreach in outback Queensland demanded to 'Rescind Adani's Unlimited Water Licence and Support Aussie farmers'. Pointing to the prolonged Queensland drought, the petition to Premier Palaszczuk asked 'nearly 90% of Queensland is currently drought declared, so why are we giving an Indian billionaire access to unlimited groundwater for a new coal mine?' (Change.org 2017).

Queensland compulsorily acquired prime agricultural land for the Adani railway, sparking concerns about the division of rural properties (Elliot 2017).[26] Another petition launched by a Galilee grazier in October 2017 attempting to stop the compulsory acquisition of grazing land for Adani's rail line drew 44,000 signatures (Smith 2018). Farmers for Climate Action (FFCA), a new national alliance between farmers and environmentalists initially formed in 2015, became a critical voice against massive water allocations to the Galilee mines and the lack of federal action on climate during this time, as one of the most severe droughts affected farmers and graziers in Central Queensland (FFCA 2018).

Despite a clear election commitment against giving handouts to Adani, the Labor government sent contradictory signals on the matter of royalties from the Carmichael mine. A new royalty deal was issued with a clarification that the Queensland cabinet had 'unanimously agreed [that]…Adani's Carmichael mine will pay every cent of royalties in full…there will be no royalty holiday' (Caldwell 2017a). News reports however contradicted this assertion, indicating a reduced holiday on royalties for six years (Robertson 2017b). An Australian Greens' analysis showed the deferral would cost A$253 million over five years (Caldwell 2017b).

Funding for the Carmichael project remained unresolved despite Adani's final investment ruling announced after the royalty deal, owing significantly to activist campaigns. The divestment campaign arm of the Stop Adani movement was led by 350.org Australia and the new divestment activism group 'Market Forces'. Market Forces and the W&J traditional owners undertook a tour of international banks and investors across the US and Europe in 2015 (Market Forces 2015). The 'National Day of Divestment Action' in October 2016 that focused on the 'Big 4 Australian Banks' to stop funding fossil fuel projects saw people from across 13 cities around Australia participating in a mass protest by cancelling their bank memberships (Market Forces 2016).[27] The divestment campaign successfully convinced all four major Australian banks to withdraw from the Carmichael project (Robertson 2017a).

The financial situation raised strong possibilities of political handouts, particularly a A$1billion Federal loan to the Adani rail project through the federally administered Northern Australian Infrastructure Facility (NAIF) fund (Koziol and Wroe 2016). Queensland Labor however repeated its previous election stand of not publicly funding the project.[28] Following re-election in December 2017, Premier Annastacia Palaszczuk vetoed the federally administered and contentious A$1 billion NAIF loan for Adani's rail project (Australian Broadcasting Corporation 2017).[29]

Queensland Labor's second term; equivocating on Carmichael

The period after re-election heralded a change in Labor's stance towards the Carmichael mine if not towards coal mining per se, even as the state Premier continued to handhold Adani such as through the reported offer to fund A$100 million for road access to the Carmichael mine (Robertson 2018b), Federal Labor leader Bill Shorten spoke about diversifying Central Queensland's economy through local infrastructures to deliver real jobs and warned against the promise of 'fake jobs' from a project that was yet to take off (Brunker 2018). Following the Paris Agreement, Federal Labor raised concerns about investment risks posed by the Galilee mines, and about the Carmichael mine being the only coal project from a greenfield region anywhere in the world (Butler 2018).

In preparation for the 2019 Federal election, Labor considered inserting a climate trigger into the EPBC Act that could be used to prevent new coal mines, or even provide a retrospective negative assessment of an existing mine, on the basis of its emissions (Murphy 2018a). This would act in a similar fashion to the water trigger that Labor inserted into the EPBC Act in 2012 to contain the effects of CSG mining on groundwater. Australia's national environmental law currently does not require the emissions (from coal-burning) of coal mines to be considered for approvals.

Labor faced pressure from the Green Party to show a stronger stand on climate change and the issue of the Carmichael mine (Wahlquist 2018).[30] The Stop Adani movement that formally started in March 2017 had also effectively made the Carmichael mine an electoral-level issue nationally. However, Trade Unions posed the biggest challenge to Labor's policy of tackling coal expansions.[31] The Construction, Forestry, Mining, Maritime, Mining and Energy Union (CFMEU), Labor's biggest internal stakeholder, warned that a hardline approach against the Carmichael coal mine could alienate Labor's blue-collar base by sending a signal that Labor is against 'any new coal' (Murphy 2018b).

Labor attempted to balance both sides through the official position that the Carmichael mine could go ahead on its own merit. However, even though at the federal level the Party equivocated on the controversial Carmichael coal mine, in Queensland, Australia's largest coal state, the Labor government quietly revived plans to expand coal mining in the adjacent Bowen Basin (Smee 2018a). This incident demonstrated the ineffectualness of Labor's balanced approach against the reality of the clout of the coal industry.

Federal Liberal–National Coalition government's political rescue for Carmichael

After prospects for a NAIF loan for Carmichael faded, the Federal Liberal government attempted to rescue the financially unviable project by attempting to broker international financing at Adani's behest. Freedom of Information documents revealed that the Department of Foreign Affairs and Trade wrote to the Chinese Embassy and met officials from Korea's Export-Import (EXIM) Bank (Slezak 2018a). Chinese banks reportedly ruled out involvement with the project owing to financial unviability and following weeks of targeted campaigning by ENGOs (Needham 2017).

In 2018 Adani proposed a separate water project through the 'North Galilee Water Scheme' to pump 12 billion litres of water from the nearby Sutton River for non-extractive activities such as washing coal (Hasham 2018). The company avoided a full impact assessment for the scheme under the water trigger of the EPBC Act, a move that angered local farmers struggling during one of the worst droughts (Lock the Gate 2018).[32] The Federal government issued a rushed approval for the water scheme weeks before the May 2019 federal elections that was challenged by water experts and ENGOs (Long and Slezak 2019). The Australian Conservation Foundation launched a fifth legal challenge against this approval process as it failed to take 'thousands of public submissions into account' in violation of the EPBC Act (Australian Conservation Foundation 2019). Although the federal government conceded that due process had not been followed (i.e. ACF won), in a similar pattern to earlier litigations against approvals issued without following due process under the EPBC Act, the Environment Minister reissued the approval at a later date stating that it had now taken public submissions into account.

By 2018 the Adani Group significantly scaled down its mine and rail projects. The mine was scaled down to a 10–15 million tonnes a year self-financed project at A\$2 billion (Talukdar 2018). The new rail proposal halved capital costs to A\$ 1 billion by reducing the railway to 200 kilometres to connect with Aurizon's existing rail to Abbott Point (Ludlow 2018). Adani's design and engineering partner AECOM disengaged from the rail project in May 2018, after it was unable to gain access to key sites to progress its design work (Smee 2018b). A volunteer-run activist outfit Galilee Blockade mounted a sustained campaign including blockades and shareholder activism against Adani's principal engineering and construction partner Downer EDI to disrupt mine construction.[33] Adani finally scrapped its agreement with the construction firm around the time when the NAIF loan was vetoed (Sydney Morning Herald 2017). Persistent direct disruptions by Galilee Blockade volunteers had also contributed to AECOM's demobilisation from the Carmichael coal project.

The Adani Group also significantly scaled down its Abbott Point expansion, proposing a conveyor and transfer tower at the existent Terminal One to increase coal throughput by 10 million tonnes a year.[34] The A\$1.8 billion investment in the

Coal Terminal is considered Adani's key asset in Australia and the reason why the conglomerate has not walked away from the Carmichael project (Smee 2018d). However, with the Australian bank Westpac declining to refinance Adani's A$2 billion loan on the Port in December 2017 as per its new climate policy, the viability of Adani's entire operations came into question (Slezak 2017b). Abbott Point also faced the risk of a stop order being issued on its operations if the concerns of the Juru traditional owners from North Queensland could not be met (Smee 2018c).[35]

The federal elections in May 2019 and its ramifications for the Labor Party in Queensland turned the tide on the prospects of the Carmichael coal mine. As opposed to federal Labor's pre-election position that the mine should proceed on its own merit, the Queensland Labor government now categorically stated it was 'fed-up of waiting' and hastened Adani's pending clearances (Bavas 2019). Final environmental approval was granted in June and Adani commenced Australia's most controversial coal mine. The multi-pronged campaign against the Carmichael mine moved toward a greater emphasis on peaceful blockades in Queensland as all other tactics failed to stop the coal mine. Camp Binbee, an 18-month long blockade camp located at 45 minutes' drive west of Bowen, began to swell with volunteers of all ages and various walks of life at the time of the final decision, in preparation for blockading the port, rail and mining activities (Krien 2019a).

The difference in approach towards the Adani mine between Liberals and Labor since 2017 was far outweighed by the consistency of their support. This can be attributed to the company's relations with leaders in both major parties (Readfearn 2015).[36] The State–corporate nexus was strengthened by the culture of revolving doors between politics and mining. Adani's lobbying firm was led by former chiefs of staff in both Labor and Liberal offices at the state and federal levels (Robertson 2017d).[37] It helped Adani get 'pretty much everything it wanted through an extraordinarily intense campaign', that made more than double the number of contacts with the government on behalf of Adani compared to any other client (Long 2017).[38] Relations were further strengthened through political donations: disclosures lodged with the Australian Electoral Commission showed that Adani donated A$49,500 to the Liberal Party and A$11,000 to Labor in 2013–2014 (Cox 2015a).

The nine-year period, from when Adani first purchased the coal tenements in the Galilee Basin to when the company could finally start the project, spanned one of the most momentous periods in Australia's coal and climate politics. The Carmichael mine both united and divided Australian politics and led to an outpouring of civil society protests.

Activisms, narratives and politics of Stop(ping) Adani

Legal challenges continued even after the project started in 2019. The sixth case by ACF in 2020 against the federal government, for Adani's new water project for the Carmichael mine not having been assessed under the water trigger in the EPBC Act for significant social and ecological impacts, was successful in May 2021 (Australian Conservation Foundation Incorporated v Minister for the Environment

2021). However, despite bringing a total of 10 legal challenges between 2014 and 2020 jointly against the Carmichael and adjoining Alpha coal mine and the expansion of the Abbott Point port related to these projects, environmental groups, and farmers were not able to stop the project from commencing or prevent its main environmental impacts.[39] It reflects how environmental laws at both the state and federal levels are structured to promote mining developments instead of being equipped to protect communities from various environmental risks and from climate change.[40] Repeated legal actions however delayed the project's timelines, established critical arguments such as the project's economic unviability, and countered false claims, particularly around its jobs. The legal challenges also challenged the moral case justifying Australian coal, by arguing for inter-generational responsibility for protecting the Reef, jobs, and future generations (Solicitor, Queensland Environment Defenders Office interview 16/09/2017).

Outside the courts, a multi-pronged anti-coal movement including divestment campaigns and mobilisations, and peaceful blockades at strategic sites was already underway by the time the Carmichael mine received environmental approvals. After the success of the divestment and 'Fight for the Reef' campaigns in arresting massive port expansions in Queensland as discussed in Chapter 6, the question on the anti-coal movement's mind was 'should we say we have won?' (Stop Adani Campaigner interview 27/10/2017). However, by 2016, subsidies and political championing by state and federal governments revived the economically unviable Carmichael project and forced a rethinking of strategies. The Stop Adani Campaigner said during the interview in Mackay, central Queensland:

> In 2016 we continued our overall strategy of divestment from fossil fuels, we were still not putting energy into organising an anti-Adani mobilisation. By the end of 2016 we started getting very scared, what if the NAIF funding came through, the W&J lost their case, and the mine went ahead?
>
> Interview (27/10/2017)

Forty environmental and climate groups from the national, state and regional levels formally united under the banner 'Stop Adani' with the aim of stopping the Adani mine as a first step in moving Australia away from exporting coal. The Stop Adani website registered over 100 new local 'Stop Adani' groups within the first 3 months of its starting in March 2017 (www.stopadani.com.au). Stop Adani's ability to respond to rapid political changes on the Carmichael issue was owing to the flexibility of volunteer-driven local action groups and new engagement models that characterised the movement. The strategist in the climate movement told me during an interview in Sydney that:

> We threw an open invitation to self-start local action groups. They did the heavy lifting in a way big groups couldn't. We stopped Westpac during the week of action in April 2017; they immediately regrouped and went after Commonwealth. Local groups are the tip of the spear and can change really quickly in a way big NGOs cannot. Volunteers have been getting on Skype

every fortnight for updates and discussions. They show a detailed understanding of the company. They discuss how Adani might do it, get money, approvals, etc, and then decide what we can do. We built a 'Directed Network Campaign'; this model of engagement was used against the Tar Sands Keystone XL Pipeline. It is very relevant for Australia. We have four elements in our engagement strategy: a shared theory of change, a shared narrative, focus and discipline, and open grassroots approach and network. Campaign actions are not top down like it used to be with big NGOs who used to say 'come to a rally'.

Interview (18/06/2018)

By mid-2017 red and white Stop Adani signs on black backgrounds had become prominent across inner city areas of the major metropolitan cities along the east coast – Melbourne, Sydney and Brisbane – where the majority of the movement's supporters and volunteers lived. Town hall meetings were organised across the major metros to bring together the local arms of the Stop Adani movement. At one meeting in Sydney that I attended in September 2017, a packed hall of more than 500 volunteers and activists cheered former Greens Senator Dr Bob Brown as he vowed that Stop Adani would grow into the 'the largest movement that Australia has ever seen, bigger than the Franklin' (Meeting notes 6/09/2017). The daylong session was dedicated to developing action strategies for local groups to pressure members of parliament.

At a Stop Adani protest in October 2017 in Sydney that I attended, paper mache heads of Prime Minister Turnbull and Gautam Adani were seen walking together

Figure 7.4 Human sign on an Australian beach on Stop Adani's 'Day of Action'.

Source: Photo by Stop Adani.

holding a bag with 'Your Taxes A$1 billion' written on it. The protest drew a couple of thousand supporters and was part of Stop Adani's 'Big Day of Action', a coordinated day of 60 anti-Adani demonstrations across Australia. Images and drone footage compilations from the national anti-Adani day showed thousands of people gathered across a variety of landscapes – beaches and coral islands, urban parklands, farmers' paddocks, and the red interiors of the continent – in 'Stop Adani' human signs (www.stopadani.com/actionday). Extensive social media broadcasts, a prominent logo and professionally managed events like the Big Day of Action proved successful in reiterating and proliferating the movement's message.

The challenge of stopping the Carmichael project in the face of deep political support compelled ENGOs in the Stop Adani movement to cooperate at a level they had not attempted before. The Climate Movement Strategist said:

> Now we know that the only way we can do this is to make governments stop it because Adani will not walk away. We have to shift the politics. Both sides are totally opposed to us. So our strategy was to put so much pressure that they have to relent. Everything we did, the government thought was impossible. They said we do not have power. We have never done things on this scale before; we have 40 NGOs on the phone every week. There are others doing things at the state level, also locally. This way of working has also helped big groups. We strategists act like the action engine room, while other groups have found their niche in the network. An oceans focused campaigner makes the links between coral deaths and coal, a conservation focused campaigner focuses on federal policies and threatened species. Youth organisations and local groups do rapid responses and snap actions; there are others working on financial strategies. GetUp! is increasing its core capacity to reach voters on a mass scale.
>
> Interview (18/06/2018)

Between 2017 and 2018, Stop Adani built a national network that exerted political pressure in strategically relevant electoral constituencies, aiming to shift the politics on the issue from the ground up. Through targeted mobilisations in Liberal Party electorates, Stop Adani and GetUp! built-up conflict within the Liberal Party on the issue. A particular case in point is the movement's mobilisation in Malcolm Turnbull's electorate of Wentworth in Sydney's eastern suburbs during the October 2018 by-election following his ousting as Prime Minister after a leadership challenge. The build up to the polling day included weekly door-knocks and GetUp! call-outs to 100,000 phone numbers in Wentworth (Stop Adani Campaigner interview 27/10/2017). Wentworth, a safe Liberal Party seat since federation, was won by an independent candidate with a climate commitment. Analysis by the Australia Institute (2018) indicates that the community campaign by Stop Adani played a role in causing a historic swing of votes away from the Liberal Party.

GetUp! started offline community organising so as to exert electoral-level pressure more effectively on the issue of the Carmichael mine, using the tactic of a 'seat-by-seat anti-Adani Pledge for Members of Parliament', where supporters were asked to meet their local representative and hold them accountable on the issue (350 Australia CEO interview 10/10/2017). It continued using its online organising strategies to generate pressure during key moments for the Stop Adani movement. The Climate Movement Strategist told me:

> When the federal government was going to give the A\$1 billion loan in 2016, GetUp! used strategy and technology to hold mass conversations with voters to shift politics and change voters. They use a calling tool – they have phone numbers of at least half the voters in the electorate. And they have a script in the calling app.
>
> Interview (18/06/2018)

Stop Adani brought consistency to various tools and tactics that had become a regular feature of a growing anti-coal movement in Australia. It established an organising model driven largely by grassroots activism and local action groups and built up a national-scale movement through it. It generated new groups that strengthened grassroots and local and constituency-based mobilisations. The Stop Adani Campaigner said:

> We had 100 registered Stop Adani groups, 65 of them were strong. The next question was 'how do we keep the momentum'? The group Tipping Point acts like an organising network; its raison d'etre was about understanding what is a grassroots model that can spread through the cities. It is a national project of the Friends of the Earth Australia. The Stop Adani campaign also seeded many constituency-based groups such as Divers for Reef Action. And then there's the Pacific Climate Warriors who are mobilising in the Western Sydney Suburbs'.
>
> Interview (27/10/2017)

The 2000-strong membership of the Farmers for Climate Action (FFCA) joined the Stop Adani alliance in 2017 to collectively protest Queensland's allocation of free water licences to Adani (Slezak 2017a). They presented petitions against the water licence and the compulsory acquisition of farmland for the Adani rail project to all sitting members of the Queensland parliament (Kippen 2018) and called for a sustainable regional plan for Central Queensland that protected against the social and environmental impacts of massive coal projects (Smith 2018). An open letter from FFCA addressed to rural and regional members of the federal parliament challenged the attack on renewable energy by the National Party, which is the regional conservative Party that represents farmers:

> The motion against renewable energy carried at the recent Nationals federal conference and ongoing political opposition to a clean energy future demonstrates that the Australian Parliament is ignoring a core conservative constituency and your long-term supporters: those of us who feed and clothe Australia and the world.
>
> www.farmersforclimateaction.org.au

Figure 7.5 Stop Adani divestment campaign action outside Commonwealth Bank in Brisbane, Queensland.

Source: Photo by Stop Adani.

Although the movement could not stop the Carmichael project, it made a national issue of the power of coal over Australian politics through exposing the nexus between Adani and state and federal governments. The Sunrise Project member said during the interview in Sydney:

> Stop Adani galvanised people in a way that has surprised us. I am continually surprised at who has come out of the woodwork to support us. We had first told a larger story of needing to move away from coal by saying mining coal at Carmichael is harmful for the Reef and is responsible for coral bleaching – but the dots had not been joined. In a way this more direct message was more effective. The Adani project became a symbol for coal and we won the narrative.
>
> Interview (20/10/2017)

Consequently, the words 'Stop Adani' became shorthand for civil society's demand to break Australian politics' affinity towards coal. Beginning with the climate, water and environmental impacts of the largest proposed coal mine in the Southern hemisphere, civil society's anti-Adani narrative grew in dimensions to reflect concerns about coal's deleterious effects on Australian democracy. The Stop Adani movement operated within two sharply contradictory realities in the post-Paris

Figure 7.6 Stop Adani Rally at Parliament House in Canberra.

Source: Photo by David McIlroy/Stop Adani.

period: the global need for a rapid phase-out of coal and the desperate politics of Australian governments to develop an entirely new coal region, despite a global withdrawal from coal. It served to delegitimise coal at various levels – the social, political, ecological and economic.

In summary, the public participation in anti-coal activism through Stop Adani's volunteer-driven actions, the formation of new and niche activisms, political and climate advocacy by Central Queensland farmers, and the movement's multi-pronged strategies offer insights into how environmentalism's anti-coal politics shifted ground from its earlier thrust of 'saving the Reef to end coal' to directly confronting Australia's 'coal business' through making Carmichael Australia's most contested coal project.

7.3 Land rights politics of the Carmichael coal mine (2010–2018)

In 2004, the Wangan and Jagalingou traditional owners' Native Title Claim was registered over 30,000 square kilometres of land in the semi-arid Galilee Basin in Central Queensland under the *Native Title Act (Cth) 1993* (NTA). Native title registration gives claimants certain procedural rights in relation to activities such as grant of mining leases that have the potential to disrupt Indigenous law and culture by affecting the land and environment. Half a decade later, the Queensland government began implementing its vision to develop the Galilee Basin for coal. The W&J's claim area enfolded lands that came under Adani's coal tenements and mining leases after 2010. The horizontally integrated Carmichael rail, port and mine proposal crossed the traditional lands of four Aboriginal nations, the W&J, the coastal Juru, the Birriah, and the Jangga people (West 2015).

Negotiations for consent for mining and an Indigenous Land Use Agreement (ILUA) commenced in 2010 between Adani and a W&J 'Applicant' that officially represented the bigger Native Title Claim Group in the negotiations.[41,42] The W&J contended that Adani negotiated in bad faith during the stipulated six-month period by taking advantage of the coercive power of the native title system.[43] They submitted to the UN Special Rapporteur on the Rights of Indigenous People that Adani used a 'divide-and-conquer' approach and mistreated Indigenous rights (Wangan and Jagalingou 2015, p. 18).

An ILUA was a prerequisite for works associated with the project, and to secure a 2750-hectare area of traditional land referred to as the surrender area, over which native title would be removed for mining related infrastructure including an airstrip, workers village and a washing plant (W&J 2018a). Without an ILUA the state government would be forced to extinguish native title and compulsorily acquire the surrender area to issue mining leases (Robertson 2016).[44] Without an ILUA Adani would be challenged to raise capital investments as most major banks comply with the Equator Principles for environmental and social risk management, and refuse to invest in projects without consent from Traditional Owners (Scambary 2013). It could also expose Adani to the financial risk of compensation claims from native title groups in the future. Adani was able to secure ILUAs with other traditional owner groups along the rail and port corridor (West 2015). However, in the case of the W&J, the ILUA was both strongly contested and created deep internal conflicts.

The W&J Family Council, a representative body that decides on all matters outside native title claim, refused consent for the Adani mine, becoming the first native title group to say an outright no to mining (Lyons et al. 2017a). Their discontent was noted as having been driven by the inadequacy of benefits and jobs proposed in the ILUA and the company's disrespectful and non-transparent dealings.[45] Independent analysis showed that the Adani deal was one of the worst for Traditional Owners (Meaton 2017). The deal offered compensation at less than half the industry average (Robertson 2017e). Three-quarters of the stated economic benefits were contingent on jobs, raising concerns about the agreement's feasibility given the negative outlook for coal's future and Adani's exaggerated jobs claim. The ILUA was struck down on three separate occasions at bona fide meetings of the W&J Native Title Claim group (Lyons et al. 2017a).[46]

Adani brought proceedings before the National Native Title Tribunal after each of the two rounds of failed negotiations in 2013 and 2015. Both times, the Tribunal ruled that the state Government could grant mining leases under the NTA (W&J 2015).[47] Queensland issued all the mining leases to Adani regardless of the W&J's lack of consent (Australian Associated Press 2016). The mining leases and the Tribunal's determinations were based on what the W&J Family Council called a 'sham agreement' and a 'fake meeting to manufacture consent' (Lyons 2017b). The

Carmichael project's political prospects became increasingly favourable by 2014 with federal approval and the Queensland Coordinator General's green signal for the project, even before the conclusion of second negotiations between Adani and the W&J. The pressure to settle an agreement or risk native title being extinguished without receiving any benefits split the W&J Applicant and caused deep divisions amongst W&J native titleholders.[48]

The W&J's experience of engagements with the native title regime during the negotiations revealed a constant prioritisation of mining and settler-state agendas over meaningful consent for Indigenous people (Lyons et al. 2017a). The conflict between the W&J people and Adani over the ILUA with the involvement of the Queensland and Federal governments was shaped through complex institutional processes involving agencies and bodies under the Native Title Act.[49] Such institutions are known to lack independence on account of being driven by state and corporate funding priorities.[50] Interactions between mining companies, Indigenous groups and the State occurring within the structures of the NTA regime have been known to favour mining interests and capitalise on divisions within Indigenous groups (Bebbington et. al 2008).

The W&J Family Council withdrew from the Tribunal proceedings in 2015 stating that:

> We cannot afford to continue a case where we do not have the resources to put our objection to the Tribunal and the cards are already stacked against us...These proceedings and the legislation under which they are held do not advance our right to live in freedom, peace, and security as distinct peoples with our own cultural values...While the legal system might weigh against us, when we say No, we mean No!
>
> Burragubba (2015, para 21)

After walking out of the NNTT, the W&J Family Council mounted legal challenges and a public and international appeal for their right to free, prior and informed consent under the United Nations Declaration of the Rights of Indigenous Peoples under the campaign slogan 'Adani, No Means No' (W&J 2015). The campaign started with the W&J Family Council presenting a 'Declaration of Defence of Country' to the Queensland government, urging against compulsory acquisition without consent (Borschmann 2015). Although Adani's 'bad faith' negotiations had triggered their resolution to fight, owing to the historic, legal, and political contexts of Indigenous rights and mining conflict, the focus of their resistance became a challenge to Australia's native title system:

> We have mounted a significant legal and political challenge to the system that enables governments and corporations to override our rights. We are taking on the racist legacy of native title and its failure to measure up to international laws that declare the rights of Indigenous people. We have sent a complaint through the UN Rapporteur on Indigenous Rights complaining about what they are doing to Indigenous people in this country.
>
> Burragubba (2016, para 12)

During a visit to Australia in September 2016 the UN Special Rapporteur Human Rights Defender singled out the case of the Carmichael coal mine as an example of poor Indigenous consultation (Lyons et al. 2017a). The Rapporteur's assessment, detailed in the *End of Mission Statement* noted that:

> Many Indigenous human rights defenders still experience severe disadvantages compared with non-Indigenous defenders. They are marginalised and unsupported by state and territory governments. This situation is compounded by the tendency of the central government to use the federal system as limitation on its ability to exercise responsibility for supporting Indigenous rights defenders. Furthermore, the right to free, prior and informed consent is not protected under Australian law, and government officials frequently fail to meaningfully consult and cooperate with Indigenous and community leaders. Indigenous rights defenders also face lack of cooperation or severe pressure from the mining industry with regard to project activities, as has been exemplified in the case of the proposed Carmichael Coal Mine in central-western Queensland.
>
> Frost (2016, para 7)

The United Nations also intervened through the Committee for the Elimination of Racial Discrimination, asking Australia to consider suspending the Carmichael project until Adani secured consent from all representative W&J claimants. Based on a submission by the W&J, the international body noted that as ILUA consultations had not been conducted in the spirit of the FPIC principles under the UNDRIP, allowing the project to proceed would violate Australia's international obligations (Robertson 2019).

However, repeated interventions from the United Nations and questions about the project's financial viability could not stop the State from paving the way for the Carmichael mine without Indigenous consent. The Australian State is considered the main driver of the Indigenous land grab behind the Carmichael coal mine (Lyons et al. 2017a). The following four subsections account for how political, parliamentary, legal and other institutional processes were used by the State to favour the Carmichael project over the free and prior informed consent of the W&J, and the significance of their resistance in this context.

State and Adani manufacturing consent (2016–2017)

The State laid the grounds for manufacturing the W&J's consent for the Carmichael mine. A meeting was organised in January 2016 of some W&J Applicants who were willing to negotiate with Queensland's Coordinator-General:

> The way I understand the meeting was: take the deal or we'll extinguish Native Title. He was very careful how he said it and he didn't say it in that way, but that's the way I took it.
>
> Craig Dallen, W&J Applicant member, quoted in Carey (2019, para 4)

Figure 7.7 Wangan and Jagalingou family members hold a demonstration in front of the Adani mine site.

Source: Photo by Anthony Esposito.

Following this, Adani organised a meeting of the W&J community in April 2016 that the five Applicant group members opposed to the mine and their extended families boycotted (Carey 2019).[51] The company claimed that the meeting achieved unanimous agreement from traditional owners, with 294 attendees having voted for and only one against the Carmichael mine (Sydney Morning Herald 2016). This contrasted with the outcomes against the mine determined in three previous meetings and one subsequent meeting (Wagner 2017).

Instead of being held on the W&J's traditional lands, the meeting was conducted in the coastal town Maryborough, where it was allegedly easier to 'rent-a-crowd' from the big Aboriginal communities in Cherbourg and Woorabinda.[52] The company also paid for transport and accommodation for 341 attendees, a figure that was a hundred more in count than the size of any previous W&J meeting (Carey 2019). The meeting had been effectively stacked with attendees who did not have the right to vote or authorise the Adani ILUA (Brigg 2018).[53] Although mining companies customarily meet all meeting expenses, the process also allows the corporations to offer inducements.[54] Ultimately this impacts outcomes from the meeting:

> You know, you come down, you get paid and why not? You get a free trip down. You get a motel, and then you go to a meeting. You get fed and then you get paid. If you're broke and you've got nothing, I'd jump into a bus too'.
>
> Craig Dallen, W&J Applicant member, quoted in Carey (2019, para 5)

This process raised questions about the validity of the ILUA certification and exposed the lack of independence of the regional Native Title Representative Body, the Queensland South Native Title Services (QSNTS), which receives most of its funding from the Federal Government. The QSNTS had helped to organise the April 2016 meeting and certified the meeting process and ILUA despite its striking anomalies, in return for a payment of A$30,000 from Adani (Brigg 2018). Adani's effort to manufacture consent for the ILUA was undermined when one of the seven individuals withdrew support at a later stage. This made it impossible for Adani to claim a majority support of the W&J Applicant (Robertson 2017c). Even pro-mine Applicants had flagged the misleading nature of the ILUA meeting.[55] Despite these contentions, and despite a clear lack of majority support, the NNTT still accepted Adani's application and registered the ILUA in December 2017 (W&J 2018b).

McGlade decision and federal native title politics (2017)

In February 2017, the McGlade decision delivered by the Western Australian Federal Court in relation to the ILUA of the Noongar People determined that the signed consent of all members of the Applicant group was required for an agreement to be valid for registration under the Native Title Act (McHugh 2017).[56] The McGlade decision came a week before a due decision by the Queensland Federal Court on the W&J's legal challenge to invalidate Adani's contested ILUA. The W&J's lawyers appealed to the Federal Court to 'strike out' Adani's claim to the ILUA's authenticity based on the McGlade ruling (W&J 2017a). However, any legal action against the disputed ILUA was withheld owing to an urgent intervention by Australia's Attorney General into the Federal Court's hearing of the W&J's appeal for a 'strike out' (W&J 2017f). The federal government also proceeded to immediately amend the Native Title Act to overturn the McGlade decision in a manner described as 'completely disrespectful' to Aboriginal people (SBS 2017).

In any event, the McGlade decision sparked a backlash from the mining sector – as it had done after the 1996 Wik Decision – with the CEO of the Queensland Resources Council warning that it threatened 126 mining projects associated with ILUAs across Australia (Parliament of Australia 2017). The federal government introduced a bill in parliament within two weeks, explaining that 'urgent amendments are imperative to preserve the operation of currently registered ILUAs and provide the sector with a prospective process for registering ILUAs which minimises the risks presented by the McGlade decision' (www.seedmob.org.au/leg al_briefing para 24). A rushed Senate Inquiry on the amendment bill allowed only two weeks for consultations with Indigenous groups and recommended that the bill be passed with minor amendments (McGlade 2018).

The W&J submission to the Senate Inquiry highlighted that Australia violated international law by not following due consultation process with Indigenous groups on matters that would affect them (W&J 2017d). The 'No Means No' campaign mobilised 6500 supporters to write letters to federal politicians on the rushed amendments (Lyons et al. 2017a). The W&J were able to influence independent Senators and secure the support of the Greens to delay the amendment bill by four months (W&J 2017e).[57]

Although the Labor opposition objected to the rushed consultations, it agreed with the spirit of the amendment bill (Hutchens 2017), reflecting a bipartisan consensus on the need to override native title considerations for mining interests (W&J 2017b). The *Native Title Amendment (Indigenous Land Use Agreements) Act 2017* came into effect once the bill passed the Senate in June 2017 (W&J 2017g).

The Adani mine was at the centre of this further delimitation of Native Title rights. Prime Minister Turnbull is reported to have delivered assurances of 'fixing native title uncertainty' to the Adani Group's CEO during a State visit to New Delhi in 2017 (Coorey 2017). On its part, Adani 'harassed' the Native Title Tribunal to breach jurisdiction and register the ILUA even as the parliamentary amendment to McGlade was ongoing, writing five letters telling the Tribunal to 'do your duty" (W&J youth spokesperson interview 20/10/2017). It was widely regarded that the government moved on native title in this precise manner to clear the way for the Carmichael project (Coorey 2017), with the Chair of the Senate Committee in federal parliament even referring to it as the 'Adani Bill' (W&J 2017g).

Failure of the legal defence against mining (2015–2018)

The W&J collectively challenged the Queensland Government, the native title institution, and Adani, through a multi-pronged legal campaign that aimed to establish that by not allowing a veto, Australia's Native Title Act fell short of complying with the UN Declaration (Lyons et al. 2017a). Their endeavour to raise questions about human rights and principles of justice in Australia's legal system and native title attracted strong support through pro-bono legal representations from senior counsels and barristers (Coyne interview in Robertson 2016) The defeat of their long drawn-out legal campaign demonstrates the inherent structural bias of the judicial and native title institutions in prioritising mining and State agendas over Indigenous rights.

There were further legal challenges. In May 2015 the W&J appealed in the Queensland Federal Court against the Native Title Tribunal's determination that mining leases could be granted, submitting that the Tribunal had been misled by Adani's fraudulent conduct in withholding expert evidence on actual jobs figures and overstating the project's economic benefits (Robertson 2016). The Court upheld the Tribunal's decision. A subsequent escalation of the challenge by the W&J before a full bench of the Federal Court was also rejected (W&J 2017c).

In April 2016, the W&J appealed to the Supreme Court for a judicial review of the State's capacity to issue the mining leases, stating that the Minister of Mines had not respected their right to 'natural justice' as per common law in issuing Adani's leases without a valid ILUA (W&J 2016b). The appeals were rejected on account of their 'narrow legal grounds'. A subsequent escalation of the appeal before the Court of Appeal in the Supreme Court was also rejected, with the judge clarifying that the State was not required by law to consult native titleholders before issuing mining leases, even when the ILUA's legal validity was undetermined (Burragubba and Ors v Minister for Natural Resources & Anor 2017).

In December 2016 the W&J Family Council challenged Adani's 'sham' agreement from the contested April 2016 meeting, alleging that a majority vote was obtained from a 'rent a crowd' gathering of Indigenous persons who had never before identified as W&J people (W&J 2016a). The federal government's intervention after McGlade delayed the Federal Court hearing for the case till March 2018. In December 2017, the W&J filed an injunction in the Federal Court to restrain Queensland from extinguishing native title, appealing that the State should not take this unprecedented step for a project in financial uncertainty (W&J 2017h). The court decision on the 'sham agreement' case in mid-2018 once again ruled against W&J, putting the risk of native title extinguishment back on the table (Robertson and Sigato 2018).

The W&J moved for a Federal Court full-bench appeal of the decision in September 2018 (W&J 2018a). Internal pressure within the Queensland Labor Party is reported to have prevented imminent native title extinguishment till the company could prove the project's financial reliability (Robertson 2018c). The W&J were ordered to pay security money for massive cost orders tallied up by Adani's lawyers against them for failed legal challenges (Archibald-Binge 2018).[58] Adani's retaliation to the W&J set a precedent as the first time an Australian traditional owner was made bankrupt by a mining company, with its severity being noted by the Federal Court (Gregoire 2019). Although the W&J were able to temporarily salvage their last legal defence against Carmichael through public donations, their appeal was finally overturned in July 2019 as Queensland removed the final roadblocks for the project's commencement (W&J 2019a).[59]

'*Adani, No Means No'!: The politics of withholding consent*

By rejecting the ILUA, withdrawing from the NNTT proceedings, and saying no to negotiating with Adani, the W&J had posed an unprecedented test for Australian native title (Borschmann 2015). The impulse for the W&J's campaign stemmed from the NTA's denial of the right to veto and consequently Free Prior and Informed Consent (FPIC) for developments on traditional lands (Howard-Wagner and Maguire 2010). The campaign narrative asserted Indigenous sovereignty against the backdrop of forced colonial dispossession and its perpetuation through today's legal and political systems that denied them the fundamental right to consent:

> The confrontation over the Galilee is the distillation of our peoples' struggle with the land grabbing and colonisation that has continued since day one of the British assertion of sovereignty over our lands and peoples – an assertion that we never ceded to and one that proceeds every day, still without our consent...We did not consent, we have not consented, we will never consent to the destruction of our country for Adani's Carmichael coal mine, or any others, on our ancestral lands. It would be against our law and order...So, we fight...We fight for our rights to free, prior, informed consent; to our own economic development; and to protection of our country and culture.
>
> Burragubba (2018, p. x)

International human rights jurists agree that extractive projects on Indigenous lands should not proceed without the affirmative consent of Indigenous peoples whose survival, rights or traditional lands can be significantly and directly harmed (Lyons et al. 2017a). The W&J appealed to the United Nations to intervene on the issue of the Adani mine proceeding without Indigenous consent. Their submission to the UN highlighted Australia's failure to fulfil its international obligation of protecting Indigenous rights:

> Our ancestral homelands in central-western Queensland, Australia, are threatened with devastation by the proposed development by a private company, Adani Mining, of the massive Carmichael Coal Mine…We exist as people of our land and waters, and all things on and in them – plants and animals – have special meaning to us and tell us who we are. Our land and waters are our culture and our identity. If they are destroyed, we will become nothing…we have not consented to the development of the Carmichael mine or any other proposed mine on our traditional lands.
>
> W&J (2015, para 4)

The W&J also appealed to international financial institutions to stop funding the Carmichael project, during a world tour in 2015 (Market Forces 2015). The W&J youth spokesperson talked about the 'tremendous task that it is in Australia and Queensland to take on a mining corporation when the laws are stacked against us' (www.adaninomeansno.com). During an interview in Brisbane, the youth spokesperson described the purpose of the tour of the international banks in May 2015 as:

> They never hear our voices or know we exist, but the decisions they make halfway across the world mean everything to us. We went around the world in 18 days', met seven banks, and directly presented our concerns as Indigenous people.
>
> Interview (20/10/2017)

By the time of the W&J's world tour of international financial institutions, 11 international banks including HSBC and Barclays had already committed to not funding Australia's largest proposed mine on grounds of respecting Indigenous rights, repeated delays in starting the project, and the effects on the Reef (Guardian 2015). During the two and a half weeks' intensive tour, the W&J met investors

Figure 7.8 Wangan and Jagalingou leaders during the world tour of banks.

Source: Photo by Anthony Esposito.

and banks in London, Zurich, New York and Hong Kong (Market Forces 2015). The London-based Standard Chartered that had previously lent to Adani eventually ended its association with Carmichael, reportedly to avoid reputational damage that the distancing of other global banks from the project had exposed it to (Rankin 2015). Following a lobbying visit to Seoul in 2018, the W&J received written commitments from South Korean banks that had been in talks with Adani to not fund the Carmichael project (Talukdar 2019).

The amendments overturning the McGlade decision and the unfavourable rulings in all legal challenges left political and legal pathways to resistance exhausted and the struggle moved into a more symbolic phase. In July 2017, the W&J held a significant 'Gathering on Country' near Clermont as an expression of their claim to ancestral lands and waters. It was the first on-Country gathering of representatives from all families since forced removals started occurring in the late 1800s (Lyons et al 2017b). It was repeated in August 2019 in anticipation of native title extinguishment (W&J 2019b) and to publicly demonstrate that 'our conflict is with the State' (W&J youth spokesperson quoted in Krien 2019b).

Despite taking a circuitous and politically circumspect path, Queensland ultimately took the unprecedented step of extinguishing W&J native title over the 1385 hectares 'surrender area' of the Carmichael mine, making it the first case of explicitly privileging private mining interests over Traditional owners (see Lyons et al. 2017a). 'No Means No' could not stop the State from allowing the Carmichael mine to proceed without the W&J's consent. However, the campaign gained recognition as a leading Indigenous rights struggle, posing a challenge to the mainstream notion of development for Indigenous people through compliance with mining (Brigg et al. 2017). The W&J's unwavering resistance exposed the limitations of Australia's native title institutions (Lyons 2019a).

Figure 7.9 Wangan and Jagalingou with North American First Nations peoples during a public event on First Nations Solidarity in Brisbane.

Source: Photo by Anthony Esposito.

Table 7.1 Timeline of the Australian anti-coal resistance

Year	Environmental politics and resistance to Carmichael	Indigenous politics and resistance to Carmichael
2004	The Wangan and Jagalingou traditional owners from central Queensland register their native title claim over ancestral lands.	
Australian labour government (federal)		
2010	Adani acquires coal assets in the Galilee Basin in central Queensland Negotiations commence between the W&J and Adani for consent for mining and an indigenous land use agreement (ILUA); the W&J contend that Adani negotiated in bad faith by taking advantage of the coercive native title system.	
2012	A Greenpeace report estimates that coal extracted and burnt at full capacity from all nine proposed mega coal mines in the Galilee would make the Basin responsible for the seventh highest emissions in the world.	The W&J refuse consent for the Adani mine due to the inadequacy of benefits and jobs in the ILUA and the company's disrespectful and non-transparent dealings; they become the first native title group to say an outright no to mining.
2013	The IESC is concerned about impacts from Carmichael's massive water use and potential contamination of the Great Artesian Basin and the Doongmabulla Springs complex that are sacred to the W&J.	Adani brings proceedings before the National Native Title Tribunal after the first round of failed negotiations with the W&J.
Liberal coalition government (federal)		
2014	Queensland approves the project with 'extensive and wide-ranging' conditions. Indian Prime Minister Narendra Modi, in Australian for the G20 Summit, reportedly signs a memorandum of understanding (MOU) for a A$1 billion loan for Adani's Carmichael mine from India's public bank the State Bank of India. The loan is (reportedly) quietly withdrawn soon after.	The W&J strike down Adani's revised ILUA.
2015	Federal Environment Minister approves the Carmichael mine with 36 'absolutely strict' conditions. In the first legal challenge the Mackay Conservation Group seeks a judicial review of the federal approval on account of the mine's impacts on two vulnerable species under the EPBC Act.	Adani brings proceedings before the National Native Title Tribunal after the second round of failed negotiations with the W&J; the Tribunal rules that Queensland can grant mining leases to Adani despite the W&J's lack of consent.

Table 7.1 (Continued)

Year	Environmental politics and resistance to Carmichael	Indigenous politics and resistance to Carmichael
	A report by Earthjustice and Environment Justice Australia highlights how Adani operates in India – through the destruction of mangroves, non-compliance with environmental conditions and illegal developments at Mundra in Gujarat.	The W&J withdraw from Tribunal proceedings, rejecting a process where the cards are already stacked against them.
	Divestment activism group Market Forces and the W&J traditional owners go on a tour of international banks and investors in the US and Europe asking them to not finance Carmichael.	The W&J appeal in the Queensland Federal Court against the Tribunal's determination that mining leases could be granted submitting that the Tribunal has been misled by Adani's overstating jobs figures and economic benefits; the Court uphold the Tribunal's decision; an escalation of the W&J's challenge before a full bench is also rejected.
	A second legal case by Coast to Country challenges Queensland's environmental approval. An Indian environmentalist from the Mumbai based Conservation Action Trust testifies in court that coal from Carmichael will not lift millions of Indians out of poverty as claimed by Australian governments. The verdict in the second legal challenge also favours Adani.	The W&J youth spokesperson describes their purpose of the tour of international banks as to 'directly present our concerns as Indigenous people'.
	Federal Environment Minister reapproves Carmichael coal mine.	The W&J build solidarities with Indigenous leaders from Standing Rock Sioux and Chickaloon Village Traditional Council in Turtle Island in the United States who resisted the Dakota Access Pipeline.
	In a third legal case following the federal reapproval the Australian Conservation Foundation challenges that the mining and burning of coal from Carmichael is inconsistent with Australia's obligation to protect the World Heritage Listed Great Barrier Reef.	The W&J launch the 'No means No' campaign by presenting a 'Declaration of Defence of Country' to the Queensland government urging against compulsory acquisition without consent
		The W&J submit to the UN Special Rapporteur on the Rights of Indigenous People that Adani used a 'divide-and-conquer' approach and mistreated Indigenous rights.

(*Continued*)

Table 7.1 (Continued)

Year	Environmental politics and resistance to Carmichael	Indigenous politics and resistance to Carmichael
2016	Following the verdict in the second court case, Queensland grants Adani the environmental licence to commence operations with 140 conditions. Queensland declares Carmichael a critical infrastructure – a status granted only four other times before and never to a private commercial development. Land Services Coast and Country launches the fourth legal challenge, against the granting of Carmichael's environmental licence, arguing that Queensland had not considered the requirement for sustainable development under the state's Environment Protection Act. With the possibility of a federal government A\$1 billion loan, GetUp! organises mass conversations with voters across electorates to build ground-level political opposition to Carmichael. On the 'National Day of Divestment Action' in October activists ask the 'Big 4 Australian Banks' to stop funding fossil-fuel projects; people from across 13 cities participate in a mass protest by cancelling their bank memberships; all four major Australian banks finally withdraw from the Carmichael project. Queensland approves Carmichael's mining licences with 200 'strict conditions'. State and federal governments clear all approvals by the end of 2016 but Adani's April 2017 deadline for starting the project appears unlikely given the pending third and fourth legal cases.	A 'self-determined' meeting of traditional owners rejects Adani's ILUA; this is the third time that the W&J reject Adani's ILUA. But Queensland issues mining leases based on a 'sham agreement' drawn up by Adani at a 'fake meeting to manufacture consent' where the company claimed it had achieved 'unanimous agreement' from traditional owners. The W&J appeal to the Supreme Court for a judicial review of the State's capacity to issue the mining leases without a valid ILUA; the Court rejects their appeal on account of its 'narrow legal grounds'; the Court also rejects a subsequent escalation of the W&J appeal. On a visit to Australia the UN Special Rapporteur Human Rights Defender singles out the Carmichael coal mine as an example of poor Indigenous consultation; the UN also intervenes through the Committee for the Elimination of Racial Discrimination asking Australia to consider suspending the Carmichael project until Adani secures consent from all representative W&J claimants and in the spirit of the FPIC principle. The W&J challenge Adani's 'sham' agreement from the contested meeting alleging that a majority vote was obtained from a 'rent a crowd' gathering of Indigenous persons who had never before identified as W&J people. But the Federal Government intervenes to delay the Court hearing for the case till 2018.

Table 7.1 (Continued)

Year	Environmental politics and resistance to Carmichael	Indigenous politics and resistance to Carmichael
2017	Queensland Premiere leads a delegation of Mayors from the state's regional centres to Mundra, Gujarat in India. A protest-delegation of Australian citizens follows the government delegation and deliver a letter signed by 90 prominent Australians and traditional owners asking Gautam Adani to abandon the Carmichael project to the corporation's headquarter in Ahmedabad, Gujarat. Various strands of environmental opposition to Adani and a collaboration of over 40 national and local ENGOs formally come together under the Stop Adani banner; there are over 100 registered Stop Adani groups across Australia. Stop Adani's 'Week of action' in April successfully makes major Australian bank Westpac withdraw from the Carmichael project. Front Line Action on Coal (FLAC) begins peaceful blockading actions at Abbott Point port and along Adani's rail corridor. Stop Adani's 'Big Day of Action' in October – a coordinated day of sixty anti-Adani demonstrations across Australia. In the lead up to the state elections GetUp! organises mass callouts to voters in marginal seats in Queensland Adani's lobbying firm is led by former chiefs of staffs from both Labor and Liberal offices at state and federal levels enabling Adani to get 'pretty much everything it wanted through an extraordinarily intense (lobbying) campaign'. Queensland grants Adani a water licence to extract unlimited groundwater with 270 conditions, in the midst of one of the worst droughts in Central Queensland. Farmers for Climate Action (FFCA) with a 2000-strong membership joins the Stop Adani alliance to protest Queensland's allocation of free water licences to Adani.	The W&J reject Adani's ILUA for the fourth time and completely reject any mining agreement with Adani. The federal government promptly amends the Native Title Act to overturn the McGlade decision – which made it essential to secure the signed consent of all members of the 'Applicant Group' for an ILUA to be valid – in a manner described as disrespectful to Aboriginal people; the amendment favours Adani and allows the 'sham agreement' to be recognised; it is even called the 'Adani bill' by some federal parliamentarians. The W&J hold a significant 'Gathering on Country' as an expression of their claim to ancestral lands and waters; this is the first on-country gathering of representatives from all families since the beginning of forced removals in the late 1800s. The W&J file an injunction in the Federal Court to restrain Queensland from extinguishing native title, appealing that the State should not take this unprecedented step for a project in financial uncertainty.

(*Continued*)

Table 7.1 (Continued)

Year	Environmental politics and resistance to Carmichael	Indigenous politics and resistance to Carmichael
2018	The Federal government issues a rushed approval for Adani's water scheme to pump to pump 12 billion litres of water from the nearby Sutton River for non-extractive activities such as washing coal before the May 2019 federal elections; the company avoids a full impact assessment for the scheme. Central Queensland farmers and environmentalists protest the impacts of the Adani mine on water through a regional road show titled 'Our Lifeblood, Our Water'. The Australian Conservation Foundation launches a fifth legal challenge against the approval process for the water scheme, which failed to take 'thousands of public submissions into account'. The Environment Minister reapproves the water scheme at a later date. Stop Adani builds pressure in 'strategic' Liberal Party electorates to shift the politics within the Liberal Party; Wentworth in NSW – a safe Liberal Party seat since federation – is won by an Independent candidate with a climate commitment during a bye-election. The Adani Group significantly scales down the Carmichael mine to a A$2 billion self-financed project and halves costs of the rail project to A$1billion. Galilee Blockade, a volunteer run local activist outfit, disrupts mine construction through blockades and shareholder activism against Adani's principal construction partner.	The court decision on the 'sham agreement' rules against the W&J, putting the risk of native title extinguishment back on the table. The W&J escalate their legal appeal not to extinguish native title and against the 'sham agreement'. Internal pressure within the Queensland Labor Party prevents imminent native title extinguishment till the company can prove the project's financial reliability. Following a lobbying tour to Seoul in 2018 the W&J receive written commitments from South Korean banks that had been in talks with Adani to not fund the Carmichael project. Adani's retaliation to the W&J sets a precedent as the first time an Australian traditional owner is made bankrupt by a mining company; the W&J are ordered to pay security money for massive cost orders tallied up by Adani's lawyers against them from failed legal challenges.
2019	Queensland grants the last remaining environmental approvals for Adani after the 2019 federal elections and paves the way for a significantly diminished coal mine to finally commence.	The W&J's legal appeal against the sham agreement and to not extinguish native title is finally overturned.

Table 7.1 (Continued)

Year	Environmental politics and resistance to Carmichael	Indigenous politics and resistance to Carmichael
	Stop Adani moves to peaceful blockades in Queensland after all other tactics fail to stop the coal mine; Camp Binbee set up near Bowen to blockade mining, rail and port activities attracts volunteers from across Australia and overseas. Queensland steps up police surveillance in anticipation of civil disobedience as the Adani mine commences; it brings in a new law banning 'locking devices' commonly used at blockades; prompting the UN to warn that Australia falls of international human rights in relation to peaceful assembly.	The W&J hold another gathering on Country in anticipation of native title extinguishment and publically send a message that 'our conflict is with the State'. After the 2019 federal elections Queensland extinguishes the W&J's native title over the 'surrender area' of the Carmichael coal mine. 'No Means No' gains recognition as a leading Indigenous rights struggle globally even though it cannot stop the State from allowing the Carmichael mine to proceed without the W&J's consent.
2020	The sixth case by ACF in 2020 against the federal government, for Adani's new water project for the Carmichael mine not having been assessed under the water trigger in the EPBC Act for significant social and ecological impacts, is successful in May 2021. ENGOs and farmers cannot stop the mine despite ten legal challenges;[a] but repeated legal actions delay timelines, expose the economic unviability, counter false claims around jobs and challenged the moral case for Carmichael.	The W&J continue their resistance through the 'Standing our Ground' campaign' through a sovereignty camp on ancestral land adjacent to the Carmichael mine and calling for a stop work order from the Queensland government on the mine.

[a] Ten legal challenges jointly against the Carmichael and Alpha coal mines and the expansion of the Abbott Point port related to these projects.

Beyond the timeline of this research, the W&J's resistance to the Carmichael coal mine has continued adjacent to the mine site through a sovereignty camp on ancestral land and calling for a stop work order from the Queensland government on the mine. The W&J also launched a new legal challenge against the Queensland government in the state's Supreme Court for its failure to protect the sacred Doongmabulla springs from contamination and drying up from mining, threatening the W&J's cultural rights that are protected under the recently legislated Human Rights Act 2019 in Queensland. With the new case, the W&J stepped outside the restrictive native title system and legally challenged the Carmichael coal mine as a risk to their human rights (see www.standing-our-ground.org).

Analysis: countering coal mining through various scales of contestations

The emergence of diverse discontents over coal mining under climate change and during the resource boom challenged the dominant narratives of economic prosperity associated with the Australian coal sector. Coal regions such as the Hunter Valley were reimagined as the 'Carbon valley' on account of the scale of transformation wrought by the mineral boom. Climate change has emerged as a transformative discourse for communities affected by intensive coal mining (Connor et al. 2009). And reflexively, coal has become the embodiment of Australia's concern for climate change (Duus 2013).

Simultaneously, social accounts such as Munro's depiction of intensive coal mining as invasion on country (Munro 2012) demonstrate a reflexive understanding of the coal boom as expressing domination by settler society, one that extends solidarity to historical Indigenous experiences of environmental loss through colonial dispossession. Anti-coal narratives during the mineral boom therefore assumed both historical and global dimensions, creating an inclusive and multi-scalar significance for resisting coal extraction. These multi-layered mobilisations against the effects of massive coal projects united various anti-coal constituents across cities and rural regions (Connor et al. 2009).

Against growing social discontent and a changing global outlook for coal, the actions of governments to develop the Galilee Basin demonstrated the effects of the resource curse in Australian politics. The efforts of the State and coal corporations to champion coal, including through coal advertisements on television, indicated that coal was losing its social licence (Federal Greens Senator Waters, notes from Brisbane Stop Adani session 29/092017). The nexus between the State and the coal sector sustained optimism for coal's outlook and false narratives about coal's economic significance. Pro-coal narratives around its 'life-saving potential' for the Southern poor indicated a concerted effort to legitimise the fossil fuel. It also rendered the narratives of the Australian government indistinguishable from the public relations campaigns of large coal corporations.

Political attempts to legitimise the Carmichael project despite its weakening economic viability and growing social discontent over coal have served to expose coal's power over Australian politics (Dennis 2018). Significantly, the political influence of the resources sector was critical in enabling the project to be rescued. This has important implications for Australian democracy.

The inability of the Stop Adani movement to stop the Carmichael project, despite causing major disruption through legal, financial and grassroots political activism, exposed the coal sector's power at various levels of politics and governance. Stop Adani's narratives and actions acquired newer layers of significance over time, from beginning as a political discourse of anti-coal activism based on environmental and climate concerns to becoming a critical test for democracy in the face of the capture of Australian politics by mining corporations with 'deep pockets' (Federal Greens Senator Waters, notes from Brisbane Stop Adani session 29/092017). In this context, the key aim and focus for the new environmentalism is to expose and shift the capture of democratic institutions by the fossil fuel sector.

The political economy of coal in Australia during the minerals boom, in part, instigated the formation of a more distributed model for effecting grassroots disruption in the Stop Adani movement, linked to an active national network of local anti-coal groups. It saw the rise of a disaffected constituency of farmers in Central Queensland who opposed both the free allocation of water to the Galilee basin coal mines and the governments' lack of climate action. These farmers' actions and politics demonstrate a leap forward from the political imperatives of the earlier Lock the Gate network. Lock the Gate had remained ambiguous on climate change, considering it an issue that 'greenies' would deal with (Friends of the Earth Australia Campaigns Coordinator Interview 20/11/2017). However, the Central Queensland farmers who simultaneously experienced political marginalisation and climate impacts through severe droughts, contested coal directly for its climate impacts.

The historical significance of the W&J's campaign as survivors of colonial dispossession who are now resisting extraction on their lands, and its formation through a dialectical process that posed a series of legal challenges and claims against the Carmichael mine, has made the land-rights politics of the Adani mine a distinct albeit critically significant part of the political conflict over the Carmichael project. Returning lands through native title and state-based land rights is meant to redress Indigenous dispossession that resulted from Australia's colonial project to secure territory and resources (Crook and Short 2014). However, the native title apparatus favoured mining interests over the informed consent and self-determination of Indigenous groups (Coyne 2017).

Consequently, the W&J experienced a reconfiguration instead of a disappearance of the settler colonial State (Lyons 2019a). Their movement challenged the highly asymmetrical power relations between the mining resources sector, the State, and Indigenous rights within the coercive native title regime (Lyons 2019a). No means No particularly helped to examine relations between the settler colonial State and Indigenous people in the context of Australia's continuing attachment to coal.

The collective resistance to the Carmichael coal mine exposed coal's power across political institutions at various scales. The Stop Adani movement operated within two sharply contradictory realities in the post-Paris period – the global need for a rapid phase-out of coal and the desperate politics of Australian governments to develop an entirely new coal region. The farmers operated within the political reality of increasing marginalisation and a growing threat to their livelihoods in part due to increased coal mining and climate change. The W&J's politicisation of the issue of Indigenous consent was based on a direct experience of institutional coercion that reaffirmed and compounded historical colonial dispossession. These various grounds for resistance exposed how coal exerts power over Australian politics and institutions – and how it can be challenged.

The Carmichael coal mine has in effect linked two landscapes – Australia where the coal is produced and India where it is sent – together, through the coal nexus.

It has shown how civil society tries to resist this coal nexus, revealed the state of democracy and what is possible as far as challenging the power of coal goes, in Australia. Through the Stop Adani movement and W&J's' No Means No' campaign, coal became a litmus test for democracy in Australia. The collective resistance politicised the contradictions of coal in Australia.

The anti-coal narratives that emerged from the Hunter Valley and Liverpool Plains and through alliances of environmental groups with farmers and Indigenous native title groups were extended in the resistance to the Adani mine. But other factors, such as explicit government alignment with a corporation with a disreputable environmental and financial record, while disregarding Indigenous consent and farmers' water concerns, generated greater public outrage, allowing anti-Adani movements to mobilise on a broader platform of public concern against coal mining. The Adani mine became synonymous with climate change and the power of the coal industry in the public imagination.

Conclusion

The collective resistance of the Stop Adani movement, the opposition of farmers from Central Queensland, and the Wangan and Jagalingou traditional owners offer insights into the political and economic context of coal in Australia. This context demonstrates the effects of the resource curse on politics and the nexus of the coal sector and the State, where the State promotes and subsidises massive coal extraction projects even though they risk becoming stranded assets owing to a global decline in coal demand. In this context, the State favours the short-term private interests of coal companies over the concerns of affected community stakeholders.

Operating within the political and economic context of coal in Australia in the post-Paris era, it became imperative for environmental activism to disrupt the power of coal over politics. The strategy, tactics and structures of the Stop Adani movement, the emergence of new and niche activisms within the movement network, and the emphasis on a distributed network of grassroots activisms, reflect this shift. The activism of Central Queensland farmers indicated a new critical politics amongst a politically conservative rural constituency. Apart from the immediate risks of coal mining to groundwater resources, it was shaped by the severe droughts that affected their livelihoods and their experience of exclusion given the clear influence of the coal sector.

Having experienced coercion within Australia's native title system, the imperative for the W&J's resistance became to expose Australia's failure to meet international Indigenous rights standards. Its campaign demonstrated how in its bid to promote coal, Australian governments failed their obligations to give Indigenous people free, prior, and informed consent. Finally, climate change emerged as a transformative argument for various discontents against coal, and for a wider transformation in environmentalism itself.

Notes

1 A Greenpeace report *Wrecking the Reef Cooking the Climate* estimated that if all nine proposals in the Galilee Basin reached production stage, they would together produce

330 million tonnes of coal per year, which when burnt could emit 705 million tonnes of carbon dioxide (Greenpeace Australia 2012b)

2 Coal from Carmichael was initially intended for Adani Enterprise's thermal power plant in Mundra. It later emerged that Adani's Australian coal would be transported to the Adani-owned Godda thermal plant in Jharkhand in eastern India, to supply electricity to Bangladesh, based on a Memorandum of Understanding (MOU) signed by the Bangladeshi government with Adani Power in 2016 (Das 2016).

3 The emissions would be primarily from the burning of coal extracted from the Hunter Valley in thermal plants at various global destinations.

4 Health impacts studies showed that rapid mining expansions caused higher incidences of respiratory disease and depression amongst those rural communities in the Hunter Valley directly exposed to coal mining (Thompson 2006). The region also recorded significantly higher levels of air pollutants than the NSW average (Connor et al. 2004).

5 The coal industry attempted to rebrand itself from being part of the problem to being part of the solution by introducing the concept of Clean Coal. It claimed that clean coal technologies could reduce emissions from thermal power generation by up to 90% (see www.newgencoal.coam.au). The Australian Coal Association (ACA) that represents coal interest from New South Wales and Queensland formed the COAL21 to raise A$1 billion for research and development for low emissions technologies in thermal power generation through a voluntary levy on coal production (Baer 2016).

6 State assistance for clean coal through Kevin Rudd's announcement of an A$100 million Global Capture and Storage Institute for research into CCS was regarded as a further subsidy for coal (Baer 2016).

7 The no-regrets approach implies climate action with no negative impact on the national economy and Australia's trade competitiveness.

8 GVK purchased most of Hancock Prospecting's coal holdings in the Galilee Basin in 2011 for A$1.26 billion, taking their ownership of the mine to 79%. The purchase occurred even as major commodity houses downgraded their forecasts for thermal coal, creating the risks of the mine becoming a stranded asset in the future (Seccombe 2014).

9 The projects in the Galilee required two rail lines, one from the north to transport coal primarily from Adani's mine and the adjoining mines belonging to Waratah Coal, and another from the South for the GVK Hancock mines.

10 The other four included Carmichael, China Stone, and Clive Palmer's Alpha North and China first mines.

11 The moral claim of helping India increase electricity generation and alleviate poverty emerged as a political argument to bolster Australia's slowing coal export industry (Hogan 2015).

12 In 2012, Newcastle coal prices that serve as the Australian benchmark fell to A$74 a tonne, well below the required A$110 price in order for mine operators to make a return on investment (Seccombe 2014). Galilee coal's high ash and low energy quality compared to the Newcastle standard – its ash content is likely double that of coal from Hunter Valley and Bowen Basin – risked further lowering Australia's coal export-prices and turning the Galilee mines stranded assets (Buckley 2016).

13 China accounts for around half the world's coal consumption, and between 2007 and 2012, China accounted for all the growth in global coal use.

14 The Newman government waived billions of dollars in royalty collections in the early stages of the Carmichael mine to stimulate the Galilee Basin's development (Dennis 2015).

15 A total of 10 legal challenges mounted between 2014 and 2020 by environmental groups against the Carmichael mine and the expansion of the Abbott Point port which is associated with the project, have either been defeated in the courts or their outcomes have been unable to stop the project from proceeding. See http://envlaw.com.au/carmich ael-coal-mine-case/. The W&J have mounted a separate and extensive legal campaign under the native title regime, which has been struck down by the courts. The W&J legal challenges are discussed in the next section. See www.adaninomeansno.com.

16 The environmental conditions can be accessed from: www.environment.gov.au/epbc/ notices/assessments/2010/5736/2010-5736-approval-decision.pdf.

17 A report by Earthjustice and Environment Justice Australia, based on an investigation of hundreds of court documents on Adani's corruption, destruction and criminal activity, highlighted the destruction of mangroves, non-compliance with environmental conditions and illegal developments at the port of Mundra in Gujarat, amongst others (adanifiles.com.au).

18 The government's legislative amendment bill was defeated in the Federal Senate during the Abbott government but subsequently revived by the Turnbull government (Hepburn 2016).

19 Section 487 pertains to the matter of standing, the legal term that decides the eligibility of a party to bring a legal case.

20 Adani advertised on regional Queensland television channels that the rail and mine would generate 10,000 direct and indirect jobs.

21 Five months later, news reports indicated that the MOU between the SBI and the Adani group had died a natural death. The MOU had stirred a controversy, with the Indian opposition dubbing this as an instance of crony capitalism (Bandyopadhyay 2015).

22 Campbell Newman's term was subjected to a Federal Senate Inquiry for corruption, human rights and environmental neglect (Dennis 2015).

23 Three out of these related to water supply infrastructures that were declared critical during record low water levels.

24 ENGOs highlighted that requiring a company that could not be trusted to monitor its own groundwater impacts and take action based on make-good agreements with affected landholders appeared risky. Landholders had been unable to arrive at make-good agreements with mining project proponents in the Galilee Basin in other instances (Extel 2016).

25 Two separate Federal Government groundwater studies conducted since Adani gained Commonwealth environmental approval in 2014 have still not been able to trace which of the two adjoining underground aquifers feeds the Springs.

26 Fragmentation of rural properties caused discontent within Queensland's Liberal National Party that had otherwise unequivocally supported and subsidised the Galilee projects. During Campbell Newman's term, Vaughan Johnson, Liberal National member for the Central Queensland electorate of Gregory that covers the Adani mine had raised the concerns of his farmer constituents (SBS News 2016).

27 The four major Australian banks are the Australian New Zealand Bank, the Commonwealth Bank, then National Australian Bank and finally Westpac.

28 NAIF is a Federal agency set up in mid-2016 to offer A$5 billion in concessional loans to projects in Queensland, the Northern Territory and Western Australia. NAIF board members' links to the mining industry represent a conflict of interest. Analysis by the Australia Institute highlighted concerns with the processes and disclosure of NAIF compared to other government organisations, its strong conflict of interest, and no scrutiny over the public interest requirement of projects (Swann 2017).

29 State governments have the final power to either authorise or reject the NAIF loan.

30 The Greens appeared poised to win the longest held federal Labor seat of Batman in Melbourne in a March 2018 by-election. The Party mounted a campaign to 'Stop Labor's Adani mine' challenging the Queensland Labor government to abandon the Carmichael project (Wahlquist 2018). The Greens had also challenged Queensland Labor on the Carmichael issue during the 2017 state elections in South Brisbane.

31 They were the factional forces behind the removal of Prime Minister Kevin Rudd from Prime Ministership in 2010 (Eltham 2012).

32 The company claimed that as the water was not required for coal extraction, it should be exempt from scrutiny under the water trigger of the EPBC Act.

33 Volunteers for Galilee Blockade became Downer shareholders and forced a special resolution to be voted in the company's annual general meeting. The resolution sought to amend Downer's constitution by inserting a clause that directors should 'ensure that the business of the company is managed in a manner consistent with the objective of holding global warming to below 2 degree Celsius above pre-industrial levels' (Robertson 2017f).

34 The proposal for terminal two had already been downsized and deferred. Terminal two would have required dredging and other works to increase its capacity to 90 million tonnes a year and had met with strong concerns from scientist and environmental groups (Smee 2018d).

35 Traditional owners were concerned that sacred sites within the Abbott Point State development area have not been properly protected. It was reported that Adani repeatedly ignored the demands of the Juru Enterprise Limited, the Indigenous business nominated to represent traditional owners on land use agreements with Adani, to conduct cultural inspections in the Abbott Point state development area, compelling them to threaten an attempted shut down of the port through a stop order (Smee 2018c).

36 Reports of Queensland and Australian trade missions to India revealed that leaders have been meeting Gautam Adani since 2010. In 2012 a 76-member business delegation led by Campbell Newman was flown in a private jet to Mundra. The delegation was given a tour of Adani's port and plant and hosted at a lavish reception at Gautam Adani's residence. The March 2017 trip of regional mayors and Premier Palaszczuk to Bhuj has also been criticised as a taxpayer funded 'junket' to court Adani's business (Readfeaern 2015).

37 Cameron Milner and David Moore headed the lobbying firm next Level Strategic Services. David Moore worked on Campbell Newman's victorious campaign in 2012 before leaving to start the lobbying services firm. Cameron Milner, a former state Labor secretary, also served as chief of staff to Federal Labor leader Bill Shorten. He played a controversial double role in Queensland by working as a lobbyist for Adani and then working on Labor Premier Palaszsczuk's re-election campaign.

38 An Australian Broadcasting Corporation analysis of the Queensland lobbyist register and found that the firm had made 33 lobbying contacts for Adani between 2015 and 2017, which was more than double the number of contacts reported for any other client. The firm's campaign was most intense when the royalty deal was being finalised in May 2017 (Long 2017).

39 Apart from the six environmental legal cases against the Carmichael coal project discussed in this section, local environmental groups also brought four challenges against the expansion of the Abbott Point coal terminal that risked harming the Great Barrier Reef. An explanation of these four cases can be found at http://envlaw.com.au/carmichael-coal-mine-case/.

40 The Queensland Land Court that held extensive hearings on environmental and water related issues with the projects did not have the jurisdiction to overturn developments and could only make recommendations.

41 Interactions between Adani and the W&J were bound by two processes in the Native Title Act. The first is the right to negotiate (RTN) process during which the proponent should negotiate in good faith with the affected Indigenous group to secure their agreement for the grant of a mining lease by the state (*Native Title Act 1993*, ss. 25, 29, 31). In case of failed negotiations after six months, the company can refer the matter to the National Native Title Tribunal (NNTT) for a determination on whether a mining lease can be granted (*Native Title Act 1993*, ss. 35, 38). The State has final discretion on whether to grant mining leases (or not) without Indigenous consent (W&J 2018a). The second involves an Indigenous Land Use Agreement (ILUA) through which the company secures the Indigenous group's agreement for mining leases as well as future acts that can impact their traditional lands (*Native Title Act 1993* ss. 24BA-24EC).

42 The W&J native title Claim Group included community members descended from the heads of the 12 different clans that together composed the W&J Traditional Owners at the time of British arrival. A Claim Group authorised the W&J 'Applicant', comprising of representatives from the 12 original families, to act on the W&J's behalf on native title claim matters under the NTA (W&J 2015).

43 The W&J submitted to the National Native Title Tribunal that Adani provided only 2 weeks for an agreement on the proposed ILUA. Aiming to deter the W&J from bargaining for benefits, Adani threatened that the State would forcefully acquire the W&J's lands if they failed to agree on the ILUA offer within the short time period. Adani also refused requests for a life of mine service agreement to be included in the ILUA and attempted to undermine the W&J's opposition to the ILUA by sidestepping the Applicant who is the official representative, during negotiations (Arnautovic 2017).

44 States gained the right to extinguish or impair native title in their jurisdictions through the 1998 amendments to the Native Title Act, particularly the right to compulsorily acquire native title land for private infrastructure (*Native Title Act 1993*, s. 24AMD(6B), arguably setting native title rights back by a decade.

45 Criticisms of the first ILUA included the mining company failing to explain its details, not allowing independent analysis of the mining deal's benefits and costs, and being unable to identify the area that would be subject to native title (Lyons 2019). A second ILUA authorised at a contested meeting in April 2016 offered a significantly reduced upfront payment to the W&J. The W&J stated the company's inability to provide transparent and honest information about the project's impacts as a challenge in providing informed consent on their part. The W&J did not trust government regulation to protect their natural and cultural values, and to maintain their country as a 'vital cultural landscape' as required by Indigenous law, in the face of mining's deleterious effects (Burragubba 2018).

46 The first ILUA was voted down at an authorisation meeting with Adani representatives in 2012. Adani's revised ILUA was struck down in October 2014 based on a decision of an authorised W&J Native Title Claim Group meeting. A third 'self-determined' meeting of Traditional Owners in March 2016 held without the company's involvement also rejected the ILUA (Lyons 2019). A fourth meeting in 2017 completely rejected any mining agreement with Adani (Wagner 2017).

47 The first negotiations between the W&J and Adani occurred between November 2011 and November 2012, and failing an agreement, Adani approached the NNTT for a determination (Adani v. Jessie Diver and Ors. 2013). The second round of negotiations

between the W&J and Adani for two mining leases lasted from October 2013 to October 2014 when the W&J Claim Group meeting rejected the ILUA for the second time and Adani took proceedings to the Tribunal (Arnautovic 2017).

48 The changing composition of the W&J Applicant reflected the divisive effect the mining company had on the W&J community through two rounds of failed negotiations (Brigg 2018). In 2014 a seven-member Applicant was replaced with a three-member group, which retained two older applicants and made a new appointment of Adrian Burragubba (W&J 2015). At an August 2015 Claim Group meeting, the three-member Applicant that had largely turned pro-mine was replaced with a 12-member Applicant comprised of a representative each from the 12 original W&J families. A Claim Group meeting in March 2016 further moved to replace four pro-Adani members from the 12 members Applicant, setting the stage for a long-drawn legal battle in Federal Court between pro-mine and anti-mine Applicants (W&J2015).

49 The National Native Title Tribunal (NTTB) registers native title claims that are then heard in the Federal Court, facilitates negotiated determinations of native title, land use agreements, and future acts. Native Title Representative Bodies (NTRB) represent native title claimants and once native title has been legally determined, Prescribed Bodies Corporates (PBC) hold this title to perpetuity (Altman 2012).

50 The Native Title Tribunal has been known to interpret the NTA to benefit mining and rarely reject projects (Corbett and O'Faircheallaigh 2006). The State also tasks project proponents to fund negotiations for future acts under the Native Title Act with affected Indigenous groups, compromising the process and undermining the latter's bargaining power (Ritter 2009). Proponents hold the power to suspend funding for native title groups' legal representations (Arnautovic 2017). Or to pressure them to accept weak agreements (O'Faircheallaigh 2006).

51 Adani managed to split the new 12-member W&J Applicant by then, and publicised the seven pro-mine applicants as 'rightful' Traditional Owners through its promotional materials, while undermining the five anti-mine Traditional Owners (Lyons 2017b).

52 Traditional Owner requests to host the meeting on W&J country were vetoed by the company (Carey 2019).

53 An analysis of the attendance register entries for the meeting by the W&J Family Council's lawyers revealed that 60% attendees had never attended a W&J Claim Group meeting before and could not be found in the members' database maintained over 12 years of the W&J's Native Title Claim (Brigg 2018). Another violation was revealed through 71 of the 341 registered attendees not recording any 'Apical Ancestors' (family members connected to Country from whom Traditional Owners traced descent, a fundamental requirement for claims under the Native Title Act) on registration forms (Carey 2019).

54 Thousands of dollars were paid to each of the seven individual applicants to recruit pro-mine attendees to the meeting, and the lawyer representing the seven was paid to engineer the ILUA (Lyons 2017b). A controversial tweet sent out by a Traditional Owner from the floor of the meeting urging 'only come meeting for money' encapsulated the concerns over the meeting process (Carey 2019)

55 The seventh Applicant who withdrew support expressed concerns about the 'pitiful nature of the agreement' and wishing Adani would renegotiate with his people (Dallen interview in Carey 2019).

56 The McGlade decision was given in an ILUA case for the Noongar people of WA, who (like the W&J) were claimants not yet granted Native Title (see McGlade v Native Title Registrar & Ors 2017).

57 Opposition to the government's mistreatment of Indigenous rights through the inadequate consultation process stalled two attempts to pass the amendment bill through the senate (W&J 2017e).
58 By the end of 2018 Adani Australia's lawyers had tallied up cost orders against the W&J from court proceedings to the tune of A$870,000. The company asked the Federal Court to direct W&J to pay A$160,000 in security money within a fortnight, failing which the W&J's appeal 'be dismissed with costs' (Robertson 2018d). The court found Adani's estimate to be 'disproportionate and revised the W&J's security money to A$50,000 (Archibald-Binge 2018).
59 They received funding from Grata Fund, a public interest litigation group (Grata Fund 2019).

References

Adani Mining Pty Ltd v. Jessie Diver and Ors on behalf of the Wangan and Jagalingou People/State of Queensland (2013) NNTA 52.

Adani Mining Pty Ltd v Land Services of Coast and Country Inc. & Ors (2015) QLC 48.

Adani Mining Pty Ltd v Land Services of Coast and Country Inc & Ors (No.2) (2016), QLC 22.

Agius, K. 2016, 'Adani's Carmichael coal mine gains final Queensland Government environmental approval', Australian Broadcasting Corporation News, 2 February, viewed 15 March 2020, <www.abc.net.au/news/2016-02-02/adanis-carmichael-mine-gains-final-state-environmental-approval/7134638>.

Altman, J. 2012, 'Indigenous rights, mining corporations, and the Australian state in the politics of resource extraction 2012', in S. Sawyer & E. Gomez (eds), *The Politics of Resource Extraction; Indigenous Peoples, Multinational Corporations, and the State*, Palgrave McMillan, pp. 46–74.

Anderson H. 1971, Coal Mining, Lothian Publishing Company, Melbourne.

Archibald-binge, E. 2018, 'Traditional owners fighting the Adani coal mine could have their legal challenge dismissed if they don't front up $50,000 by the end of January', SBS News, 18 December, viewed 20 September 2020, <www.sbs.com.au/nitv/article/2018/12/18/traditional-owners-fighting-adani-coal-mine-ordered-pay>.

Arnautovic, K. 2017, Resources, Race and Rights: A Case Study of Native Title and the Adani Carmichael Coal Mine, PhD Thesis, viewed 15 March 2020, <https://ro.ecu.edu.au/theses_hons/1503/>.

Australian Associated Press. 2016, 'Adani's Carmichael coalmine leases approved by Queensland', The Guardian, 3 April, viewed 20 March 2020, <www.theguardian.com/environment/2016/apr/03/adanis-carmichael-coalmine-leases-approved-by-queensland>.

Australian Broadcasting Corporation. 2012, 'UN report scathing of Barrier Reef plan', ABC News, 2 June, viewed 20 March 2020, <www.abc.net.au/news/2012-06-02/un-report-scathing-of-barrier-reef-plan/4048498>.

Australian Broadcasting Corporation. 2016, 'Carmichael coal mine: mining leases approved for $21 billion project in Queensland's Galilee Basin', ABC News, 3 April, viewed 15 March 2020, <www.abc.net.au/news/2016-04-03/mning-leases-approved-carmichael-mine-qld-galilee-basin-adani/7295188>.

Australian Broadcasting Corporation. 2017, 'Adani: Premier Annastacia Palaszczuk withdraws government involvement in mine funding', ABC News, 3 November, viewed

15 March 2020, <www.abc.net.au/news/2017-11-03/premier-annastacia-palaszczuk-veto-qld-government-adani-brisbane/9117594>.

Australian Broadcasting Corporation. 2018, 'The big dry: 'See us, hear us, help us'', ABC News Rural, 29 July, viewed 15 March 2020, <www.abc.net.au/news/rural/2018-07-29/the-big-dry-see-us-hear-us-help-us/10030010?nw=0>.

Australian Conservation Foundation Incorporated v Minister for the Environment and Energy (2017) FCAFC 134.

Australian Conservation Foundation Incorporated v Minister for the Environment (2021) FCA 550.

Australian Conservation Foundation. 2019, 'ACF wins legal challenge to Adani's water scheme approval as Federal government concedes', Media Release, Australian Conservation Foundation and Stop Adani, Melbourne, 12 June, viewed 20 June 2020, <www.acf.org.au/acf_wins_legal_challenge_to_adanis_water_scheme_approval_as_fed eral_govt_concedes_case>.

Australian Government, Department of Climate Change, Energy, the Environment and Water. n.d., Energy Data: Electricity Generation, viewed 20 December 2017, <www.ene rgy.gov.au/energy-data/australian-energy-statistics/electricity-generation>.

Baer, H. 2016, 'The nexus of the coal industry and state in Australia: historical dimensions and contemporary challenges', *Energy Policy*, vol. 99, pp. 194–202.

Bandyopadhyay, T. 2015, 'The quiet death of the SBI-Adani loan agreement', *Live Mint*, 20 April, viewed 20 September 2020, <www.livemint.com/Opinion/PSAjYMTctZg144d W94ODUL/The-death-of-a-1-billion-loan-agreement.html>.

Bavas, J. 2019, 'Adani delays lead Annastacia Palaszczuk to ask Coordinator-General to intervene', Australian Broadcasting Corporation, 22 May, viewed 20 March 2020, <www.abc.net.au/news/2019-05-22/adani-approvals-removal-environment-department/11138140>.

Bebbington, A., Hinojosa, L., Bebbington, D. H., et al. 2008, 'Contention and ambiguity: mining and possibilities of development', *Development and Change*, vol. 39, no. 6, pp. 887–914.

Borschmann, G. 2015, 'Wangan and Jagalingou people reject $16 billion Carmichael mine to be built in Central Queensland', ABC News, 26 March, viewed 20 September 2020, <www.abc.net.au/news/2015-03-26/wangan-jagalingou-people-say-no-to-16-billion-car michael-mine-q/6349252>.

Brigg, M. 2018, 'Killing Country (Part 5): native title colonialism, racism and mining for manufactured consent', New Matilda, January 30, viewed 20 September 2020, <https://newmatilda.com/2018/01/30/native-title-colonialism-racism-adani-and-the-manufact ure-of-consent-for-mining/>.

Brigg, M., Quiggin, J. & Lyons, K. 2017, 'The last line of defence: indigenous rights and Adani's land deal', *The Conversation*, 19 June, viewed 20 October 2020, <https://theconve rsation.com/the-last-line-of-defence-indigenous-rights-and-adanis-land-deal-79561>.

Brunker, M. 2018, ''ALP to pay' for Adani sell-out', *The Australian*, 7 February.

Buckley, T. 2016, 'Adani's Carmichael coal project remains unbankable', *Institute of Energy Economics and Financial Analysis*, August 29, viewed 20 March 2020, <http://ieefa.org/ieefa-australia-adanis-carmichael-coal-project-remains-unbankable%E2%80%A8/>.

Buckley, T., Nicholas, S., & Shah, K. 2018, 'New South Wales Thermal Coal Exports Face Permanent Decline', *Insititute of Energy Economics and Financial Analysis*, October 2018, viewed 20 March 2020, <http://ieefa.org/wp-content/uploads/2018/10/NSW-Coal-Exports-November-2018.pdf>.

Buckley, T. & Nicholas, S. 2019, 'Conflating Queensland's Coking and Thermal Coal Industries: Thermal Coal Adds little to Queensland's State Budget', Institute of Energy Economics and Financial Analysis (IEEFA), June 2019, viewed 20 March 2020, <http:// ieefa.org/wp-content/uploads/2019/05/Conflating-Queenslands-Coking-and-Thermal-Coal-Industries_June-2019.pdf>.

Burragubba, A. 2015, 'Statement by the Wangan and Jagalingou People on the Carmichael mine', Wangan and Jagalingou Family Council, 26 March, viewed 20 September 2020, <https://wanganjagalingou.com.au/stories-two/>.

Burragubba, A. 2016, 'Adrian Burragubba: the struggle to save country from Adani', Green Left Weekly, no. 1091, 15 April, viewed 20 March 2020, <www.greenleft.org.au/content/ adrian-burragubba-struggle-save-country-adani>.

Burragubba, A. 2018, 'High noon in the Galilee: Wangan and Jagalingou law and order', in *The Coal Truth: The Fight to Stop Adani, Defeat the Big Polluters and Reclaim Our Democracy*, University of Western Australia Publishing, Australia, pp. vii–xiv.

Burragubba and Ors v Minister for Natural Resources & Anor (2017) QCA 179, viewed 20 June 2020, <https://archive.sclqld.org.au/qjudgment/2017/QCA17-179.pdf>.

Butler, M. 2018, *Managing Climate Related Financial Risk; Lessons from Adani*, Speech at the Sydney Institute, 18 May.

Caldwell, F. 2017a, 'Palaszczuk rules out royalty holiday for Adani', *Sydney Morning Herald*, 26 May, viewed 20 March 2020, <www.smh.com.au/business/companies/palaszc zuk-rules-out-royalty-holiday-for-adani-20170526-gwe937.html>.

Caldwell, F. 2017b, 'Deferred royalties on Adani could hold back $253 million from government', Sydney Morning Herald, 1 June, viewed 20 March 2020, <www.smh.com.au/ business/companies/deferred-royalties-on-adani-could-hold-back-253m-from-governm ent-20170601-gwi4cn.html>.

Carey, M. 2019, 'An inside look at how Adani dealt with Traditional Owners', NITV News, 16 May, viewed 20 September 2020, <www.sbs.com.au/nitv/nitv-news/article/2019/05/ 16/inside-look-how-adani-dealt-traditional-owners>.

Change.org. 2017, *Rescind Adani's Unlimited Water Licence and Support Aussie Farmers!*, October 2017, viewed 15 March 2020, <www.change.org/p/premier-annastacia-palaszc zuk-rescind-adani-s-unlimited-water-license-and-support-aussie-farmers>.

Cleary, P. 2012, *Mine-Field: The Dark Side of Australia's Resource Rush*, Black Inc., Collingwood, Melbourne.

Cole, J. 2014, 'Australia's biggest coal state plans for life beyond coal', *The Conversation*, 8 April, viewed 20 March 2020, <https://theconversation.com/australias-biggest-coal-state-plans-for-life-beyond-coal-24673>.

Coorey, P. 2017, 'Malcolm Turnbull tells Adani native title issues will be fixed', *Australian Financial Review*, 11 April, viewed 20 March 2020, <www.afr.com/politics/malcolm-turnbull-tells-adani-native-title-issues-will-be-fixed-20170411-gvi6i3>.

Comerford, J. 1997, 'Coal and colonials: the founding of the Australian coal mining industry', United Mineworkers of Australia (NSW Northern District Branch of the CFMEU Mining and Energy Division), Aberdare, New South Wales, Australia.

Connor, L. G., Albrecht, G., Higginbotham, N., et al. 2004, 'Environmental change and human health in Upper Hunter communities of New South Wales, Australia', *Ecohealth*, vol. 1, no. Suppl. 2, pp. 47–58.

Connor, L. G., Freeman, S. & Higginbotham, N. 2009, 'Not just a coalmine: shifting grounds of community opposition to coal mining in Southeastern Australia', *Ethnos*, vol. 74, no. 4, pp. 490–513.

Corbett, T., & O'Faircheallaigh, C. 2006, 'Unmasking the politics of native title: the national native title tribunal's application of the NTA's arbitration provisions', *University of*

Western Australia Law Review, vol. 33, no. 1, pp. 153–172, viewed 14 June 2020, <www.austlii.edu.au/au/journals/UWALawRw/>.

Corrighan, T. 1980, 'The political economy of minerals', *Journal of Australian Political Economy*, vol. 7, pp. 28–40.

Coyne, B. 2017, *Re-greening Rights Indigeneity, Climate Change, and a Timely Reconfluence of Human Rights and the Environment*', keynote address, 2017 Environment and Planning Law Association (NSW) Annual Conference.

Cousins, G. 2018, 'An environmentalist first and foremost', in D. Ritter (ed.), *The Coal Truth: The Fight to Stop Adani, Defeat the Big Polluters and Reclaim Our Democracy*, University of Western Australia Publishing, Australia, pp. 147–152.

Cox, L. 2015a, 'Uncertainty over massive Queensland mine after election shock and concerns over Indian company', *Sydney Morning Herald*, 6 February, viewed 20 March 2020, <www.smh.com.au/business/uncertainty-over-massive-queensland-mine-after-election-shock-and-concerns-over-indian-company-20150206-137mbi.html>.

Cox, L. 2015b, 'Adani's Carmichael mine is unbankable says Queensland Treasury', *Sydney Morning Herald*, 30 June, viewed 20 March 2020, <www.smh.com.au/business/companies/adanis-carmichael-mine-is-unbankable-says-queensland-treasury-20150630-gi1l37.html>.

Cox, L. 2015c, 'Gautam Adani makes special request to Malcolm Turnbull over $15b deal', *Sydney Morning Herald*, 9 December, viewed 20 March 2020, <www.smh.com.au/business/companies/adani-demanded-certainty-from-turnbull-20151209-gliuk8.html>.

Crook, M. & Short, D. 2014, 'Marx, Lemkin and the genocide-ecocide nexus', *International Journal of Human Rights*, vol. 18. no. 3, pp. 298–313.

Crothers, T. 2013, Draining the lifeblood – Galilee Basin at risk, videorecording, Youtube, 22 September, viewed 20 March 2020, <www.youtube.com/watch?v=bqwIpgoJysY>.

Crough, G. J. & Wheelwright, E. L. 1983, 'Australia: Client State of International Capital: a case study of the mineral industry', in E. L. Wheelwright & K. Buckley (eds), *Essays in the Political Economy of Australian Capitalism*, Australia and New Zealand Book Company, Sydney, pp. 15–42.

Crowe, D. 2019, 'Democracy for sale: what did Clive Palmer get for his $50m-plus?', *Sydney Morning Herald*, 25 October, viewed 20 June 2020, <www.smh.com.au/politics/federal/democracy-for-sale-what-did-clive-palmer-get-for-his-50m-plus-20191024-p533vv.html>.

Cubby, B. & Environment Reporter. 2009, 'Coal Group coy about port exposure to rising seas', *Sydney Morning Herald*, 15 June, viewed 20 March 2020, <www.smh.com.au/environment/climate-change/coal-group-coy-about-port-exposure-to-rising-seas-20090614-c7g3.html>.

Das, R. K. 2016, 'Adani to supply 1600-Mw power to Bangladesh from Jharkhand project', Business Standard, 12 July, viewed 20 July 2020, <www.business-standard.com/article/economy-policy/adani-to-supply-1600-mw-power-to-bangladesh-from-jharkhand-project-116071200589_1.html>.

Davidson, S. & de Silva, A. 2013, *The Australian Coal Industry – Adding Value to the Australian Coal Economy*, Report prepared for the Australian Coal Association, April 2013.

Davison, H. 2017, '"Irreversible consequences': Adani coalmine granted unlimited water access for 60 years', *The Guardian*, 5 April, viewed 15 March 2020, <www.theguardian.com/environment/2017/apr/05/irreversible-consequences-adani-coalmine-granted-unlimited-water-access-for-60-years>.

Dennis, R. 2015, 'Why was Newman handing out billions to an Indian coalmining company that didn't need it?', The Australia Institute, 9 February, viewed 20 March 2020,

<www.tai.org.au/content/why-was-newman-handing-out-billions-indian-coal-mining-company-didnt-need-it>.

Dennis, R. 2018, 'Adani coal mine. Matt Canavan's symbolic war that went wrong', *Sydney Morning Herald*, 9 February, viewed 12 July 2020, <www.smh.com.au/opinion/adani-coal-mine-matt-canavans-symbolic-war-that-went-wrong-20180125-h0oizp.html>.

Duus, S. 2013, 'Coal Contestations: learning from a long broad view', *Rural Society,* vol. 22, no. 2, pp. 96–110.

Elliot, T. 2017, 'Carmichael coalmine magnate Gautam Adani: from school drop out to $12bn empire', *Sydney Morning Herald*, 6 November, viewed 20 March 2020, <www.smh.com.au/lifestyle/carmichael-coal-mine-magnate-gautam-adani-from-school-drop out-to-12bn-empire-20171106-gzfobl.html>.

Eltham, B. 2012, 'Faceless and facepalm: the ALP, factions and us', Australian Broadcasting Corporation, The Drum, 14 February, viewed 20 March 2020, <www.abc.net.au/news/2012-02-14/eltham--/3829416>.

Energy Minerals Branch. 1999, *Australia's Export Coal Industry*, Department of Industry, Science and Resources, Canberra.

ENS Economic Bureau. 2014, 'Adani Group's Australia deal sealed, SBI to give him $1-billion loan', Indian Express, 18 November, viewed March 20 2020, <https://indianexpr ess.com/article/business/economy/adanis-australia-deal-sealed-sbi-to-give-him-1-bill ion-loan/>.

Environment Defenders Office Queensland. 2016a, *Legal Implications of the Declarations of Adani's Carmichael Combined Project as a 'Prescribed Project' and 'Critical Infrastructure'*, 25 October, viewed 20 March 2020, <https://d3n8a8pro7vhmx.cloudfr ont.net/edoqld/pages/376/attachments/original/1499240420/Legal-implications-of-the-declarations-of-Adanis-Carmichael-Combined-Project-October-2016.pdf?1499240420>

Environment Defenders Office Queensland. 2016b, *Water Reforms Passed – Exemption from Public Scrutiny for Adani*, 10 November.

Environment Justice Australia. 2014, *Indian Conservation Group Launches Legal Challenge to Adani's Carmichael Coal Mine*, 8 October.

Environment Justice Australia. 2015, *A Review of the Adani Group's Environmental History in the Context of the Carmichael Coal Mine Approval*, 15 March

Environmental Law Australia. 2016, Carmichael Coal ('Adani') Mine Cases in Queensland Courts, viewed 15 March 2020, <http://envlaw.com.au/carmichael-coal-mine-case/>.

Evans, G. 2010, 'A rising tide: linking local and global climate justice', *Journal of Australian Political Economy*, vol. 66, pp. 199–221.

Extel, C. V. 2016, 'Meet the landholders of the Galilee Basin', Radio National Breakfast, Australian Broadcasting Corporation, 29 January, viewed 15 March 2020, <www.abc. net.au/radionational/programs/breakfast/meet-the-landholders-of-the-galilee-basin/7096062>.

Fagan, B. & Bryan, D. 1991, 'Australia and the changing global economy: background to social inequality in the 1990s', in J. O. O'Leary & R. Sharp (eds), *Inequality in Australia: Slicing the Cake*, The Social Justice Collective, William Heinemann, Melbourne, pp. 7–31.

Farmers for Climate Action. 2018, 'Graziers gather in Mackay to oppose extractive industries', Media Release, 25 November..

Fitzgerald, R. 1984, *From 1915 to the Present: A History of Queensland*, University of Queensland Press, St Lucia.

Frost, M. 2016, End of Mission Statement by Michael Frost, United Nations Special Rapporteur on the Situation of Human Rights Defenders, visit to Australia, 18 October, viewed 20 March 2020, <www.ohchr.org/en/NewsEvents/Pages/DisplayNews.aspx?New sID=20689&LangID=E>.

Frydenberg, J. 2015, 'Josh Frydenberg puts "strong moral case" for coal exports to prevent deaths', *The Guardian*, 18 October, viewed 20 May 2021, <www.theguardian.com/austra lia-news/2015/oct/18/josh-frydenberg-puts-strong-moral-case-for-coal-exports-to-prev ent-deaths>.

Galligan, B. 1989, *Utah and Queensland Coal: A Study in the Micro Political Economy of Modern Capitalism and the State*, Queensland University Press, St Lucia.

Gartry, L. 2019, 'Mega mine next to Adani quietly put on hold, thousands of promised jobs in doubt, Australian Broadcasting Corporation', ABC News, 23 May.

Ghoukassian, A. & Crook, A. 2015, 'What Abbott has delivered (or promised) to his IPA mates', *Crikey*, 14 April, viewed 20 June 2020, <www.crikey.com.au/2015/04/14/what-abbott-has-delivered-or-promised-to-his-ipa-mates/>.

Grant-Taylor, T. 2011, 'India secures $1.8b Abbott Point coal terminal deal', Courier Mail, 4 May. .

Grata Fund. 2019, 'W&J appeal will hold Adani accountable to Australian law', Media Release, Grata Fund, Sydney, 25 January, viewed 20 March 2020, <www.gratafund.org. au/media_release_to>.

Greenpeace Australia. 2012a, *Boom Goes the Reef: Australia's Coal Export Boom and the Industrialisation of the Great Barrier Reef*, March 2015, viewed 20 June 2020, <www.gre enpeace.org.au/news/boom-goes-the-reef/>.

Greenpeace Australia. 2012b, *Cooking the Climate Wrecking the Reef: The Global Impacts of Coal Exports from Australia's Galilee Basin*, December 2015, viewed 20 June 2020, <https://fdocuments.in/document/cooking-the-climate-wrecking-the-reef-the-global-impacts-of-coal-exports-from-australias-galiliee-basin.html>.

Gregoire, P. 2019, 'In Defence of Country: An Interview with Wangan and Jagalingou Council's Adrian Burragubba', Sydney Criminal Lawyers, 14 September, viewed 20 October 2020, <www.sydneycriminallawyers.com.au/blog/in-defence-of-country-an-interview-with-wangan-and-jagalingou-councils-adrian-burragubba/>.

Guardian. 2015, Australia, the New Coal Frontier, 15 May, viewed 20 March 2020, <www. theguardian.com/environment/ng-interactive/2015/may/15/carbon-bomb-australia-the-new-coal-frontier>.

Hamilton, C. 2017, 'That lump of coal', *The Conversation*, 15 February, viewed 20 February 2020, <https://theconversation.com/that-lump-of-coal-73046>.

Hasham, N. 2015, 'Adani Carmichael: Australia's largest coalmine free to proceed after Greg Hunt gives approval', *Sydney Morning Herald*, 15 October, viewed 20 March 2020, <www.smh.com.au/politics/federal/adani-carmichael-australias-largest-coal-mine-to-proceed-after-greg-hunt-gives-approval-20151015-gk9wof.html>.

Hasham, N. 2018, 'Adani shuns water trigger despite droughts', Sydney Morning Herald, 12 June, viewed 20 March 2020, <www.smh.com.au/politics/federal/adani-shuns-water-trig ger-despite-drought-20180612-p4zkz1.html>.

Hepburn, S. 2015, 'Brandis' changes to environmental laws will defang the watchdogs', *The Conversation*, 19 August, viewed 20 October 2020, <https://theconversation.com/bran dis-changes-to-environmental-laws-will-defang-the-watchdogs-46267>.

Hepburn, S. 2016, 'Turnbull wants to change Australia's environment act: here's what we stand to lose', *The Conversation*, 31 October, viewed 20 October 2020,

<https://theconversation.com/turnbull-wants-to-change-australias-environm ent-act-heres-what-we-stand-to-lose-67696>.

Hepburn, J., Burton, B. & Hardy, S. 2011, 'Funding proposal for the Australian anti-coal movement', Media Watch, Australian Broadcasting Corporation, 1 November.

Hepburn, S., Lucas, A., Froome, C., et al. 2015, 'Greg Hunt approves Adani's coal mine again: experts respond', The Conversation, 16 October, viewed 20 September 2020, <https://theconversation.com/greg-hunt-approves-adanis-carmichael-coal-mine-again-experts-respond-49227>.

Hannam, P. 2017, 'Barbaric: Adani's giant coalmine granted unlimited water licence for 60 years', *Sydney Morning Herald*, 4 April, viewed 20 March 2020, <www.smh.com.au/ environment/barbaric-adanis-giant-coal-mine-granted-unlimited-water-licence-for-60-years-20170404-gvd41y.html>.

Hogan, B. 2015, 'The life saving potential of coal: how Australian coal could help 82 million Indians access electricity', *Institute of Public Affairs,* 22 June, viewed 20 March 2020, <https://ipa.org.au/wp-content/uploads/archive/22Jun15-BH-Report_The_life_saving_ potential_of_coal.pdf>.

Howard-Wagner, D. & Maguire, A. 2010, 'The holy grail or the good, the bad and the ugly?: a qualitative investigation the ILUA agreement making process and the relationship between ILUAs and native title', *Australian Indigenous Law Review,* vol. 14, no. 1, pp. 71–85.

Huleatt, M. B. 1991, *Handbook of Australian Black Coals: Geology, Resources, Seam Properties, and Product Specifications*, Department of Primary Industries and Energy, Bureau of Mineral Resources, Geology and Physics, Resource Report 7, Australian Government Publishing Service, Australian Government, Canberra.

Hutchens, G. 2017, 'Labor to support native title changes to protect mining deals', The Guardian, 21 March, viewed 20 March 2020, <www.theguardian.com/australia-news/ 2017/mar/21/labor-to-support-native-title-changes-to-protect-mining-deals>.

Hutton, D. 2013, *Mining: The Queensland Way*, At A Glance Pty Ltd, Queensland.

International Energy Agency (IEA). 2017, *World Energy Outlook 2017*, Paris.

Independent Expert Scientific Committee. 2013, *Advice to Decision Maker on Coalmining Project. Proposed Action: Carmichael Coal Mine and Rail Project, Queensland (EPBC 2010/5736) – New Development*, 4 March, viewed 20 March 2020, < www.iesc.gov.au/ projectadvice/advice-decision-maker-coal-mining-project-iesc-2013-034>.

Kelly, J. 2015, 'Strong moral case for Adani's Carmichael coal mine: Josh Frydenberg', *The Australian*, 18 October.

Kenny, M. 2014, 'Tony Abbott recounts backpacking days in India', Sydney Morning Herald, 4 September, viewed 20 March 2020, <www.smh.com.au/politics/federal/tony-abbott-recounts-backpacking-days-in-india-20140904-10cd3d.html>.

Kenworthy, T. & Gordon, K. 2011, 'Enabling exports clouds environmental, economic goals', *Centre for American Progress*, April 2011.

Khadem, N. 2019, 'Glencore moves to cap global coal output after investor pressure on climate change', Australian Broadcasting Corporation, 20 February, viewed 18 June 2020, <www.abc.net.au/news/2019-02-20/glencore-moves-to-cap-global-coal-output-post-investor-pressure/10831154>.

Kippen, T. 2018, 'More pressure on the government to pull support for Adani', Daily Mercury, 5 March, viewed 20 March 2020, <www.dailymercury.com.au/news/more-pressure-on-the-government-to-pull-support-fo/3352583/>.

Koziol, M. & Wroe, D. 2016, 'Turnbull government eyes $1 billion Adani loan backed by new infrastructure fund', Sydney Morning Herald, 4 December, viewed 15 March 2020,

<www.smh.com.au/politics/federal/turnbull-government-eyes-1-billion-adani-loan-bac ked-by-new-infrastructure-fund-20161204-gt3joz.html>.

Krien, A. 2019a, 'Part one: inside the Adani blockade', The Saturday Paper, 31 August, viewed 20 October 2020, <www.thesaturdaypaper.com.au/news/environment/2019/08/ 31/part-one-inside-the-adani-blockade/15671736008677#hrd>.

Krien, A. 2019b, 'Part two: second Adani blockade established', The Saturday Paper, 7 September, viewed 20 September 2020, <www.thesaturdaypaper.com.au/edition/2019/ 09/07>.

Lee, M. & Draper, S. 1988, 'The coal industry: the current crisis and the campaign for a National Coal Authority', *Journal of Australian Political Economy*, vol. 23, pp. 45–60.

Lock the Gate. 2013, *Draining the Life-Blood: Groundwater Impacts of Coal Mining in the Galilee Basin*, 23 September, viewed 20 March 2020, <www.lockthegate.org.au/drainin g_the_lifeblood>.

Lock the Gate. 2018, 'Adani to take 10 billion litres river water without Federal environ-mental assessment', Media Release, Lock the Gate, 13 June, viewed 20 March 2020, <www.lockthegate.org.au/adani_to_take_10_billion_litres_river_water_without_feder al_environmental_assessment>.

Long, S. 2017, 'The Labor Insider who lobbied for Adani', *Australian Broadcasting Corporation*, 23 November, viewed 20 March 2020, <www.abc.net.au/news/2017-11-23/ the-labor-insider-who-lobbied-for-adani/9181648>.

Long, S. & Slezak, M. 2019, 'Inside Melissa Price's decision to approve Adani's ground-water plan', Australian Broadcasting Corporation, 11 April, viewed 20 September 2020, <www.abc.net.au/news/2019-04-11/adani-damning-assessment-turned-into-approval/ 10990288?nw=0>.

Ludlow, M. 2015, 'Adani distorted jobs from Carmichael mine', Australian Financial Review, 22 November, viewed 20 March 2020, <www.afr.com/politics/adani-distorted-jobs-from-carmichael-mine-20151123-gl5bhq>.

Ludlow, M. 2018, 'Adani moves to slash costs to get Carmichael mine across the line', *Australian Financial Review*, 13 September, viewed 20 March 2020, <www.afr.com/ news/politics/adani-moves-to-slash-costs-to-get-carmichael-mine-across-the-line-20180 912-h15b2l>.

Lyons, K. 2017a, 'The Queensland government is the real driver in Adani's dirty land grab', New Matilda, 22 November, viewed 20 March 2020, <https://newmatilda.com/2017/11/ 22/the-queensland-government-is-the-real-driver-in-adanis-dirty-land-grab/>.

Lyons, K. 2017b, 'Traditional Owners expose Adani's relentless pursuit of W&J country', *New Matilda*, November 23, viewed 15 March 2020, <https://newmatilda.com/2017/11/ 23/traditional-owners-expose-adanis-relentless-pursuit-of-wj-country/>.

Lyons, K. 2019, 'Securing territory for mining when Traditional Owners say 'No': The exceptional case of Wangan and Jagalingou in Australia', *The Extractive Industries and Society*, vol. 6, pp. 756–766.

Lyons, K., Brigg, M. & Quiggin J. 2017a, *Unfinished Business: Adani, the State, and the Indigenous Rights Struggle of the Wangan and Jagalingou Traditional Owners Council*, University of Queensland, June 2017 and Earthjustice, Brisbane, Queensland, viewed 20 June 2020, <https://earthjustice.org/sites/default/files/files/Unfinished-Business.pdf>.

Lyons, K., Brigg, M. & Quiggin J. 2017b, *Return to Country: Researcher's Report to Families*, Wangan and Jagalingou Traditional Owner's Council, July,

Market Forces. 2015, 'Funding the Galilee coal mines just got harder', Market Forces, June 15, viewed 20 March 2020, <www.marketforces.org.au/funding-the-galilee-coal-mines-just-got-harder/>.

Market Forces. 2016, 'Divestment Day Live', Market Forces, 7 October, viewed 15 March 2020, <www.marketforces.org.au/divestmentdaylive/>.

Martin, C. H, Hargraves, A. J., Kininmonth, R. J. & Saywell, S. M. C. 1993, *History of Coalmining in Australia*, Con Martin Memorial Volume, Monograph series no. 21, Australasian Institute of Mining and Metallurgy, Melbourne.

Massola, J., Kerr, P. & Cox, L. 2014, ' 'Coal is good for humanity', says Tony Abbott at mine opening', *Sydney Morning Herald*, 13 October, viewed 20 January 2020, <www.smh.com.au/politics/federal/coal-is-good-for-humanity-says-tony-abbott-at-mine-opening-20141013-115bgs.html>.

McGlade v Native Title Registrar & Ors (2017) FCAFC 10.

McGlade, H. 2018, 'The McGlade case: a Noongar history of land, social justice and activism', *Australian Feminist Law Journal*, vol. 43, no. 2, pp. 185–201.

McHugh, B. 2017, 'Turmoil over Indigenous land use ruling', *Australian Broadcasting Corporation Rural*, 8 February, viewed 20 March 2020, <www.abc.net.au/news/rural/2017-02-08/turmoil-over-indigenous-land-use-ruling/8250952>.

McKenna, M. 2016, 'Gautam Adani's dream to light India's darkened nights', *The Australian*, 4 June.

Mckeown, M. 2018, 'Our Water Our Lifeblood', *Mackay Conservation Group*, 16 April, viewed 20 October 2020, <www.mackayconservationgroup.org.au/water_forum>.

McKibben, B. 2017, 'In the battle for the planet's climate future, Australia's Adani mine is a line in the sand', The Guardian, 27 March, viewed 20 February 2020, <www.theguardian.com/commentisfree/2017/mar/27/in-the-battle-for-the-planets-climate-future-australias-adani-mine-is-the-line-in-the-sand>.

Meaton, M., 2017, 'Adani Carmichael coal mine ILUA assessment', Economic Consulting Services, December 2017, viewed 20 March 2020, <http://wanganjagalingou.com.au/wp-content/uploads/2017/12/Adani-Coal-presentation-2December2017.pdf>.

Milman, O. 2015, 'Mining industry's new 'coal is amazing' TV ad labelled desperate', *The Guardian*, 6 September, viewed 20 June 2020, <www.theguardian.com/environment/2015/sep/06/mining-industrys-new-coal-is-amazing-tv-ad-slammed-as-desperate>.

Milne, C. 2008, 'Coal is on the nose, no matter the branding', The Greens, 12 November, viewed 20 March 2020, <https://mail.greensmps.org.au/articles/coal-nose-no-matter-branding>.

Mitchell-Whittington, A. 2016, 'Adani Carmichael coal mine rail line and camp approved', *Sydney Morning Herald*, 5 December, viewed 15 March 2020, <www.smh.com.au/business/companies/adani-carmichael-coal-mine-rail-line-and-camp-approved-20161205-gt3vgu.html>.

Munro, S. 2012, *Rich Land, Waste Land: How Coal is Killing Australia*, Pan Macmillan, Sydney.

Murphy, M. 2010, 'Adani seals deal with Linc Energy', Sydney Morning Herald, 4 August, viewed 20 March 2020, <www.smh.com.au/business/adani-seals-deal-with-linc-energy-20100803-115k4.html>.

Murphy, K. 2018a, 'Labor weighs Adani options as Canavan says Australia needs to get these jobs going', The Guardian, 5 February, viewed 20 March 2020, <www.theguardian.com/environment/2018/feb/05/labor-weighs-adani-options-as-canavan-says-australia-needs-to-get-these-jobs-going>.

Murphy, K. 2018b, 'Labor shouldn't toughen its stance on Adani coalmine CFMEU head warns', The Guardian, 15 February, viewed 20 March 2020, <www.theguardian.com/australia-news/2018/feb/15/labor-shouldnt-toughen-its-stance-on-adani-coalmine-cfmeu-head-warns>.

Murray, C., Browne, B. & Campbell, R. 2018, 'The Impact of Galilee Basin Development on employment in existing coal regions', *The Australia Institute*, 15 July, viewed 20 January 2020, <www.tai.org.au/content/impact-galilee-basin-development-employment-existing-coal-regions>.

Muttitt, G. 2016, 'The Sky's Limit: Why the Paris Climate Goals Require a Managed Decline of Fossil Fuel Production', *Oil Change*, September 2016, viewed 20 March 2020, <http://priceofoil.org/2016/09/22/the-skys-limit-report/>.

Needham, K. 2017, 'Chinese money will not fund 'dirty' Adani mine says embassy', Sydney Morning Herald, 5 December, viewed 20 March 2020, <www.smh.com.au/world/chinese-money-will-not-fund-dirty-adani-mine-says-embassy-20171205-gzzewy.html>.

Nicholas, S. 2022, 'Carmichael coal is not reducing poverty in South Asia', *Institute for Energy Economics and Financial Analysis*, 14 December, viewed 20 January 2024, <https://ieefa.org/resources/carmichael-coal-not-reducing-poverty-south-asia>.

O'Brien, C., & Mellor, L. 2016, 'Adani's $22-billion Carmichael coal mine to be headquartered in Townsville', ABC News, 5 December, viewed 15 March 2020, <www.abc.net.au/news/2016-12-05/adani-carmichael-coal-mine-to-be-headquartered-in-townsville/8092896>.

O'Faircheallaigh, C. 2006, 'Aborigines, mining companies and the state in contemporary Australia: a new political economy or 'business as usual'?', *Australian Journal of Political Science*, vol. 41, no. 1, pp. 1–22.

Owen, A.D. 1988, 'Australia's role as an energy exporter Status and prospects', *Energy Policy*, vol. 16, no. 2, April, pp. 131-151.

Parliament of Australia. 2017, 'Report: Native Title Amendment (Indigenous Land Use Agreements) Bill 2017 [Provisions]', Commonwealth of Australia, 20 March 2017, viewed 20 March 2020, <www.aph.gov.au/Parliamentary_Business/Committees/Senate/Legal_and_Constitutional_Affairs/NativeTitleILU2017/Report>.

Pearse, G. 2009, 'Quarry vision: coal, climate change and the end of the resources', Quarterly Essay, vol. 33, Black Inc., Melbourne.

Pearse, G., McKnight, D., Burton, B. 2013, *Big Coal Australia's Dirtiest Habit,* New South, University of New South Wales, Sydney.

Parker, M. W. & Chang, F. 2014, 'Less, Less, Less: The Beginning of the End of Coal. Asian Coal, Power and Renewables', Bernstein Research, 17 March, viewed 20 January 2020, <https://policyintegrity.org/documents/PARKERPanel1_2014.pdf>.

Paton, J. 2011, 'Australian State Targets A$6.2 Billion Coal Port Expansion', Bloomberg, 1 June, viewed 20 March 2020, <www.bloomberg.com/news/articles/2011-06-01/australian-state-targets-a-6-2-billion-coal-port-expansion-1->.

Readfeaern, G. 2015, 'Private dinners, lavish parties and shoulder rubbing: how coal giant Adani charmed Australia's political elite', *The Guardian*, 21 August, viewed 15 March 2020, <www.theguardian.com/environment/planet-oz/2015/aug/21/private-dinners-lavish-parties-and-shoulder-rubbing-how-coal-giant-adani-charmed-australias-political-elite>.

Ritter, D. 2009, *Contesting Native Title: From Controversy to Consensus in the Struggle Over Indigenous Land Rights*, Allen and Unwin, Crows Nest, Sydney.

Robertson, J. 2016, 'Carmichael coalmine appeal says Adani 'misled' Native Title Tribunal over benefits', *The Guardian*, 8 September, viewed 20 March 2020, <www.theguardian.com/environment/2016/sep/08/carmichael-coalmine-appeal-says-adani-misled-native-title-tribunal-over-benefits>.

Robertson, J. 2017a, 'Big four banks distance themselves from Adani coalmine as Westpac rules out loan', The Guardian, 28 April, viewed 20 June 2020, < https://www.theg uardian.com/environment/2017/apr/28/big-four-banks-all-refuse-to-fund-adani-coalm ine-after-westpac-rules-out-loan#:~:text=Australia's%20big%20four%20banks%20h ave,the%20resources%20minister%2C%20Matthew%20Canavan>.

Robertson, J. 2017b, 'Adani Carmichael mine to get six-year holiday on royalties, report says', The Guardian, 26 May, viewed 20 March 2020, <www.theguardian.com/envi ronment/2017/may/26/adani-carmichael-mine-to-get-six-year-holiday-on-royalties-rep ort-says>.

Robertson, J. 2017c, 'Adani mine loses majority support of traditional owner representatives', The Guardian, 15 June, viewed 20 March 2020, <www.theguardian.com/environment/ 2017/jun/15/adani-mine-loses-majority-support-of-traditional-owner-representatives>.

Robertson, J. 2017d, 'Queensland Labor strategist announces he will stop lobbying for Adani', The Guardian, 16 October, viewed 20 March 2020, <www.theguardian. com/business/2017/oct/16/queensland-labor-strategist-announces-he-will-stop-lobby ing-for-adani>.

Robertson, J. 2017e, 'Adani's compensation well below industry standard, report finds', Australian Broadcasting Corporation, 1 December, viewed 20 June 2020, <www.abc. net.au/news/2017-12-01/adani-compensation-well-below-industry-standard-report-finds/ 9212058>.

Robertson, J. 2017f, 'Adani parts way with mining services company Downer', Australian Broadcasting Corporation, 18 December, viewed 20 March 2020, <www.abc.net.au/ news/2017-12-18/adani-parts-way-mining-services-company-down-carmichael-mine- qld/9267778>.

Robertson, J. 2018a, 'Adani groundwater plans risks permanent damage to desert springs', Australian Broadcasting Corporation, 21 March, viewed 20 March 2020, <www.abc.net. au/news/2018-03-21/adani-groundwater-plan-risks-permanent-damage-to-desert-spri ngs/9569184>.

Robertson, J. 2018b, 'Queensland government considering funding $100 million road for Adani mine, documents show', Australian Broadcasting Corporation, 6 June, viewed 15 March 2020, <www.abc.net.au/news/2018-06-06/adani-mine-qld-government-consider ing-funding-$100m-road/9837042>.

Robertson, J. 2018c, 'Adani site to remain under native title until finance confirmed, min- ister says', Australian Broadcasting Corporation, 15 September, viewed 20 March 2020, <www.abc.net.au/news/2018-09-15/adani-site-to-remain-under-native-title-until-fina nce-confirmed/10249692>.

Robertson, J. 2018d, 'Adani aims to quash traditional owner challengers, tells court they're impecunious', Australian Broadcasting Corporation, 14 December, viewed 20 March 2020, <www.abc.net.au/news/2018-12-14/adani-aim-quash-traditional-owner-challeng ers-over-money/10616732>.

Robertson, J. 2019, 'Adani coal mine should be suspended, UN says, until all traditional owners support the project', Australian Broadcasting Corporation, 25 January, viewed 20 March 2020, <www.abc.net.au/news/2019-01-25/adani-mine-should-be-suspended-un- traditional-owners/10686132>.

Robertson, J. & Sigato, T. 2018, 'Adani Indigenous challenge dismissed by Federal Court, Government could cancel mine native title', Australian Broadcasting Corporation, 17 July, viewed 20 September 2020, <www.abc.net.au/news/2018-08-17/adani-federal- court-traditional-owners-native-title/10131920>.

Queensland Government. 2013, *Plan to Develop Galilee Basin Unveiled*, Media Statement, 7 November, viewed 20 March 2020, <http://statements.qld.gov.au/Statement/2013/11/7/plan-to-develop-galilee-basin-unveiled>.

Queensland Government. 2014, *Carmichael Coal Mine and Rail Project Report*, May 2014, viewed 20 September 2020, <www.statedevelopment.qld.gov.au/resources/project/carmichael/carmichael-coal-mine-and-rail-cg-report-may2014.pdf>.

Queensland Government. 2016, *Premiere Secures Adani Commitment on Regional Queensland Jobs*, Media Statement, 6 December, viewed 15 March 2020. <http://statements.qld.gov.au/Statement/2016/12/6/premier-secures-adani-commitment-on-regional-queensland-jobs>.

Rankin, J. 2015, 'Standard Chartered quits controversial Queensland coal mining project', *The Guardian*, 10 August, viewed 20 March 2020, <www.theguardian.com/business/2015/aug/10/standard-chartered-quits-controversial-queensland-coal-mining-project>.

Ray, G. 2005, 'Valley's 'lunar landscape' prompts demand for mine free zone', *Newcastle Herald*, 8 February.

Readfeaern, G. 2015, 'Private dinners, lavish parties and shoulder rubbing: how coal giant Adani charmed Australia's political elite', *The Guardian*, 21 August, viewed 15 March 2020, <www.theguardian.com/environment/planet-oz/2015/aug/21/private-dinners-lavish-parties-and-shoulder-rubbing-how-coal-giant-adani-charmed-australias-political-elite>.

Remeikis, A. 2014, 'Queensland wedded to coal and proud: Campbell Newman', Brisbane Times, 24 September, viewed 20 February 2020, <www.brisbanetimes.com.au/national/queensland/queensland-wedded-to-coal-and-proud-campbell-newman-20140923-10l39c.html>.

Rolfe, J. 2014, 'Carmichael mine is a game changer for Australian coal', The Conversation, 29 July, viewed 20 July 2020, <https://theconversation.com/carmichael-mine-is-a-game-changer-for-australian-coal-29839>.

SBS News. 2016, 'Adani's Carmichael mine: The long controversial road to approval', SBS News, 6 December, viewed 20 October 2020, <www.sbs.com.au/news/adani-s-carmichael-mine-the-long-controversial-road-to-approval>.

SBS News. 2017, 'Labor slams government over native title move', SBS News, 16 February, viewed 20 March 2020, <www.sbs.com.au/news/article/2017/02/16/labor-slams-govt-move-native-title>.

Scambary, B. 2013, *My Country, Mine Country: Indigenous People, Mining and Development Contestation in Remote Australia*, Australian National University Press, Canberra.

Seccombe, M. 2014, 'The End of Coal', The Saturday Paper, 26 April, viewed 20 March 2020, <www.thesaturdaypaper.com.au/news/resources/2014/04/26/the-end-coal/1398434400#.VDaEJfmSySr>.

Slezak, M. 2017a, 'Farmers join fight against Adani coalmine over environmental concerns', The Guardian, 30 June, viewed 20 March 2020, <www.theguardian.com/environment/2017/jun/30/farmers-join-fight-against-adani-coalmine-over-environmental-concerns>.

Slezak, M. 2017b, 'Abbot Point coal terminal: Westpac may not refinance Adani loan', The Guardian, 12 October, viewed 20 March 2020, <www.theguardian.com/environment/2017/oct/12/abbot-point-coal-terminal-westpac-may-not-refinance-adani-loan>.

Slezak, M. 2018a, 'Adani asked coalition to help secure funding from China FOI shows', *The Guardian*, I March, viewed 20 March 2020, <www.theguardian.com/environment/2018/mar/01/adani-asked-coalition-to-help-secure-funding-from-china-foi-shows>.

Slezak, M. 2018b, 'Clive Palmer seeks approval for 'monster mine' next door to Adani', Australian Broadcasting Corporation, 26 April, viewed 20 March 2020, <www.abc.net.au/news/2018-04-26/clive-palmer-seeks-approval-for-monster-mine-near-adani/9698680>.

Smee, B. 2018a, 'Plans to mine 62 billion tonne Queensland coal deposit quietly revived', The Guardian, 27 March, viewed 20 March 2020, <www.theguardian.com/environment/2018/mar/27/plans-to-mine-62bn-tonne-queensland-coal-deposit-quietly-revived>

Smee, B. 2018b, 'Carmichael rail line engineering firm says it has demobilised from Adani project', The Guardian, 16 May, viewed 20 March 2020, <www.theguardian.com/business/2018/may/16/carmichael-rail-line-engineering-firm-says-it-has-demobilised-from-adani-project>.

Smee, B. 2018c, 'Adani coal port under threat of stop order amid concern for sacred sites', The Guardian, 28 June, viewed 20 March 2020, <www.theguardian.com/environment/2018/jun/28/adani-coal-port-under-threat-of-stop-order-amid-concern-for-sacred-sites>.

Smee, B. 2018d, 'Adani lodges slimmed down plan to expand Abbot Point coal terminal', The Guardian, 1 August, viewed 20 March 2020, <www.theguardian.com/business/2018/aug/01/adani-lodges-slimmed-down-plan-to-expand-abbot-point-coal-terminal>.

Smith, L. 2018, 'Mixed feelings over constructing a Galilee Basin rail line', *The Morning Bulletin*, 8 November, viewed 15 March 2020, <www.themorningbulletin.com.au/news/mixed-feelings-over-constructing-a-galilee-basin-r/3333392/>.

Smyth, J. 2017, 'World's biggest coal port looks to life after fossil fuels', Financial Times, 18 December, viewed 15 June 2020, <www.ft.com/content/e1453830-e2f4-11e7-97e2-916d4fbac0da>.

South Asia Times. 2014, 'Setback for Adani, GVK as 'land Lease' quits Abbott Point project', South Asia Times, February 27, viewed 20 March 2020, <www.southasiatimes.com.au/news/?p=4441>.

Swann, T. 2017, 'Redirecting Adani's NAIF loan into other industries', *The Australia Institute,* 16 November, Canberra, viewed 20 September 2020, <www.tai.org.au/content/redirecting-adani's-naif-loan-other-industries>.

Sydney Morning Herald. 2016, 'Traditional owners split over Adani mine', *Sydney Morning Herald*, 17 April, viewed 15 March 2020, <www.smh.com.au/business/companies/traditional-owners-split-over-adani-mine-20160417-go8780.html>.

Sydney Morning Herald. 2017, 'Adani scraps Downer deal after loan veto', *Sydney Morning Herald*, 18 December, viewed 15 March 2020, <www.smh.com.au/business/companies/adani-scraps-downer-deal-after-loan-veto-20171218-p4yxsq.html>.

Talukdar, R. 2018, 'Adani is Byword for Government's Climate Inaction as Australia Gears for Elections', *Newsclick,* 19 December, viewed 20 March 2020, <www.newsclick.in/adani-byword-governments-climate-inaction-australia-gears-elections>.

Talukdar, R. 2019, 'No means no: Adani's land woes in Australia', Newsclick, 11 February, viewed 12 March 2020, <www.newsclick.in/no-means-no-adanis-land-woes-australia>.

Taylor, L. 2015, 'Coal from Carmichael mine 'will create more annual emissions than New York'', *The Guardian*, 11 November, viewed 20 September 2021, <https://www.theguardian.com/environment/2015/nov/12/coal-from-carmichael-mine-will-create-more-annual-emissions-than-new-york>.

Thompson, F. 2006, 'Communities resist mines', *Herald*, 30 October, p. 13.

Turner, J. & Blyton, G. 1985, The Aboriginals of Lake Macquarie: a brief history, Lake Macquarie City Council, Speers Point, Australia.

Van Vonderen, J. 2015, 'Australian Conservation Foundation challenges Adani's Carmichael coal mine in Federal Court', *Australian Broadcasting Corporation*, 9 November, viewed

15 March 2020, <www.abc.net.au/news/2015-11-09/adani-mine-australian-conservation-foundation-court-challenge/6923598>.

Wagner, M. 2017, 'Australia's Ongoing Violation of the Rights of the Wangan and Jagalingou People to be Consulted in Good Faith About the Development of the Native Title Amendment (Indigenous Land Use Agreements) Bill 2017 (Cth) and its Impacts on the Wangan and Jagalingou', Earth Justice, 1 March, viewed 20 March 2020, available online at www.wanganandjagalingou.com.au.

Wahlquist, C. 2018, 'Batman by-election: Adani casts long shadow over former Labor stronghold', *The Guardian*, 3 February, viewed 15 March 2020, <www.theguardian.com/australia-news/2018/feb/03/batman-byelection-adani-casts-long-shadow-over-former-labor-stronghold>.

Wangan and Jagalingou. 2015, *Submission to the Special Rapporteur on Indigenous Peoples by the Wangan and Jagalingou People,* 2 October, viewed 20 September 2020, <http://wanganjagalingou.com.au/wp-content/uploads/2015/10/Submission-to-the-Special-Rapporteur-on-Indigenous-Peoples-by-the-Wangan-and-Jagalingou-People-2-Oct-2015.pdf>.

Wangan and Jagalingou. 2016a, 'We stand in the way' of Adani mine say Traditional Owners', *Wangan and Jagalingou Family Council,* 5 December, viewed 20 September 2020, <https://wanganjagalingou.com.au/we-stand-in-the-way-of-adani-mine-say-traditional-owners-seek-urgent-meeting-with-gautam-adani-after-filing-objection-to-carmichael-mine-land-use-agreement/>.

Wangan and Jagalingou. 2016b, 'Traditional Owners construct 'legal line of defence' against Adani and Qld Govt', *Wangan and Jagalingou Family Council*, 7 December, viewed 20 September 2020, <https://wanganjagalingou.com.au/traditional-owners-construct-legal-line-of-defence-against-adani-and-qld-govt/>.

Wangan and Jagalingou. 2017a, 'W&J resist mining industry push to amend Native Title Act to secure Carmichael mine proposal', Wangan and Jagalingou Family Council, 12 February, viewed 20 September 2020, <https://wanganjagalingou.com.au/wj-resist-industry-push-for-amended-native-title-act-to-secure-carmichael-mine-proposal/>.

Wangan and Jagalingou. 2017b, 'Senate inquiry report tabled: labor support for Native Title Bill profoundly disappointing', Wangan and Jagalingou Family Council, 21 March, viewed 20 September 2020, <https://wanganjagalingou.com.au/senate-inquiry-report-tabled-labor-support-for-native-title-bill-profoundly-disappointing/>.

Wangan and Jagalingou. 2017c, 'Adani faces strong Indigenous fight despite s66B court outcome', *Wangan and Jagalingou Family Council*, 12 April, viewed 20 September 2020, <https://wanganjagalingou.com.au/adani-faces-strong-indigenous-fight-despite-court-outcome/>.

Wangan and Jagalingou. 2017d, 'W&J call on Labor, Greens and Xenophon Senators to hold fast against government push on native title vote', *Wangan and Jagalingou Family Council,* 11 May, viewed 20 September 2020, <https://wanganjagalingou.com.au/wj-call-on-labor-greens-and-xenophon-senators-to-hold-fast-against-govt-push-on-native-title-vote/>.

Wangan and Jagalingou 2017e, 'Senate frustrates Government's push to pass Native Title Bill', *Wangan and Jagalingou Family Council,* 11 May, viewed 20 September 2020, <https://wanganjagalingou.com.au/senate-frustrates-governments-push-to-pass-native-title-bill/>.

Wangan and Jagalingou. 2017f, 'Brandis intervenes in W&J court action against Adani', *Wangan and Jagalingou Family Council*, 18 May, viewed 20 September, <https://wanganjagalingou.com.au/brandis-intervenes-in-indigenous-court-action-against-adani/>.

Wangan and Jagalingou. 2017g, 'Native title law now tainted by Adani', *Wangan and Jagalingou Family Council*, 15 June, viewed 20 September 2020, <https://wanganjagalin gou.com.au/native-title-law-now-tainted-by-adani/>.

Wangan and Jagalingou. 2017h, 'W&J claimants again vote down Adani deal, seek injunction', *Wangan and Jagalingou Family Council*, 3 December, viewed 20 September 2020, <https://wanganjagalingou.com.au/wj-claimants-again-vote-down-adani-deal-seek-inj unction/>.

Wangan and Jagalingou. 2018a, 'The Path of Resistance', Wangan and Jagalingou Family Council, 11 July, viewed 20 March 2020, <http://wanganjagalingou.com.au/the-path-of-resistance/>.

Wangan and Jagalingou. 2018b, 'Request to the United Nations Committee on the Elimination of Racial Discrimination for Urgent Action under the Early Warning and Urgent Action Procedure', Wangan and Jagalingou Family Council, 31 July, viewed 20 September 2020, <http://wanganjagalingou.com.au/wp-content/uploads/2018/08/Requ est-for-Urgent-Action-by-Wangan-and-Jagalingou-People-to-CERD-31-July-2018.pdf>.

Wangan and Jagalingou. 2019a, 'Federal Court Adani decision: W&J rights fight will continue', Wangan and Jagalingou Family Council, Media Statement, 12 July, viewed 20 September 2020, <https://wanganjagalingou.com.au/federal-court-adani-decision-%EF%BB%BFwjs-rights-fight-will-continue>.

Wangan and Jagalingou. 2019b, 'Defending our people, our rights, and our country', *Wangan and Jagalingou Family Council*, 1 September, viewed 20 September 2020, <https://wanganjagalingou.com.au/defending-our-people-our-rights-and-our-country/>.

Waters, L. 2015, 'Senate to vote on investigation into Adani's ownership and tax arrangements for Abbott Point port and Carmichael mine', Media Release, 22 June, viewed 15 March 2020, <http://larissa-waters.greensmps.org.au/content/media-releases/senate-vote-invest igation-adani's-ownership-and-tax-arrangements-abbot-point>.

West, M. 2015, 'Adani shown the door by traditional owners', *Sydney Morning Herald*, 4 July, viewed 20 March 2020, <www.smh.com.au/business/adani-shown-the-door-by-trad itional-owners-20150702-gi3y2h.html>.

West, M. & Cox, L. 2015, 'Adani and Commonwealth Bank part ways, casting further doubt on Carmichael coal project', *Sydney Morning Herald*, 5 August, viewed 15 March 2020, <www.smh.com.au/business/companies/adani-and-commonwealth-bank-part-ways-cast ing-further-doubt-on-carmichael-coal-project-20150805-gisd11.html>.

Westpac. 2017, 'Westpac launches updated Climate Change Action Plan', Media Release, 28 April, viewed 18 March 2020, <www.westpac.com.au/about-westpac/media/media-releases/2017/28-april/>

Wilson, J. 2018, 'Townsville isn't an outlier: It shows with clarity what Australia could become', The Guardian, 21 February, viewed 20 February 2020, <www.theguardian.com/ commentisfree/2018/feb/21/townsville-isnt-an-outlier-it-shows-with-clarity-what-austra lia-could-become>.

Legislations

Commonwealth Coal Industry Act 1946
Environment Protection Act 1994, Queensland
Minerals Resources Act 1989, Queensland
Native Title Act (Cth) 1993
Native Title Amendment (Indigenous Land Use Agreements) Act 2017

Websites

https://littleblackrock.com.au
https://adanifiles.com.au
www.adaninomeansno.com
http://envlaw.com.au/carmichael-coal-mine-case/
www.environment.gov.au/epbc/
www.farmersforclimateaction.org.au
www.standing-our-ground.org
www.stopadani.com

8 Resistances from coal's new frontier in the Galilee Basin in Central Queensland

Nothing has changed for our people except the way in which our lands are appropriated and our people subjugated…So, we fight. We fight to protect our connection to Country and prevent damage to our ancient heritage.

Wangan and Jagalingou Elder Adrian Burragubba (2018)

A national anti-coal environmental movement was consolidated from resistances across key sites of extraction during Australia's resource boom. It ranged from anti-coal contestations in the Hunter Valley and Liverpool plains, opposition to coal seam gas in New South Wales and Victoria, and a smaller extent in Southern Queensland, and opposition to coal-port developments on the Great Barrier Reef. The movement was shaped through various anti-coal socio-political dynamics related to these sites at the local, regional, and national levels. In the Hunter Valley, the environment movement joined local and grassroots networks against coal while in the Liverpool Plains alliances were built between environmentalists, farmers, and native titleholders. Conservative rural communities and farmers were politicised through resisting coal seam gas extraction, and the deteriorating Reef became the symbol of Australia's climate-destroying economic pathway of coal.

Compared to the previous resistances, challengers to the Galilee coal mines were faced with the spatial factor of sparse human geography and dominant pro-coal rhetoric from Central Queensland. The Galilee Basin region is sparsely populated by rural towns and scattered properties of pastoralists and graziers. According to the 2011 census, an estimated 20,000 people live in the region, and agriculture is the main industry, employing one-third of the population. Community activity is mostly concentrated around the main towns – Alpha, Jericho, Barcaldine, Aramac, Tambo, Blackall, Charleville, Richmond, Augathella and Hughenden – where 75% of the region's residents live (Australian Bureau of Statistics 2013). The closest large economic centres lie on the central coast, at Townsville to the northeast and Mackay to the east. Mackay serves as a mining services town for the coal-producing Bowen and Surat Basins.

Australian governments have defended the logic of opening up the greenfield Galilee Basin for coal mining on account of the region's seeming insignificance as

DOI: 10.4324/9781003410416-8

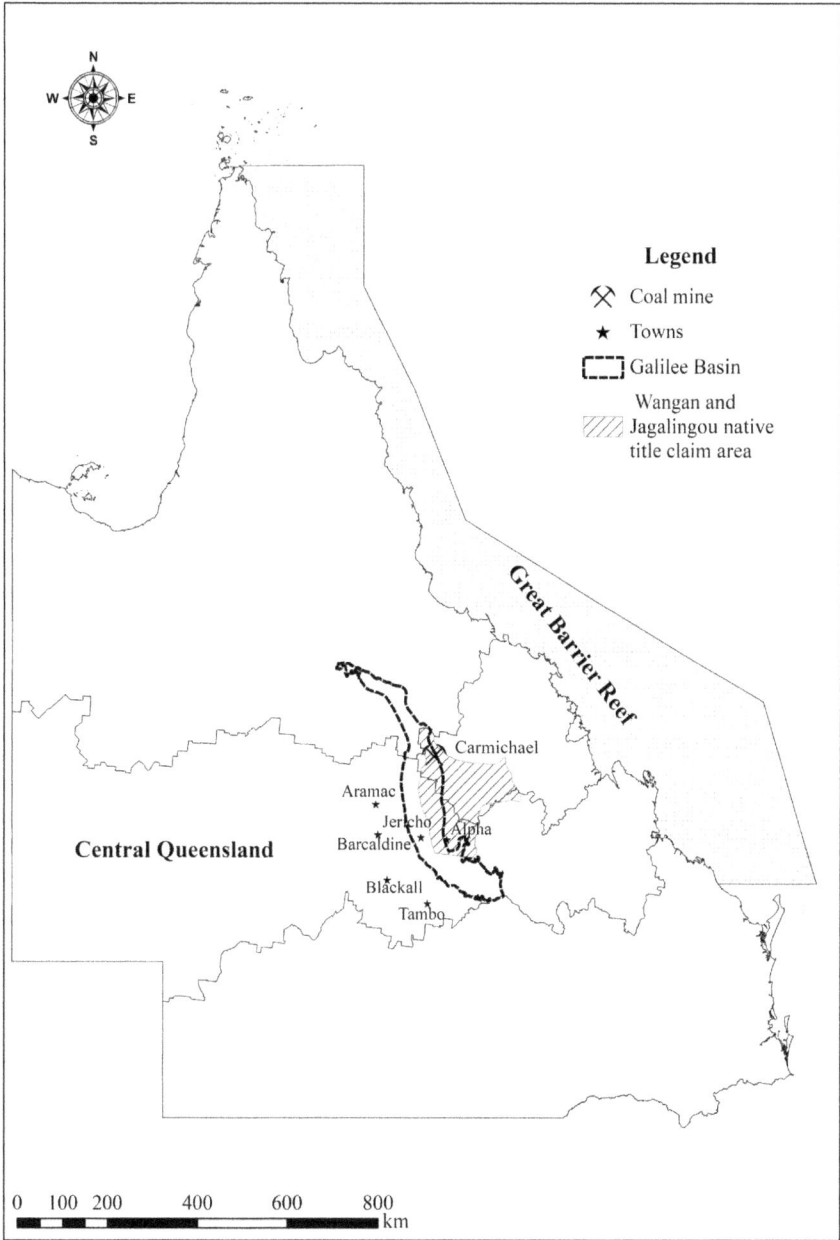

Figure 8.1 Galilee Basin geographic area and Wangan and Jagalingou native title claim area.

a place owing to its remoteness. Greg Hunt, the federal environment minister in the Tony Abbott government, described the Carmichael project as a 'mining operation in the deep outback of Queensland...[in] one of the most remote areas...an enormous distance from any significant town' (cited in Sturmer 2014, para 5). Apart from ignoring climate change and environmental concerns around the region's unique biodiversity, this understanding disregards the sacredness of Country and dreaming places for the Wangan and Jagalingou (W&J) people and the critical value of groundwater for farmers. Jolley and Richards (2020) argue that such an understanding demonstrates what Howlett and Lawrence (2019, p.1) describe as settler colonialism within mineral governance in Australia.

At the regional scale, a nexus of politics, the coal industry, and the Newscorp media that enjoys primary readership in Central Queensland, promised that the Galilee Basin coal mines would be the 'next big thing' for the region. The economic benefits of the resource boom had been felt unevenly across Central Queensland. While some economic centres had prospered, others had been left behind. Mackay, previously a base for Queensland's sugar industry, had flourished during the mining boom with an expanded coal-export trade through the Hay Point Terminal south of the city. In contrast, Townsville had reeled from the closure of a nickel refinery owned by mining magnate Clive Palmer and faced a high rate of local unemployment. The Galilee coal mines were promised as the economic solution that Townsville needed.

The pro-coal narrative constructed spatial identities in the national debate by representing the issue of the Carmichael mine as a cultural friction between Australia's north and south and as an urban–rural binary that is fundamental to the mainstream Australian identity (Jolley and Richards 2020). This construct was made possible in part through a combination of a dislike for influences from Sydney and Melbourne and largely Southern Australia, conservative politics, and a rural populism characterised by a frontier ethos that built up over time owing to Queensland's developmental history (Stuart 1985; Duus 2015). This construct created a generic imagery of an inner-city 'greenie' who is emotionally and spatially removed from the social and economic questions of Central Queensland as the face of the anti-Adani opposition (Jolley and Richards 2020). It disregarded the class struggle implicit in farmers' discontent over coal mining, the historic grounds of the W&J's challenge, and the concerns of local residents opposed to the Galilee mines.

With a focus on the politics of coal in relation to Central Queensland, this chapter analyses the tactics and pathways to the resistance of the three socially and historically distinct anti-coal constituents – local environmental groups, farmers in the Galilee Basin, and the W&J traditional owners – who now faced the common risk of coal mining. It answers the third research question of this book from an Australian perspective, about the discourses, tactics, and relations of anti-coal activisms; and their significance for environmentalism.

The background summarises the political economic history of the Galilee Basin from the beginning of White settlement and colonial dispossession of Indigenous people. It traces the changing power of pastoralists from the colonial period till the present times. The first, second and third sections look at these three streams of mobilisations against the Carmichael project. For the environment movement, the local tactics of challenging coal in Central Queensland have been considered critical for challenging Australia's coal exports at the grassroots level. For the farmers of the Basin, a sustained disaffection owing to the State's neglect toward them has turned into a conflict over artesian water that has been put at risk by the Galilee mines. The W&J's campaign has focused on the inadequacy of Australia's native title regime. They have forged international Indigenous solidarities and strategic Indigenous–green relations that have situated their story within a global context of fossil fuel extraction on Indigenous lands.

The Analysis discusses how the three campaign streams built a matrix of opposition to coal mining that both grounded the issue of climate change within the local conflict as well as foregrounded climate change as an overarching concern for their disaffections. The conclusion highlights the significance of the collective resistance to Australia's new anti-coal environmentalism. My fieldwork in Central Queensland a month before the 2017 state election offered an opportunity to observe the full extent of the social and political conflict over the Carmichael mine that had become a cardinal election issue in the Central Queensland region.

Background

Although coal mining occupies a dominant economic position in Central Queensland, it is new to the Galilee Basin, which lies toward the central-west of the broader region. The State's approach of deeming the Galilee Basin a suitable sacrifice zone for coal extraction ignores the region's political and economic history. A historical perspective can offer insights into the significance of collective anti-coal opposition from disparate regional communities in the Galilee who share a conflicted past.

Queensland's history of frontier violence is considered the grimmest of all the Australian colonies. The scale of violence is partly attributable to the relatively short period of time in which Queensland was colonised and the simultaneous advance of the multiple frontiers of pastoralism, mining, maritime and plantation on Indigenous lands (Evans 2004). From the 1860s, the expansion of agriculture in the new colony was driven by settler society's ambitions to extend both Christianity and 'civilisation' to Aboriginal land (Evans 2007). Through a process of regional transformation that included extensive ecological changes resulting in the disruption of Indigenous food and water sources, the pastoral industry removed Indigenous people from Country within one generation in most cases (Duus 2015).

In the essay, *The Place of Many Waters* author Sujatha Fernandes (2022) offers an account of the brutality of conflicts on Country over a period of four decades from the time of the advent of British colonisation in the Central Queensland Hinterland in the 1840s:

There were several battles with Aboriginal Groups between 1857 and 1861 as white settlers set up stations on Aboriginal lands, bringing herds of invasive species like sheep and cattle. In a series of incidents at Hornet Bank station and Cullin-la-ringo, the Iman and Gayiri people fought the settlers with nulla-nullas, killing a few dozen. In retaliation, colonial police and death squads hunted them down and massacred them in the hundreds. From the 1860s to the 1880s, there were further massacres in Belyando and elsewhere.

Fernandes (2022, para 10)

The conflicts, massacres, and consequent removals and forced relocations from Country – there were waves of forced removals including in 1918 when communities and families were taken to the Cherbourg Mission and forced relocations including to the Woorabinda Mission in 1927 – set the historical socio-political backdrop against which the establishment of pastoralism was progressed in Central Queensland beginning with securing water supplies.

As a permanent and exhaustive water source in Australia's dry interiors, groundwater from the Great Artesian Basin held great significance for settler society and for pastoralism that was central to the Basin's economy. Artesian water was the first 'resource' to be 'discovered', utilised and valued in the area (Duus 2015). Often referred to as 'liquid gold', Queensland's first groundwater was 'extracted' at Barcaldine in 1887. Hundreds of flowing bores quickly followed in western Queensland and helped to secure water supplies for the outback pastoral industry (Hoch 1992, p. 29). The colonial government supported the expansion of pastoralism in the semi-arid central-west through funding explorations and the establishment of artesian bores (Duus 2015).

In the early years of White occupation, economic output in the Galilee Basin was dominated by wool production for export. Apart from displacing the original people, the wool industry was also characterised by conflicts between the classes of pastoral workers, landholders and governments, leading to the prolonged Shearer's Strike in 1891 (Svensen 2008). The power of pastoralists was built through a combination of preferential legislations and the dominance of pastoral interests in the colonial government. Starting with the *Regulating the Occupation of Unoccupied Crown Lands in the Unsettled Districts Act 1860*, Queensland's earliest land acts were designed to incentivise and facilitate the occupation of Country by pastoralists (Duus 2015).

The livestock industry continued to change the ecology of the Galilee Basin through broad-scale vegetation clearing and planting of exotic grass for pasture. The state incentivised mechanised clearing after World War II through tax deductions; these measures increased the pace and scale of vegetation clearing (Wear 2010). Queensland had one of the highest land clearing rates in the world, and within the Galilee Basin, places such as Jericho Shire became the most extensively cleared sites, in pursuit of pasture for livestock (Cooper 2005). Land clearing is a source of historic conflict between environmentalists and graziers in Queensland. Queensland passed successive legislations to tackle land clearing from the 1990s; these measures were met with opposition from agricultural interests. Broad-scale clearing was finally phased out through legislation passed in 2004 that aimed to harness the benefits of conservation values and reduce carbon emissions (Duus 2015).

In 1991, Queensland's Labor government passed Indigenous land rights legislation through the *Aboriginal Land Rights Act 1991*. It was regarded as a weak legislation enacted to ameliorate the racist legacy of the previous Joh Bjelke-Peterson government's Queensland Coastal Islands Declaratory Act 1984 which aimed to block land claims by Torres Strait Islanders (Tatten and Djnnbah 1991). The State has however processed Indigenous claims to Country and returned lands under the *National Native Title Act 1993* (NTA) since its inception. As per government records, over 26 years of the NTA, Queensland has resolved 148 native title claims and opposed fewer than one in 12 claims (Robertson 2019). The W&J claim area is currently bounded by other native title holdings or claims such as the Jangga to the northeast, Barada Barna Kabalbara and Yetimarla to the west, Kangoulo to the southwest, and the Bidjara people to the south (National Native Title Tribunal (NNTT) 2020).

Since the beginning of White settlement in the Galilee Basin, industry and trade have resulted in waves of social and ecological disruptions, with the initial period being characterised by one of Australia's grimmest frontier wars. For Indigenous communities, surviving colonial dispossession is an ongoing process that continues. Despite its restrictive scope, returning lands through the native title regime has begun to redress some historical injustices of colonial dispossession toward Indigenous peoples. Industry and trade have also been characterised by alliances between governments and private interests (Duus 2015). However, for farmers who had historically enjoyed political power, the pre-eminence of coal exports in the Australian economy from the 1980s has resulted in a progressive marginalisation of their influence over governments.[1] The nexus of the State and coal mining companies now directly affects the interests of the agriculture sector.

In the 21st century, communities in the Galilee Basin have encountered another wave of transformation through the prospect of massive coal mining projects. Even as climate change has posed risks to the region's primary agricultural sector through water insecurity and increased droughts (Queensland Government 2019), high influx of foreign capital and record prices of Australia's resource exports during the minerals boom have made it viable to develop the remote Galilee basin for coal.

8.1 Tactics of anti-coal environmentalism

The 'Fight for the Reef campaign' had made Queensland's coal exports a national issue through highlighting the risk to the Reef from port expansions and exacerbation of climate change through increased burning of coal. The divestment arm of the anti-coal environmental movement made a significant dent to the economic prospects of coal port expansions and the Carmichael project. In addition, Labor's win in the 2015 state election made the anti-coal movement feel that 'a bit of heat was taken off the issue' of the Carmichael mine (Stop Adani Campaigner interview 27/10/2017). However, sustained and strong political support for the Adani project at both the state and federal levels meant that the earlier success of campaign tactics for financially disrupting the project alone could not stop the development of the Galilee Basin. The coordinator of the Mackay Conservation Group (MCG) told me during an interview in the ENGO's office in Mackay that:

> The Newman government had a lot of hubris. 2013–2014 was one of the worst times, with both Abbott and Queensland advocating coal, cracking down on activists, and giving a A\$300 million loan so Adani won't have to pay royalties for 5 years. No other coalmine has been given that. And then Labor won unexpectedly. But we soon realised that did not change things. [Premier] Palaszchuk does not want to be seen as anti-jobs like greenies. The government is working on the China Stone approval now while seeing Adani through; the idea is it will be ready to go when Adani starts.
>
> Interview (28/10/2017)

My interactions and interviews with volunteers and coordinators of Central Queensland-based groups and listening to campaign discussions over a one-week period at the MCG's office revealed the structural nature of the challenge to resist Adani in Central Queensland. For the MCG, a peak ENGO in Central Queensland, the priority issues in its 30 years of operations had shifted from tackling land clearing and privatisation of National Parks to now finding themselves 'in the frontline of coal, a fight we did not choose to fight' (MCG Coordinator interview 28/10/2017). Central Queensland was transformed through the expansion of coal mining in the Bowen Basin over 15 years of the minerals boom, to become Australia's largest coal-producing region.

Owing to the scale of its transformation during the minerals boom, the region's current political economic context posed strategic challenges to attempts by local and regional environmental groups to challenge the Carmichael coal mine. The coordinator of the MCG told me:

> The Bowen Basin has thick seams of coal, so plenty of coal to be extracted, 65% is coking coal and 35% thermal coal. Most of Queensland's coal comes from there. There's been mining in the Bowen Basin from the 1970s. Mackay is different from everyone else; it has gone from boom to bust. Ask anyone born in the 1980s; an entire generation has grown up with the boom. During the boom, mine jobs generated big incomes for Mackay, but ultimately they weren't sustainable. The bust came, and then the promise of Adani followed.
>
> Interview (28/10/2017)

The risk of the Galilee Basin's development became evident to Central Queensland groups when protected nature reserves began to be cleared for the first mega mines, long before the national movement took cognisance of the issue. The former coordinator of the MCG told me during the interview in the ENGO's office in Mackay that:

This all goes back to 2007 when Clive Palmer wanted to mine Bimblebox Reserve that had been protected by Federal and state legislation. We went out to Bimblebox and spotted 146 threatened species. Then we did a survey of impacted wildlife for all the leased mine sites in Galilee – Palmer, Adani, Rhinehart.

Interview (28/10/2017)

The extent of coal mining and a corresponding weakening of environmental regulations during Queensland's coal rush significantly affected the groups' ability to respond to environmental destructions. The former coordinator of the MCG told me:

In the 1970s we had strong regulation that followed on from the UN. The boom presented a very different picture for environmental regulation. During the boom you could have multiple Environment Impact Assessments dumped on communities with only 20 days to have a say. The size of projects used to be 1 to 8 million tonnes of coal (per annum), then during the boom it became 20 to 30 million, and then with the Galilee mines it became 60 million tonnes. That kind of increase in scale is unbelievable; it leaves communities without the ability to cope. Groups are disempowered.

Interview (28/10/2017)

The process of disempowerment of communities also involved significant funding cuts from ENGOs and Queensland's Environmental Defenders legal network. Funding cuts reduced the MCG to a very small operation with only five full-time staff, and it had to rely on its committed group of volunteers. One of Stop Adani's first priorities in Queensland was to build the capacity of ENGOs and local community groups. The strategist in the climate movement told me during the interview that:

New South Wales and Victoria-based groups on coal seam gas had been there much longer. But groups in Queensland were weaker from last two decades. We spent blood, sweat, and money to build capacity with them. Giving capacity to the legal track of campaigning in Queensland was critically important. Conservation groups in Northern Queensland mostly did legal actions and freedom of information work. Lock the Gate's major focus has been on mine rehabilitation, to make sure Adani provides for the cleanup. And they have also led advocacy on the water allocation issue. Farmers for Climate Action have given commentary on water and climate.

Interview (18/06/2018)

Central Queensland groups like the MCG felt the double pressure from facing both the hostility of local conservative politicians and a hostile press. The domination of the pro-coal narrative in the Murdoch-owned Newscorp local newspapers amplified the challenges of local activists and added to the polarising social effect on the Carmichael issue by prioritising the message about jobs from coal over other concerns surrounding the issue. The coordinator of the MCG told me:

> We are at a point where coal is seen as toxic. There is some sort of an upheaval. People want jobs but are still uncomfortable with Adani. People in Townsville are used to big men coming and saying they will save the world. And then things go bust! Townsville Bulletin makes the Adani mine seem like manna from heaven. It is Murdoch media, as is the Courier Mail, Cairns Post, and Daily Mercury. We did a random survey of 220 residents in Mackay. 90% don't want Adani to get free water. Almost 50% don't want any new coalmines. You won't hear that in the media. The local media is pro-Adani. So now George Christensen, our MP (Federal Member for Dawson on the coast of Central Queensland), is calling us anti-jobs, anti-coal terrorists, a threat to sovereignty!
>
> Interview (28/10/2017)

The period between the Queensland elections in 2015 and 2017 was critical for the Stop Adani movement to build resistances in Central Queensland. The Stop Adani Campaigner told me during the interview in Mackay:

> We thought of three things. The first was to go off after the banks again. We thought of reviving direct action to help the financial campaign. As Adani got ready to build the rail corridor FLAC prepared for direct action to show investors we are disrupting the project. We won against Westpac in the 'Week of Action' in April 2017, and went after Commonwealth Bank. The second was to build local support in Bowen. We had to mitigate effects of possible crackdowns on our blockades near Abbott Point. The third was to build capacity in Queensland. The intention was to have thousands of people day after day to blockade.
>
> Interview (27/10/2017)

Front Line Action on Coal (FLAC) began peaceful blockading actions at Abbott Point and along Adani's rail corridor in 2017 (Gregoire 2019b). The grassroots direct actions led to the idea of a permanent activist camp in the Galilee Basin. Binbee, a permanent anti-Adani base camp was set up by October 2017 at a location that is a 45-minute drive from Bowen. Located on private land in the Birri Indigenous people's Country through which Adani's rail corridor passes, Binbee, which means good in the Birri language, served as a learning space for peaceful civil disobedience and collective social disruption to address the climate emergency (Gregoire 2019b). The culture at Binbee included an acknowledgment of the long Indigenous history and its continuing presence on the land. Activists at Binbee assumed a practical approach towards local sustainability through growing a kitchen garden on the campground.

Organised by FLAC and other volunteer-based grassroots networks against coal such as Reef Defenders and the more recently formed Galilee Rising,

Camp Binbee attracted volunteers from all around as well as outside Australia. Participants peacefully blockaded the port, rail corridor and Adani's mining activities, and were prepared to face arrests for doing so. The Camp also registered the presence of other grassroots civil disobedience activist networks including Extinction Rebellion and was seen by the collective as an organic template for social change at a time when politics could not be trusted to deliver climate justice (Daley 2019).

When the federal Labor opposition started to address the climate and financial risks around coal after the Paris Agreement, for activists, it indicated a small and temporary shift in the political ground on the Carmichael coal mine between the two major parties. GetUp led a months-long electoral mobilisation on the Carmichael issue in the lead-up to the 2017 Queensland election. It combined offline organising along with the standard online mobilisation tactics to achieve maximum impact by building electoral power. The strategist in the climate movement said:

> They would organise a gathering of calling volunteers, to talk about what do you care about? And then they go and call. 130,000 calls went out in two weeks to voters in marginal seats in QLD during elections, half of them were by Stop Adani volunteers, the other half by GetUp! Coordinators.

> Interview (18/06/2018)

Figure 8.2 A meeting in progress at activist Camp Binbee in Central Queensland.

Source: Photo by Ricky Carioti/The Washington Post.

The earlier Fight for the Reef campaign had generated support from select groups of environmentally supportive communities along Queensland's central coast and from Reef Tourism operators. The campaign had played out in a highly politically conflicted social environment. The long existent tension between Queensland's coal and Reef-dependent tourism industries could now be observed through the deepening social divide on the issue of the Carmichael mine between towns such as Bowen and Airlie Beach along the Central Queensland coast. The coordinator for the Airlie Beach area for the Australian Marine Conservation Society told me during the interview in Mackay:

> Arlie beach is the biggest jumping off point for the Reef on the Central Coast. We were badly affected by cyclone *Debbie*, we haven't had a storm like this before. Airlie beach knows what is going on with the Reef; they don't deny climate change. But inland in Bowen they do! It was initially planned as the capital of North Queensland, but then Townsville took over. They were the fruit bowl of Australia, and many including the Mayor made money off it. They had prawn trawling and meat works. Merinda meatworks was the main employment source, but it closed down. Fruit business suffered, people lost jobs, and the government took back fishing licences. So they put their faith in coal projects. Adani won Bowen over with a sausage sizzle. Bowen sees Airlie beach as full of fortunate wealthy greenies, and generally feel hard done up by.
>
> Interview (30/10/2017)

All regional councils on the Central Coast – Mackay, Rockhampton and Whitsundays – and Townville in North Queensland supported the project (Krien 2017). Central and North Queensland became the political battleground during the lead-up to the 2019 federal elections, with both federal political parties targeting the region's voters. Pro-coal politicians gathered to show their support for the Adani mine in the remote town of Clermont on W&J Country (Lyons 2019b).

On the other hand, federal Labor, which appeared to have shifted its stance toward the Carmichael mine, if not towards coal mining per se, targeted marginal seats in coastal electorates with a plan to diversify the regional economy, appealing to blue-collar constituents to choose local infrastructure over the financially unviable Carmichael project (Murphy 2018). Mining magnate Clive Palmer's election intervention polarised the electorate with an anti-Labor advertising campaign that he called a 'service to the nation' (Howells 2019). The success of the mining magnate's misleading campaign indicated the high moral ground coal enjoys in Central Queensland.

The resistance to Adani in Queensland did not dissipate after the mine officially commenced in September 2019. Grassroots disruption of the mine and rail construction assumed greater significance during this period. Camp Binbee started to grow in numbers after Queensland issued Adani's last approvals and extinguished the W&J's native title over the 'surrender area' (Krien 2019). The Queensland government responded by stepping up police surveillance of protestors in anticipation of civil disobedience when the Adani mine commenced (Smee 2019a). It also criminalised protests through a new law banning activists from using 'locking devices'. 'Locking on' to machinery using these devices had become a common activist tactic at peaceful blockades of coal mine and rail construction. This move prompted the UN to warn that Australia fell short of international human rights obligations in relation to peaceful assembly (Smee 2019b).

Outside the timescale of this research, since 2019, the resistance also extended to other Galilee coal mines, particularly mining magnate Clive Palmer's Waratah, against which a youth climate justice group brought a human-rights-based legal objection. Youth verdict alleged that by contributing to climate change, the mine would infringe on their right to life, the protection of children, and the right to culture as protected by the newly legislated Queensland Human Rights Act 2019.[2] Following the directive of Queensland's Land Court where the case was heard, the state government denied an environmental licence for Waratah because of its 'unacceptable' climate impacts.

8.2 Rural discontent over coal and Farmers for Climate Action

The idea that began with the 'Lock the Gate' initiative in the coal seam gas mining impacted farming regions of New South Wales and Victoria – farmers locking their gates to hydraulic fracturing on their properties – expanded in Central Queensland through farmers opposing Galilee coal mines to also include demands for climate action from governments. Several local farmers joined the new national alliance, called the Farmers for Climate Action. Farmers in Central Queensland were fighting multiple threats from coal mining in the Galilee Basin that were a result of decades of a structural shift in government's priorities toward favouring mining corporations over the farming sector. A Longreach-based farmer who led the national farmer's call to climate action told me during a telephonic interview that:

> The Lake Eyre Basin Advisory Committee is fighting Queensland and Federal Governments on water issues. We are worried that 350 km long and 50 km wide coalmines will be dug right on the main recharge zone of the GAB over the long term. None of the coalmines in the Bowen Basin sit on recharge zone of the GAB nor have free licences. It is a huge risk to water in the outback. Then they give free licences to Adani but not to farmers. Governments are batting for companies who are putting in the money, not for people. The government has become anti-people. Farmers are the community, mining companies are corporations; this is a direct clash of the interests of the Liberal National Party. This is a huge strike at Australia's sense of fair go.
>
> Interview (23/08/2017)

As in the case of Lock the Gate, environmental activists who operate under the title of organisers, working to knit together communities facing environmental risks, helped bring together previously unconnected landowners in Central Queensland under the umbrella of the Farmers for Climate Action (FFCA), indicating another alliance forged amongst two traditionally hostile groups. While Lock the Gate had avoided discussing the issue of climate change, FFCA unambiguously called for climate action. The coordinator of Farmers for Climate Action told me during an interview that:

> Greenies have always pointed fingers at them and not understood them. I usually work with graziers. They always talk in generations. Farmers for Climate Action was formed one year ago by farmers. Our focus is on projects that are climate killers and affect policy. It is an advocacy group for unsustainable land use. We are advocates for farmers' rights and for long-term sustainability for land use and groundwater. We are becoming a voice for the industry's sustainability.
>
> Interview (30/10/2017)

The area surrounding the Carmichael mine, the electoral seat of Gregory, is considered a safe seat for the Liberal National Party. However, the sense that the elected political representative does not represent the interests of farmers is acute amongst the small and scattered community, and was reflected in what the third-generation cattle farmer from Longreach told me:

> We are formed from conservative voting populations. We have always been the core constituents of right wing politicians. And now new alliances are forging outside of politics.
>
> Interview (23/08/2017)

The transformation of Central Queensland through mining, the increasing marginalisation of farmers, and risks to farming's future viability in the region have become foremost concerns for the electorate. A cattle farmer from the rural town of Jericho told me during an interview that:

> Coalmining in Clermont used to be low key and part of the landscape. It used to be agriculture with a little bit of mining, and now it seems like a lot of mining with little agriculture, and you wonder about the long-term sustainability. Increased automation means only down to three people in a mine and you feel very vulnerable about job security. We are basically left without options in this region and governments just do not care about sustainability. We have some work to do at our end to work with Greenies; we need a cultural change to form alliances with them.
>
> Interview (01/11/2017)

This farmer's cattle property near Jericho, which I visited during my field trip in Central Queensland, lies adjacent to GVK-Hancock's proposed Alpha coal mine. Alpha poses a direct water risk to the property. The cattle property's water bores dip into aquifers adjoining the Alpha's underground sources; the property's water supply was at risk of being contaminated by coal mining. The farmer was part of the legal challenges to the Alpha and Kevin's Corner coal mines in the Land Court.

Being unable to afford legal fees, he was forced to make his own legal representation on an issue critical to his livelihood. He was also part of the citizens' delegation to India that apprehended the Queensland Premier's entourage to Mundra in March 2017. The Jericho farmer told me that seeing the condition of local communities in Mundra left him shocked as to what Adani can do to people.

Having lost faith in the Liberal National Party, the lack of power amongst ordinary people to influence political decisions, and wanting a sustainable future for Central Queensland, he has run as an independent candidate for the electorate of Gregory in successive state elections. The Jericho farmer said:

> In the Alpha case, where we won, mining companies did not acknowledge that landholders are a legitimate voice. They always tell the story of greenies stopping investment and stopping jobs. The companies refuse to put all the problems into the make good agreements, and they ask you to sign a confidentiality agreement, signing a make good agreement. We need an independent voice. They all ignore you when they get into power. One of things that happened since we started at this property at the end of 2005 is that they are weakening the right of people to object to mines unless the mine is on your property. Newman changed legislation later to give statutory water licence to mining corporations. Core issue for my election campaign is transparency and accountability of government. Coal is a classic example for these problems.
>
> Interview (01/11/2017)

Figure 8.3 A cattle farm near Jericho in Central Queensland.

Source: Photo by author.

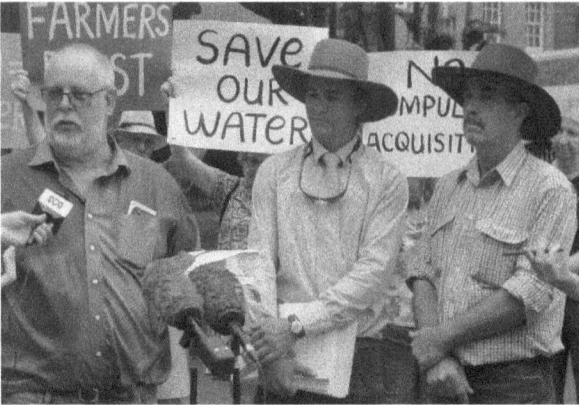

Figure 8.4 Graziers from Central Queensland as a part of a Farmers for Climate Action
delegation delivering a signature petition to Queensland Parliament against
Adani's water licence.

Source: Photo by Michael Kane/Farmers for Climate Action.

As one of its first politically focussed activities since its formation, Farmers
for Climate Action brought all the candidates for the seat of Gregory together
for a forum before the Queensland elections in November 2017 (Slezak 2017).
Attended by all candidates contesting the seat of Gregory, including those of the
One Nation, Green and Labor Parties, and the sitting LNP member, the forum
debated the issues of mining impacts, water, and climate change for the region.
During my drive from Rockhampton to Jericho along with the Farmers for Climate
Action Coordinator who was preparing the forum, we stopped at several properties
encouraging farmers to attend the forum. Strong political disaffection was evident
through farmers talking about having voted for the populist 'One Nation Party' as
a protest response during the 2015 state election.

Farmers from Central Queensland joined environmental groups to protest
the impacts of the Adani mine on water through a regional roadshow titled 'Our
Lifeblood, Our Water' in April 2018 (www.lockthegate.org.au/watermackay). They
became a vocal constituency for climate action in the wake of a severe drought and
criticised the federal Coalition government's failure to develop a long-term climate
response (Cox 2018).

8.3 'We meet at the crossroad': Wangan and Jagalingou's alliances

Conflict over fossil fuel extraction on Indigenous lands has emerged as a critical intersectional justice issue in the present era. Indigenous groups have found
themselves both on the frontline of climate impacts and of fighting fossil fuel
projects. While the tactics and strategies of environmental activisms and farmers'

mobilisations in Central Queensland were shaped in response to coal's regional politics and power, the W&J pitched their campaign at an international scale and established themselves as part of a global Indigenous solidarity against fossil fuel mining on First Nations land.

The W&J's international campaign coincided with significant movements in North America against oil extraction on Indigenous lands such as the protests in Canada against the Keystone Pipeline and in the United States against the Dakota Access Pipeline (Talukdar 2019). The W&J youth spokesperson told me during an interview in Brisbane that during the international advocacy tour in 2015, the W&J had a firsthand experience of the devastation from large-scale fossil fuel projects when they flew over the tar sands on the land of the Athabasca people in Canada (interview 20/10/2017). The W&J also built solidarities with Indigenous leaders from Standing Rock Sioux and Chickaloon Village Traditional Council in Turtle Island in the United States, who had resisted the Dakota Access Pipeline (W&J 2018). The forging of international solidarities between the W&J's resistance and the North American Indigenous struggles signified a resurgence of original sovereign rights over settler-colonial State formation (Lyons 2019a).

The intersection of climate change and fossil fuel extraction on Indigenous lands found strategic solidarity from international environmental activism that aimed to 'keep fossil fuels in the ground'. This included prominent legal activism such as undertaken by the US-based Earth Justice, which prepared the W&J's appeals to the United Nations (www.earthjustice.org). Earth Justice's pro-bono legal representations of Indigenous fossil fuel fights have also included the Standing Rock Sioux Tribe's resistance to the Dakota Access Pipeline and a Chickaloon Native Village's fight against a coal mine in Alaska (www.earthjustice.org).

Global environmentalism's new approach was drawing attention to sites of fossil fuel extraction on Indigenous lands; implying that Indigenous land justice was essential for climate justice. Global organisations such as 350.org connected stories of Indigenous resistances to protect water, land, culture and heritage from the impacts of large-scale fossil fuel projects across global locations, creating a new human rights-centric narrative for environmentalism.

In Australia, the W&J collaborated with Stop Adani while reiterating the distinctiveness of their struggle on account of its historic dimensions, rights-centric agenda and the disproportionate vulnerability of Indigenous groups in a mining conflict with the State and corporation. Spokesperson Adrian Burrugabba articulated the critical differences of the W&J's fight in a few different ways. At the 2016 Beyond Coal and Gas Summit near Newcastle in New South Wales that I attended, he emphasised the higher stakes for the W&J in the fight:

> While it will benefit people in general and the environment more widely, this is not an environmental campaign. The Wangan and Jagalingou people have joined with the environmental movement but we are running our own campaign, based on the singular act of self-determination and our right to say 'no' as the Traditional Owners and custodians of our ancestral lands where our ancestors still reside. It is possible to succeed in getting rid of this mining company. Even

the economics of global energy alone could stop it. But for us it is about self-determination. If we lose this battle we lose our right to defend our law and custom and culture. It will mean disaster for our people.

Burragubba speech, from summit notes (04/04/2016)

Burragubba described how the W&J are continuing to struggle with historic injustice and how that makes their struggle unique from other anti-Adani campaigns:

The way I see it, I was born Aboriginal. We had nothing. My parents had to leave the mission station because they had no money...this is the history of this country... All I have is my land, my lore and my culture. And that is what I have been fighting for. No one else can fight that. No one else can explain that.

Burragubba interview in Gregoire (2019a, para 35)

The W&J's campaign and the credibility of its spokesperson Adrian Burragubba were attacked by politicians, the coal mining industry, as well as prominent Indigenous intellectuals who advocated mining for Indigenous prosperity. Adani and Liberal Party politicians labelled Burragubba as an activist who did not represent the interests of the entire native title Claim Group (Brigg et al. 2017). Indigenous leader Noel Pearson and historian Marcia Langton alleged that the 'No Means No' was a campaign of a minority W&J faction that was doing the bidding of the Greens (Murphy 2017).

Members of the W&J's campaign were of the view that being perceived as subsumed within the Stop Adani environmental movement by both Indigenous and non-Indigenous civil society stakeholders, politicians and Adani, served to weaken and possibly even delegitimise the sovereign assertion of historic land justice by the W&J's campaign. How 'No means No' could be represented alongside Stop Adani in the narration of a broader story of resistance to the Carmichael coal mine therefore proved to be an ongoing point of tension between the two movements, with the cards being entirely stacked against the W&J. The campaign manager and strategic adviser to the W&J is of the view that the Stop Adani campaign, on the other hand, gained social legitimacy in being regarded as extending solidarity to the Indigenous cause (interview 05/011/2018). The tension between the Indigenous and mainstream environmental streams of the resistance to Carmichael serves as a reminder of the highly uneven socio-political terrain even in Australia within which movements operate.

Speaking at a Stop Adani Summit in Brisbane in September 2017 that I attended, Burragubba explained the fallacy of an Indigenous elder being labelled an activist by pro-mining critics:

I try not to see myself as an activist. I see my role as a water protector. Every Aboriginal person will tell you that a natural spring is a sacred site with a creation story, dreaming story, a Rainbow Serpent.

Notes from Summit (03/09/2017)

At an Indigenous Climate Justice Summit at the University of Technology Sydney in 2018, the W&J youth spokesperson said that the W&J have rejected the 'centering and normalising of the Black corporate identity' inherent in the criticism of their resistance by Marcia Langton and reiterated the criticality of their 'Indigenous rights driven work that is anchored in programs on Country' The W&J youth spokesperson talked about preparations for a 'Wangan and Jagalingou republic soon, so there is something specific that people can rally around, and counter the Marcia Langton kind of position' that argues for mining as the path to Indigenous prosperity (notes from the Summit, 05/07/2018).

Speaking at a Sydney Stop Adani Climate Summit in August 2017, Adrian Burragubba explained how mining without consent repeats colonial dispossession for Indigenous people, and why their struggle cannot be the same as that of environmentalists:

> When an Indigenous elder talks to an audience, the message is still one appealing to White people that we can learn from one another. In the end, we might have to sacrifice more than the others. So, appreciate what we do, and we meet you at the crossroads. You are now fighting extractive Adani; but we have been fighting since day 1, 1788.
>
> Notes from summit (06/08/2017)

The W&J have expressed mixed opinions about the alignment of narratives of their human rights-centric campaign with the climate change narrative of the environment movement, and the possibility of the latter obfuscating their message. According to the W&J Youth spokesperson, the focus on climate risks from fossil

Figure 8.5 Launch of the 'No Means No' campaign in Brisbane.

Source: Photo by Wangan and Jagalingou Council.

Figure 8.6 Wangan and Jagalingou spokespersons at a press conference after a Federal
Court hearing of their case.

Source: Photo by Anthony Esposito.

fuel projects could be a helpful approach for drawing attention to Indigenous
justice:

> Some climate stories make our fight more relevant. Many people have come
> from the climate story. And got to know about our issues. It has not been inten-
> tional whom and how we link to groups. Generally speaking there is a tendency
> to push our stories to the back. But some climate stories act as a gateway point
> for our stories, since they point to those of us who are on the frontline due to
> extraction on our lands.
>
> Notes from Indigenous Climate Justice Summit,
University of Technology Sydney (05/07/2018)

This agreement on a shared vision with an environmental movement reflects
a new Indigenous–green politics that has formed since a strategic shift in
environmentalism's focus on 'keeping fossil fuels in the ground'. The Australian
Indigenous youth organisation Seed articulates a specific political narrative of climate justice, which although based on the shared approach, connects the historical
reality of colonisation, the present reality of Indigenous struggles against fossil fuel
projects, and the future impacts of climate change into a continuous arc (Baldwin
and Copland 2017).

Green-Black solidarity based on a narrative of land rights and historic justice
for Indigenous people is evident within the Stop Adani movement. At Power
Shift, an annual environment and climate movement conference, organised by

the Australian Youth Climate Coalition that I attended in Melbourne in 2017, discussions foregrounded and emphasised solidarity, diversity and inclusiveness of climate stories and decolonisation. Banners hung in the seminar hall of La Trobe University, the venue for the conference read 'Song Lines not Coal Mines' and Land Rights, Not Mining Rights'. Opening the Conference, the National Directors of Seed said:

> What about all the cultural stories from our country, places that are burning, bleaching? That is what I want to talk about. I don't want to talk just about the science, that kind of story is not relevant for me culturally. But I don't know if they will understand or care about my stories, so it is important for me to tell them.
>
> Summit notes (16/08/2017)

The other W&J perspective takes a cautious approach to working with Stop Adani as the W&J's main focus is to reform Australia's Indigenous land rights regime rather than stop coal exports. The campaign manager and strategic advisor to the W&J Council said during the Indigenous Climate Justice Summit at the University of Technology Sydney in 2018:

> 'No means No' is a rights-based campaign that only tangentially connects to any environmental claims. Stop Adani is a fraction of the W&J's focus, since an environmental focus can obscure the rights issue. The W&J's economic interests are not represented by the system and legislation. Bulk of the 'No Means No' campaign is focussed on [the failures of the] native title legislation; it is about Truth Telling.
>
> Notes from the Summit (05/07/2018)

Apart from a strategic environmental interest in Indigenous sovereignty in relation to fossil fuel extraction on ancestral lands, the new solidarities of Indigenous anti-extractive struggles have also received collaborative assistance from academic projects. A team of researchers from the University of Queensland's Global Change Institute chronicled the W&J's resistance through academic and non-academic publications, helped to document responses of families on Country, and in workshops to strategically envision a climate future without mining (see Lyons et al. 2017). This strategic collaboration built up towards a symposium and public forum in Brisbane in July 2018, where a global alliance of Indigenous resistances against fossil fuel extraction called for a rethinking of development so that the rights of Indigenous people and the realities of a climate-constrained world are both kept at the centre (W&J 2018).

Time and again the W&J held the last line of legal defence against the Carmichael mine as compared to the professional environmental movement (Brigg et al. 2017). Despite this, their efforts have received relatively little national media coverage compared to the Stop Adani campaign, except where they have been falsely alleged to be puppets on the strings of the environment movement, demonstrating

Figure 8.7 Wangan and Jagalingou families meet on their ancestral land.

Source: Photo by Anthony Esposito.

Figure 8.8 Wangan and Jagalingou families perform a ceremony on their land to register their resistance as the Carmichael mine formally opens.

Source: Photo by Mark Doyle.

an obvious inequality in social and structural power between the White and Blak streams of the resistance to the Carmichael coal mine (Lyons 2016).

Outside the timescale of this research, the W&J's resistance has continued even after opening up of the Carmichael coal mine in 2019, albeit through a different approach from 'No means No'. The new 'Standing our Ground' campaign involved a permanent camp on their traditional lands adjacent to the mine site. In early 2024, the W&J launched yet another legal challenge alleging that the State is failing to protect their sacred Doongmabulla Springs from irreversible harm from open-cut coal mining at Carmichael, threatening their cultural survival and human rights.[3]

During the anti-coal and anti-coal seam resistances in the Liverpool Plains in New South Wales, some relations had been forged between farmers, environmentalists, and the Gomeroi traditional owners. However, no such direct ties were forged between the Galilee farmers and the W&J. This highlights differences within present social dynamics between graziers and Indigenous people across various coal mining regions in Australia. It reflects a spatial difference between the two regions where coal was contested and differences in the extent to which the two groups share a fraught history across the two regions.

Analysis: the significance of countering Adani from Central Queensland

The pro-coal coalition consisting of the industry, parts of government, and the Newscorp media has cast the Carmichael issue as a spatial conflict of identities between elite and urban 'greenies' from the south and blue-collar regional Queenslanders in the north of Australia (Colvin 2020). Coal mining corporations have leveraged the stereotypical anti-coal identity to ignore farmers' concerns over water allocations and dubious make-good agreements. Indigenous intellectuals and leaders have also dismissed the W&J's historical justice claims and labelled them as pawns for 'greenies'. This stereotyping has ignored concerns brought by local environmental and resident groups about the ecology and economy of Central Queensland and concerns from other coal regions that could face job losses if the Galilee mines opened up at a time of declining global demand for coal. The Galilee Basin was deemed a suitable sacrifice zone for the expansion of coal in Central Queensland, which became Australia's largest coal-producing region.

The dominance of the regional scale in the pro-coal advocacy that justified the project on the grounds of economic justice in Central Queensland made it imperative for the national Stop Adani movement to mobilise in the region with the aim of disrupting coal's power. According to Jolley and Richards (2020), the Carmichael controversy demonstrates a conflict between an old politics that attempts to retain coal's power and a new politics that disrupts it. While the pro-coal coalition asserted coal's significance along the vertical scalar geographies of the Central Queensland region, the state of Queensland and the Australian nation, the mobilisations reconfigured the politics by building networks and alliances that cut across territorially bounded electoral politics. The inclusion of climate change and climate justice in the contestations of environmentalists, farmers and the W&J

played a central role in re-scaling the Carmichael debate by connecting the local with the global.

At the local scale, the mobilisations challenged coal mining based on livelihoods, water, sacred land and culture, and ecology. From a historical perspective, the local politics over the Carmichael mine served as a reminder of the long arc of political transformation from the beginning of settler colonialism, through the State's approach toward groundwater. From the mid-1800s, colonial governments incentivised pastoralism for White settlement by securing artesian water supplies amongst other measures. By the 21st century, as new spaces were drawn into the global extractive economy, Australia's 'client State' incentivised mega coal mines through unlimited and free water allocations that risked water supplies for agriculture.

Although historically responsible for their dispossession through settler colonialism, graziers are now facing a common risk of disruption of Country and water alongside Indigenous groups (Mayes 2018). Coal mining is casting both groups as dispensable and invisible (Jolley and Richards 2020). Exacerbation of coal mining in Central Queensland from the 1980s also shifted farmer–environmentalist hostilities from historic conflict over land clearing towards collaborations for protecting groundwater from the impacts of coal mining. As opposed to conservative politics' ideological scepticism towards climate science, farmers' concerns about climate impacts in the Basin reflect pragmatism on account of the biophysical reality of intensifying droughts in Australia's interiors.

A national-scale mobilisation through the Farmers for Climate Action indicates farmers' political response to structural marginalisation by governments that promote intensive coal mining and fail to act on climate change. At a local scale, a radicalisation against the Basin's farmers' historical political alignment has emerged as a meaningful resistance to their marginalisation. For farmers from the Galilee Basin, addressing the unsustainable practices of the agricultural sector has become a necessary area of focus through the process of challenging their structural marginalisation by the State.

The challenges encountered by Stop Adani's mobilisations highlight critical spatial differences between Central Queensland and coal-impacted areas in New South Wales, where the national anti-coal movement was forged. Weakened environmental regulations, reduced capacity and attack on environmental groups, and the lack of alternative viewpoints in regional publications have left few avenues for anti-coal advocacy in regional Queensland. These factors compounded the inability to register concerns about coal mining projects, leaving communities feeling overwhelmed and environmental groups unable to challenge the multiple risks posed by the massive coal projects. It became imperative for Stop Adani to mobilise in Central Queensland, particularly when it became evident that the project could not be stopped despite the success of the divestment campaign in defunding the project, owing to its strong political support.

Stop Adani's environmentalism in Central Queensland reflects the new elements of the national anti-coal movement – a strong grassroots approach, niche organisations performing specialised activist tactics, and the deployment of new tools and technologies for 'scaling up' mass mobilisations and electoral engagement. The political rescue of Carmichael left direct non-violent disruption of mine-and-rail construction as the only effective mechanism for civil resistance, in a repetition of a pattern seen in New South Wales. Camp Binbee's vision of collective disruption as essential for social change reflects the politics of earlier blockade camps in the Hunter region and Liverpool Plains. As in the earlier cases, non-violent direct action in the Galilee Basin has embedded a global call for climate justice. Further, with the environment movement significantly relying on mass actions and grassroots disruptions against coal mining for over a decade, such actions have come to signify necessary civil society tactics for democratising Australia's coal-driven economic growth. Grassroots disruptions have continued to be relevant forms of resistance to the Carmichael project even after its commencement.

The W&J have internationalised their call for self-determination through establishing solidarities with other Indigenous struggles against fossil fuel extraction. Indigenous climate justice is now understood as both the remediation of historic dispossession and resistance to fossil fuel extraction on Indigenous lands. This understanding has allowed a resurgence of a call for sovereign rights by Indigenous peoples. The W&J have emerged as leaders in the global climate change and human rights movement and challenged Australia to meet its international responsibilities (Lyons 2016). They have exposed coal's institutional power in Australia as embedded within the structures of the native title system. The W&J's international alliances and solidarities made an assertive call for strengthening Indigenous rights for self-determination and free, prior and informed consent as a necessary step towards climate and energy justice.

The W&J's tactical relations with the Stop Adani movement underscore the distinctiveness and the sovereignty of their campaign. As a land rights movement, 'No Means No' stands out on account of its outright rejection of mining and its direct challenge to Australia's native title system. Owing to the nature of its resistance, 'No Means No' has received critical support from the legal and academic community, and developed strategic international relations with environmental legal networks. The W&J have maintained a sovereign distinction between their campaign for historic justice, a fight no one else can fight for them, and the ahistorical environmental movement that aimed to stop the coal mine. This dynamic between 'No Means No' and 'Stop Adani' also signifies a new dimension for Indigenous-green relations in Australia predicated on environmentalists being able to meet traditional owners 'at the crossroads'.

The three anti-coal contestations used climate change in distinctive ways to politicise their resistances to the controversial Carmichael coal mine and to re-scale the pro-coal claim. While Stop Adani and farmers localised and particularised the problem of climate change through attempting to shift coal's political power, the W&J internationalised their struggle for self-determination based on a political understanding of Indigenous climate justice that asserts Indigenous sovereignty. Through this process of distinct and intersecting rescaling of a coal mining conflict in Central Queensland, the mobilisations allowed alternative social, economic and environmental visions to emerge.

Conclusion

The shift in environmentalism's strategic focus allowed a new political and social understanding of climate action to emerge, one that connects the local, social and political dynamics of stopping coal extraction, with the global need to reduce greenhouse gas emissions. At the same time, Indigenous narratives put the historical dimension of colonial dispossession and present injustices of fossil fuel extraction on Indigenous lands at the centre of an understanding of global climate justice. The mobilisations from Central Queensland reflect these new understandings and demonstrate political turns that are strategically relevant for climate action's new imperatives.

Like previous anti-coal grassroots environmentalisms in New South Wales, Stop Adani's activism in Central Queensland attempted to disrupt coal's dominance in a region of intensive coal extraction. Although it could not stop the coal mine from commencing, it had a democratising effect on the region's dominant economic narrative of coal-led growth. The farmers' mobilisation reflected the long arc of political and economic transformation in the Galilee Basin. The crisis of political identity experienced by the farmers indicates a formative moment for pathways for sustainability for the Basin. The internationalisation of their assertion of sovereignty and self-determination added new dimensions to the national-scale land rights campaign of the W&J people. In solidarity with other Indigenous narratives against fossil fuel extraction and climate justice, it cast a global spotlight on the conflict over the Carmichael coal mine that is centred on historic and present Indigenous concerns.

The differences between the three anti-coal constituents reflect the significance of stopping coal and the need for climate justice in building a shared imperative across critical and historical divides. The collective resistances from Central Queensland demonstrate a relational politics that is characteristic of new anti-coal activism in Australia. The specificity of their collaborations reflects the realities of space and history and adds new elements to the understanding and dynamics of Indigenous–green relations in Australia.

Notes

1 The value of mineral exports has been greater than that of agricultural and cattle exports in the Queensland economy since 1982–1983, with coal primarily contributing to this change (Queensland Government 2009).
2 See https://climatecasechart.com/non-us-case/youth-verdict-v-waratah-coal/.
3 See https://standing-our-ground.org/.

References

Australian Bureau of Statistics. 2013, Census *Data* by *Location*, ABS, Canberra, viewed 20 March 2020, <www.abs.gov.au/websitedbs/D3310114.nsf/Home/2016%20search%20by%20geography>.

Baldwin, L. & Copland, S. 2017, 'Protecting country: first nations people and climate justice', Green Agenda, 13 April, viewed 20 October 2020, <https://greenagenda.org.au/2017/04/protecting-country-first-nations-people-climate-justice/>.

Brigg, M., Quiggin, J. & Lyons, K. 2017, 'The last line of defence: indigenous rights and Adani's land deal', *The Conversation*, 19 June, viewed 20 October 2020, <https://theconversation.com/the-last-line-of-defence-indigenous-rights-and-adanis-land-deal-79561>.

Burragubba, A. 2018, 'High noon in the Galilee: Wangan and Jagalingou law and order', in *The Coal Truth: The Fight to Stop Adani, Defeat the Big Polluters and Reclaim Our Democracy*, University of Western Australia Publishing, Australia, pp. vii–xiv.

Colvin, R. 2020, 'Our social identity shapes how we feel about the Adani mine – and it makes the energy wars worse', *The Conversation*, 25 March, viewed 20 October 2020, <https://theconversation.com/our-social-identity-shapes-how-we-feel-about-the-adani-mine-and-it-makes-the-energy-wars-worse-133686>.

Cooper, J. 2005, *Sufficient for Living: A History of Pastoral Industries in the Alpha District*, Alpha Historical Society, Alpha, Qld.

Cox, L. 2018, 'Drought-stricken farmers challenge Coalition's climate change stance in TV ad', *The Guardian*, 16 September, viewed 20 October 2020, <www.theguardian.com/environment/2018/sep/16/drought-stricken-farmers-challenge-coalitions-climate-change-stance-in-tv-ad>.

Daley, M. 2019, 'From Lismore to Camp Binbee fighting Adani every step of the way', *Echo Daily*, 22 August, viewed 20 October 2020, <www.echo.net.au/2019/08/from-lismore-to-camp-binbee-fighting-adani-every-step-of-the-way/>.

Duus, J. 2015, *Unearthing the Sun: Making Sense of the Proposed Coal Developments in the Galilee Basin*, PhD thesis, Australian National University, unpublished.

Evans, R. 2004, '"Plenty Shoot 'Em": the destruction of aboriginal societies along the Queensland Frontier', in A. Dirk Moses (ed.), *Genocide and Settler Society: Frontier Violence and Stolen Indigenous Children in Australian History*, Berghahn Books, USA, pp. 150–173.

Evans, R. 2007, *A History of Queensland*, Cambridge University Press, New York.

Fernandes, S. 2022, 'The place of many waters: for two Indigenous communities maintaining presence is power', Orion Magazine, viewed 20 December 2023, <https://orionmagazine.org/article/coal-mining-indigenous-land-rights/>.

Gregoire, P. 2019a, 'In defence of country: an interview with Wangan and Jagalingou Council's Adrian Burragubba', *Sydney Criminal Lawyers*, 14 September, viewed 20 October 2020, <www.sydneycriminallawyers.com.au/blog/in-defence-of-country-an-interview-with-wangan-and-jagalingou-councils-adrian-burragubba/>.

Gregoire, P. 2019b, 'Join the Adani Blockade: an interview with Frontline action on coal's Andrea Valenzuela', *Sydney Criminal Lawyers*, 19 September, viewed 20 October 2020, <www.sydneycriminallawyers.com.au/blog/join-the-adani-blockade-an-interview-with-frontline-action-on-coals-andrea-valenzuela/>.

Hoch, I. 1992, *To the Setting Sun: A History of Railway Construction, Rockhampton to Longreach 1865–1892*, Robert Brown & Associates, Buranda, Qld.

Howells, M. 2019, 'Clive Palmer says big election spend 'more effective' than charity donations', *Australian Broadcasting Corporation*, 21 May, viewed 20 March 2020, <www.abc.net.au/news/2019-05-21/clive-palmer-55m-federal-election-ad-spend/11135636%3Fsmid=Page:%20ABC%20Australia-Facebook_Organic%26WT.tsrc=Facebook_Organic%26sf213048920=1%26fbclid=IwAR2BnRZPOli3-pXrjTbBeucgT2CYBztIf1FXIWwaJFQ>.

Howlett, C. & Lawrence, R. 2019, 'Accumulating minerals and dispossessing Indigenous Australians: native title recognition as settler-colonialism', *Antipode*, vol. 51, no. 1, pp. 1–20.

Jolley, C. & Richards, L. 2020, 'Contesting coal and climate change using scale: emergent topologies in the Adani mine controversy', *Geographical Research*, vol. 58, no. 1, pp. 6–23.

Krien, A. 2017, 'Revealed: Gautam Adani's coal play in the state facing global-warming hell', *Sydney Morning Herald*, 25 July, viewed 15 March 2020, <www.smh.com.au/lifest yle/adani-how-we-got-conned-by-coal-20170525-gwcw5h.html>.

Krien, A. 2019, 'Part one: inside the Adani blockade', *The Saturday Paper*, August 31, viewed 20 October 2020, <www.thesaturdaypaper.com.au/news/environment/2019/08/ 31/part-one-inside-the-adani-blockade/15671736008677#hrd>.

Lyons, K. 2016, 'Australia's coal politics are undermining democratic Indigenous rights', *The Conversation*, 26 October, viewed 20 October 2020, <https://theconversation.com/ australias-coal-politics-are-undermining-democratic-and-indigenous-rights-66994>.

Lyons, K. 2019a, 'Securing territory for mining when Traditional Owners say 'No': the exceptional case of Wangan and Jagalingou in Australia', *The Extractive Industries and Society*, vol. 6, pp. 756–766.

Lyons, K. 2019b, 'Wangan and Jagalingou country the Frontline in Adani's Federal Election', *New Matilda*, 7 May, viewed 6 Jan 2020, <https://newmatilda.com/2019/05/07/wangan- and-jagalingou-country-the-frontline-in-adanis-federal-election/>.

Lyons, K., Brigg, M. & Quiggin, J. 2017, *Unfinished Business: Adani, the State, and the Indigenous Rights Struggle of the Wangan and Jagalingou Traditional Owners Council*, University of Queensland, June 2017 and Earthjustice, Brisbane, Queensland, viewed 20 June 2020, <https://earthjustice.org/sites/default/files/files/Unfinished-Business.pdf>.

Mayes, C. 2018, *Unsettling Food Politics: Agriculture, Dispossession and Sovereignty in Australia*, Rowman and Littlefield, London.

Murphy, K. 2017, 'Indigenous people victims of 'green' fight against Adani mine, says Marcia Langton', *The Guardian*, 7 June, viewed 20 October 2020, <www.theguardian. com/australia-news/2017/jun/07/indigenous-people-victims-of-green-fight-against- adani-mine-says-marcia-langton>.

Murphy, K. 2018, 'Labor MP says Adani mine would displace jobs and sabotage Paris targets', *The Guardian*, 19 February, viewed 20 March 2020, <www.theguardian.com/ australia-news/2018/feb/19/labor-mp-says-adani-mine-would-displace-jobs-and-sabot age-paris-targets>.

National Native Title Tribunal (NNTT). 2020, Native Title Claimant Applications and Determination Areas, 30 September, viewed 30 October 2020, <www.nntt.gov.au/Maps/ QLD_NTDA_Schedule.pdf>.

Queensland Government. 2009, *Historical Tables, Economy, 1860–2008*, Queensland Treasury.

Queensland Government. 2019, *Climate Change in the Central Queensland Region*, version 1, Department of Environment and Science, published by the State of Queensland.

Robertson, J. 2019, 'Traditional owners driven apart by the Adani issue unite to fight Queensland Government', *Australian Broadcasting Corporation*, 6 December, viewed 20 September 2020, <www.abc.net.au/news/2019-12-06/adani-native-title-claim-wangan- jagalingou-galilee-basin-court/11770936>.

Slezak, M., 2017. 'Queensland farmer raises $25,000 to run ad opposing Adani water licence', *The Guardian*, 9 November. Available from: www.theguardian.com/australia-news/2017/nov/09/queensland-farmer-raises-25000-to-run-ad-opposing-adani-water-lice nce. Accessed 15 March, 2020.

Smee, B. 2019a, 'Queensland police to get new powers to search climate change protestors', *The Guardian*, 20 August, viewed 20 March 2020, <www.theguardian.com/australia-news/2019/aug/20/queensland-police-to-get-new-powers-to-search-climate-change-pro testers>.

Smee, B. 2019b, 'Queensland anti-protest laws inherently disproportionate UN human rights experts say', *The Guardian*, 13 December, viewed 20 March 2020, <www.theguard ian.com/australia-news/2019/dec/13/queensland-anti-protest-laws-inherently-dispropor tionate-un-human-rights-experts-say?utm_term=Autofeed&CMP=twt_gu&utm_med ium=&utm_source=Twitter#Echobox=1576214747>.

Stuart, R. 1985, 'Resource development policy: the case of Queensland's Export Coal Industry', in P. Allan (ed.), *The Bjelke-Petersen Premiership, 1968–1983: Issues in Public Policy*, Longman Cheshire, Melbourne, Australia, pp. 53–80.

Sturmer, J. 2014, 'Environmentalists concerned about go-ahead of QLD coal-mine, PM', Australian Broadcasting Corporation [database], 28 July, viewed 6 September 2020, available at: ProQuest.

Svensen, S. 2008, *The Shearers' War: The Story of the 1891 Shearers' Strike*, rev. ed., Hesperian Press, Western Australia.

Talukdar, R. 2019, 'No Means No: Adani's land woes in Australia', *Newsclick*, 11 February, viewed 12 March 2020, <www.newsclick.in/no-means-no-adanis-land-woes-australia>.

Tatten, R. & Djnnbah. 1991, 'Queensland land rights – an illusion floating on rhetoric', *Aboriginal Law Bulletin*, vol. 51, [online], viewed 20 September 2020, <www.austlii.edu. au/au/journals/AboriginalLawB/1991/51.html>.

Wangan and Jagalingou (W&J). 2018, *The Path of Resistance*, Wangan and Jagalingou Family Council, 11 July, viewed 20 March 2020, <http://wanganjagalingou.com.au/the-path-of-resistance/>.

Wear, R. 2010, 'Johannes Bjelke-Petersen: straddling a barbed wire fence', Queensland Historical Atlas: Histories, Cultures, Landscapes, 2009–2010.

Legislations

Aboriginal Land Rights Act 1991
National Native Title Act 1993

Websites

www.earthjustice.org
www.lockthegate.org.au/watermackay
https://standing-our-ground.org/

9 A global outlook for anti-coal climate justice activism

Community struggles at sites of fossil fuel extraction have been an ongoing feature in both the global North and global South. However, since 2009, environmental and climate groups in the global North and international activist networks have been increasingly politicising fossil fuel extraction. Through its new approach of cutting carbon from the ground up, environmental activism's politics and narratives have become entangled with the struggles of other environmental actors – Indigenous groups, subsistence-based communities in the global South, and farmers – on the frontline of extractive projects who emphasise the deleterious effects of coal extraction on their lands, livelihoods, culture and survival.

Through the Australian and Indian case studies, this book has explored numerous dimensions of this new anti-coal environmentalism. One the one hand, climate activism has politicised coal based on the risk of climate change caused by carbon emissions. Coal is now seen as dirty and risky. The so-called demonisation of coal by climate activists has motivated the industry and pro-coal politicians to defend coal, as noted through advertising campaigns claiming coal to be good for humanity, and measures to promote 'clean coal' in Australia.

On the other hand, challenging coal extraction has brought a tangible account-ability to the issue of climate change for environmentalists as opposed to abstract carbon emissions. The role played by environmentalists in linking multiple grievances against coal by various actors in anti-coal resistances has entangled abstract carbon emission in material and historic concerns over land and natural resources, making environmentalism of the global North acknowledge various human relations with nature that it has historically excluded. Engaging with dis-tributive and procedural injustices of those on the frontlines of coal extraction has made it possible for environmentalism to address additional facets of climate risk beyond the generic 'melting of the ice', and additional facets of climate justice by considering 'how and for whose concern justice might be applied' (Forsyth 2014, p. 232). Such an engagement helps ensure that some people's concerns are not overlooked or their problems exacerbated.

Politicisation of coal mining by alliances of communities and environmentalists has brought together intersecting claims from various stakeholders that simultan-eously encompass multiple scales of place – the land, local and the global – and

DOI: 10.4324/9781003410416-9

time –the historic, present and future. This can be clearly seen in the collective process of politicisation of multiple grievances against the Carmichael coal mine in Australia, as noted in Chapters 6, 7 and 8. Multi-pronged and multi-scalar anti-coal politics are enabling a turn towards an interconnected politics of justice, as opposed to a previous Northern environmentalism that, as Jedadiah Purdy (2015) notes about the United States in the 1970s, while taking on big new scientific challenges, ended up excluding certain kinds of people.

Impacted communities in the Australian case – the Wangan and Jagalingou (W&J) and Farmers for Climate Action – countered coal on multiple grounds including the risk of climate change. But the same cannot necessarily be said about India, where climate change has not been a prominent issue in people's movements and the Mahan case, as seen in Chapters 3, 4 and 5. Looking at the Australian and Indian cases, it could be wagered that a global North–South difference persists in climate justice activism focussed on 'keeping coal in the ground'. There is a tendency in activism and research in climate justice to transplant Northern environmental discourses onto the South, which can impede building a contextual understanding and effectively responding to issues there. To build a global outlook for climate justice activism, it is imperative for activists and scholars to pay attention to narratives around which intersecting social and environmental justice claims come together in a place like India, and reflexively interpret their significance for climate justice.

Through a comparison of the politics of coal in Australia and India, this book essentially links the two landscapes together by revealing similar patterns of power and a State-corporate nexus in the coal sector. In comparing how civil society counters coal, this book exposes the state of democracy and what is possible as far as challenging the coal nexus goes in the respective countries. I have argued that the new focus of global climate activism, predicated on forming solidarities with local struggles, necessitates a deep rethinking of the nature of alliance-making across global North–South intersectional differences. To draw these two landscapes of resistance to coal together in solidarity, anti-fossil fuel environmentalism will need to carefully consider the similarities and differences between the two cases in this book and find an intersectional outlook.

This chapter discusses the final questions of this book, about the critical similarities and differences between the two anti-coal resistances and their contexts, and what global outlook for environmentalism emerges from a comparison of the Australian and Indian cases. It discusses the various climate justice activisms emerging from the two cases in Section 9.1, and relations formed by environmentalists with other actors in the anti-coal resistances in Section 9.2. It compares the W&J and Mahan struggles for land justice and self-determination in Section 9.3, and the politics, narratives, tactics and risks associated with the Stop Adani and Greenpeace anti-coal activisms under Section 9.4.

Table 9.1 summarises the varieties of climate justice and relations forged by environmentalists with other environmental actors from Sections 9.1 and 9.2. Table 9.2 summarises the similarities and differences across the Australian and Indian cases from Sections 9.3 and 9.4. Drawing on previous sections, Section 9.5

makes recommendations for how climate activism can assume a global North–South intersectional outlook. Section 9.6 reflects on this book's contributions to political ecology and outlines areas for further research in global climate justice activism and comparative environmentalism that can build on its findings. The Conclusion reemphasises the purpose of this book and its findings – to offer insights for activists, practitioners and researchers in the field of energy transition and social movements across the global North and global South.

9.1 Varieties of climate justice

Here, I delineate various approaches to climate justice seen through the two cases and propose how a North–South difference in climate justice activism focussed on keeping coal in the ground can be bridged through an intersectional global outlook.

Climate justice, livelihood and sovereignty in the Galilee Basin

As seen in the case of movements emerging from coal mining regions in New South Wales in Australia, the common risk of coal mining has linked anti-coal actors together into unlikely alliances, particularly farmers with environmental and Indigenous groups. Concerns about the impact of mining on water, the loss of farming land and health impacts from pollution near mine sites have united groups across urban and rural regions (Connor et al. 2009). What has emerged even more distinctly through the Stop Adani movement whose supporters are primarily based in the largest cities – Sydney, Melbourne and Brisbane – is a dominant concern for climate change. Stop Adani has made the issue of the so-called 'climate-wrecking' Carmichael mine a shorthand for the government's failed commitments on climate (Murphy 2017).

But even though the three strands of the resistance – environmentalists, farmers and Indigenous groups – have expressed concern about climate change, their specific imperatives for contesting coal mining, and consequently what climate justice has meant to them, have been different.

For the farmers in the Galilee Basin, climate justice signified protecting water sources and security for agriculture. Their imperative for demanding climate action was shaped to a certain extent by hardships faced during the 2018 Queensland drought. The Queensland government's allocation of free and unlimited water for the Carmichael coal mine during the drought exacerbated their concerns over water security. The weakness of the federal government's drought response measures, along with its politics of climate denialism, heightened their concerns about secure livelihoods. As opposed to the earlier "Lock the Gate" movement, which did not explicitly advocate for climate policy action, these experiences shaped the Basin's farmers' stand against the coal mines as a call for climate action.

For the W&J people, against a history of Indigenous dispossession in Australia's settler colonial society, climate justice came to signify the need for sovereignty, particularly through the native title legal system. The imperative for the W&J to challenge the Carmichael coal mine was based on their experience of the repressive native title system that favoured mining corporations and did not allow Indigenous native title groups to exercise their free, prior and informed consent on mining operations, an internationally recognised Indigenous human right. Their international links with Indigenous fossil fuel struggles in North America served to strengthen and collectivise their claim for sovereignty over their traditional lands.

Although 'No Means No' was a land rights campaign that aimed to expose the failure of Australian native title, it both acknowledged the climate change problem, and articulated a form of Indigenous climate justice based on the demand to redress historic injustices. Climate injustice through fossil fuel projects on Indigenous lands formed the basis for their international activism, such as through their alliance with the environmental legal network Earth Justice.

The various imperatives, politics and concepts of climate justice in the Australian movement were articulated against the context of a pro-coal and climate-denialist national politics. In addition to the environmental movement, the anti-coal arguments of the farmers and the W&J were recast within a common frame of climate justice in response to this context. Both the W&J and farmers articulated climate injustice as an extension of their grievances against coal mining in the Galilee Basin, reflecting a dialectical process of meaning-making in the collective movement.

Forest rights and democracy in Mahan

The Indian case also reflects a dialectical process of meaning-making by the movement in response to the State's bias towards the corporation and its withholding of forest rights and self-determination. But it cannot be said that like the W&J and the farmers in central Queensland, the Mahan movement also linked their grievances to climate change of their own accord. Overall, the narrative of forest and democratic rights dominated the local and broader civil society discourse on anti-coal activism, with concerns over climate change not being strongly or directly articulated. Movement claims collectivised around a call for democracy against a context of attacks by the government on civil society groups and democratic institutions and mechanisms.

Climate justice can assume a derived rather than directly articulated significance for subsistence-based Southern communities such as Mahan. And this takes us into a discussion of certain elements that are fundamental to the formulation of climate justice in a global South context, which is characterised by social unevenness, a developing political economy and a postcolonial climate politics of governments.

Daily struggle for survival

Fellow climate justice researchers have asked me whether the people of Mahan talked about climate change. Failure to link their everyday struggles with climate change issues amongst India's livelihood-focussed people's movements can be both on account of socio-economic and political factors which I discuss in various places throughout this chapter. This phenomenon must not be mistaken as the environmental parochialism of the Southern poor. Instead, it signifies imminent risks from industrialisation to their survival and security, as noted in discussions on Indian environmentalism in Chapters 2 and 3. Being overwhelmed with the daily struggle for survival, as also seen in India's so-called energy capital of Singrauli in Chapter 5, environmentalism of the poor's assertions remains grounded in immediate injustices.

Unlike educated, middle-class urban activists, environmentalism of the poor movements are composed of largely rural populations that can often lack a scientific understanding of the issue of climate change, even though they are attuned to changing weather patterns. The South Asian People's Action on Climate Crisis (SAPACC), formed in 2019, is a recent and unique collaboration between livelihood-focussed people's movements, Indigenous groups, trade unions and farmers across South Asia. It marks an emergent space in mass activism in South Asia that makes climate change central to people's movements and attempts to link existent struggles and grievances of ecosystems-dependent subsistence communities with the broader problem of climate change (Adve 2020).

However, on the whole, being overwhelmed with the daily struggle and a lack of connection with the scientific language of climate change are likely to make rural communities, as opposed to urban middle-class activists and environmental nongovernmental organisation (ENGO) members, articulate a different narrative of environmental resistance; one that is squarely situated in the immediate injustices they are facing.

Role of interlocutors

Anti-coal narratives emerging from global South environmentalisms are often re-interpreted as climate justice narratives by other globally oriented actors, even when the livelihood-focussed movement itself might not directly express climate change as a concern. In India, Greenpeace connected the Mahan struggle with the issue of climate change by linking the anniversary celebrations of victory over coal mining with the 'Break Free From Fossil Fuel's' global event in 2017. As a global ENGO, it re-interpreted the movement's significance as a quest for climate justice through its campaign (Talukdar 2019).

Distinctions between issues of local identity and politics versus globally oriented activism can be seen in resistances in both the global North and South. The global cause may serve to undermine or complement the local issue. In this case, Greenpeace and MSS's activisms proved complementary and effective in their respective socio-political spheres. However, in a global South context, as noted in the case of Mahan, global–local relationships can raise stronger structural questions about the agency of subaltern actors, including who speaks for local people and how their narratives are re-interpreted at the global level.

Owing to persistent structural divisions in global South societies, the question of agency, access to various platforms, and visibility for communities points to a North–South difference in climate justice struggles even in the case of Indigenous resistances. I discuss this in Section 9.3 by comparing Mahan and W&J's resistance.

While, on the one hand, the indispensability of an interlocutor in global South campaigns raises questions about structural justice in the process of including the voices of communities in global activism, on the other it points to the responsibility that comes with this role. As discussed in Chapter 2, Naomi Klein sees the task of the climate activist as 'overcoming the various disconnections and connecting our various movements' (Klein 2016). But there is a tendency in global activism and even climate justice research to transplant Northern environmental discourses onto the South, which can, as Lawhon (2013) puts it, hamper a contextualised understanding of the relation between poverty and environmental justice (amongst other things) to emerge. The role of an interlocutor extends to being able to hold all the interconnected threads of discourses emerging from the ground in the global South and resisting such transplantings. I draw on my own experience as an interlocutor to discuss the significance and challenges of this role in Section 9.5.

Precarious rights and justice

The rejection of coal mining by the people of Mahan was based on an assertion of their newfound forest rights, and signified shifting ground within India's environmental movement. Against the historical context of the dispossession of Adivasi and forest-dependent communities under colonial-era laws, and its continuation during India's post-independence industrial development, the Forest Rights Act and other related legal mechanisms brought a new language of rights into environmental justice movements. Despite having formal constitutional rights over their lands in Scheduled areas, the rights of Adivasis were superseded in practice by the priority given to coal production in government policies in postcolonial India. Through legal tools like the FRA, forest-dependent communities could claim rights over their forests, reject coal mining, and protect and enhance their forest-dependent livelihoods, effectively democratising development from the ground up.

However, low levels of implementation by state governments and low awareness amongst forest-dependent communities of their legal rights have remained as barriers to community empowerment in central India. The FRA has also been undermined by India's neoliberal development model: provisions within the FRA for community participation in resource development including requiring consent for coal mining and construction of thermal plants have been eroded since 2014 under the Narendra Modi government (Sethi 2019). Finally, violation of the consent mechanism in the FRA and denying communities self-determination in matters of coal mining and other extractive uses of Indigenous lands keep recurring. Against this fraught political–economic context, climate justice for forest-dwelling communities impacted by coal mining signifies both being able to save their forests and livelihoods and protect their legal rights under the FRA.

The vulnerability of their newfound rights poses a challenge for theorising climate justice in a transitional Southern context as O'Neill (2012) points out, particularly under neoliberalism in a developing political economy such as India today. With the inclusion of Mahan in the global climate justice matrix, climate justice can be expanded to signify forest rights for communities in central India. Indigenous ownership and governance of forests and land is increasingly being recognised as crucial both for climate justice and better climate outcomes (Agarwal and Dash 2022). Therefore, global climate justice advocacy can call for strengthening and securing forest rights and forestland tenures for communities.

Subaltern climate justice: countering a postcolonial climate justice narrative

The Mahan struggle unfolded against the backdrop of climate politics of a post-colonial developmental State in the global South. Unlike in Australia, the Indian government carefully positions itself as a supporter of climate action. India points to the West's historic responsibility for climate change, and calls for climate justice by asserting the carbon space for India to grow. Even though this so-called moral position generates a deep contradiction, particularly around climate's worsening effects on India's vulnerable, ecosystems-dependent communities, it still does not generate the specific political imperative such as in Australia to mobilise against the government on the issue of climate change, even for urban-based civil society organisations (CSOs). Overall, the postcolonial contradictions of the Indian government generate a different paradox from that of coal-driven Northern economies such as Australia's where climate denialism exists as a disruptive political force.

The injustice of expectations on India to reduce its carbon footprint while the West does not do its fair share has also preoccupied several globally oriented elite Indian civil society actors, especially in the first two decades of international climate negotiations since 1992. The Mahan struggle signifies a subaltern climate justice politics and narrative that is not preoccupied with abstract carbon but with the direct disruption of people's livelihoods and forests. Arguments by elite CSOs and the Indian government around a fair share of carbon space to grow are underpinned by the historical injustice of India's colonisation. It articulates the dominant postcolonial aspiration and imperative to catch up with the development of the West. But the subaltern Mahan struggle signifies a climate justice that investigates past, present and future injustices towards communities on whose lands and in whose forests the story of India's colonial extraction, postcolonial development, and present neoliberal economic growth is being wrought.

A global outlook for climate justice needs to recognise that the Mahan struggle signifies a systemic critique of the extractive economic development model, one which bypasses them while dispossessing them of their *jal, jangal, jameen.* Mahan's struggles for claiming forest rights and asserting democracy on the ground were their means for claiming a stake in an extractive economy. But Mahan's subaltern climate justice approach essentially implies that alternatives be considered that do not sacrifice the lands, forests and livelihoods of communities, and instead secure them as a means for ecologically and socially sustainable futures in a climate-impacted world. The inclusion of the Mahan struggle in a global matrix of climate justice activism implies that climate justice questions the extractive economic models within which global energy systems are built, not just coal and other fossil fuels but also so-called clean and renewable energy. I discuss this further in Section 9.3.

A global outlook

Table 9.1, which is found after Section 9.4, summarises the politics and narratives of climate justice of the mobilisations against coal mining in Australia and India from this section. The table demonstrates what climate justice signifies through the inclusion of various situated imperatives, logics and modes of resistance to coal mining, and acknowledges that multiple scales and layers are involved in a global approach to climate justice activism focussed on keeping coal in the ground.

In Australia, the abstract notion of carbon emissions is made tangible through the national Stop Adani movement, and local impacts and historic dimensions of injustice are connected to the overarching climate change issue through local ENGOs, farmers' protests, and the W&J's campaign. In India, through the interlocutor Greenpeace, the local and historical dimensions of the struggle are 'tied' to the global story of climate activism. This implies a responsibility for the globally oriented actor, an activist or researcher in climate justice, who acts as the interlocutor to hold all the threads from the ground up and overcome challenges in scaling local concerns to the global arena. A global outlook for climate justice must take cognisance of this full picture of climate justice across the North and South.

9.2 Green relations with Indigenous and farmers' groups

Naomi Klein (2014) envisions environmentalism's new approach as a global-scale 'blockadia' of activists working in solidarity with local struggles against fossil fuel projects. In the Australian case study, I have discussed two sets of relations of environmental activists: with farmers in the Galilee Basin and with the W&J traditional owners. In the Indian case study, I have discussed the relationship between environmental activists from Greenpeace and the community at Mahan. Together, these relations of environmental activists offer perspectives about a new relational politics in global environmentalism today.

Environmentalist–farmer relations in Australia

Both Lock the Gate (LTG) and the Farmers for Climate Action (FFCA) demonstrated a new approach for Australian environmentalism, which is based on finding common ground with farming communities affected by fossil fuel mining. FFCA is a national alliance of farmers concerned about climate change that focusses on issues of sustainability, land management, agriculture and the use of natural resources. The Galilee farmers opposed to coal mining, who are also FFCA members, raised systemic concerns about mining that were both ecological and economic, and envisioned a future for the Basin beyond coal mining. Groundwater allocations to the Galilee coal mines emerged as a crucial link in their alliance with Stop Adani; they campaigned collaboratively against free and unlimited water allocations and mounted joint legal challenges against environmental approvals for the Carmichael and Alpha coal mines. Groundwater issues also dominated the narrative and the politics of the state-level Stop Adani movement in Queensland and generated mass demonstrations in which farmers participated.

The issue of groundwater allocation for coal mining and the risk of groundwater contamination from coal seam gas extraction emerged as a significant political issue at the peak of Australia's coal boom in 2011. It sparked environmentalist–farmer alliances and compelled the federal government to take legislative measures to ameliorate concerns over coal projects by adding a water trigger into the EPBC Act (see Chapter 6). The interviews in this research revealed that environmentalists formed collaborations based on how farmers approached sustainability and their vision for change. While environmentalists were compelled to transform their ideology and politics due to the twin risks of climate change and coal mining, farmers too had to recalibrate their approach toward land and water management and advocate for sustainability within their own industry. This direction is reflected in the aims and objectives of the FFCA, and through the Galilee Basin's farmers' articulations about the need for sustainability in agriculture.

Historical political power structures in Queensland have been transformed through the increasing influence of the coal and coal seam gas mining industries on governments. The marginalisation of farmers by governments, particularly in the Galilee Basin where the State had supported the establishment and development of pastoralism since settler colonialism, led to a new approach of forming alliances outside formal politics and collaborating with environmentalists despite a history of antagonistic relations over issues like land clearing. Forging alliances with environmentalists assumed a critical significance for the Basin's farmers due to the marginalisation of their interests within the National Party that normally represented them (see Chapter 8).

The relationship between the Galilee farmers and the Stop Adani movement has added another dimension to environmentalist–farmer relations in Australia, by making ground water security in the Great Artesian Basin through stopping the Galilee coal mines synonymous with climate action. Due to the presence of neoliberal logic in community discourses in Australia, an environmentalist–farmer alliance is unlikely to fundamentally question structural inequality, or translate into a resistance against capitalism, as compared to the developing world where these themes can be central to mining conflicts (Arashiro 2017). However, in the absence of government action, the environmentalist–farmer relations, forged at the brink of a crisis in the agriculture sector and rural Australia, and against the expansion of coal mining on agricultural lands, hold possibilities for forcing long-term sustainability in Australia's political economy.

The farmer–environmentalist relations in this ethnography raise a question that can be answered through further research on anti-fossil fuel solidarities:

How can environmentalist–farmer relations in Australia move beyond relations of convenience against the common risk of coal mining, towards a shared vision for a future beyond coal?

Green–Black relations in Australia

Although the W&J people formed a tactical alliance with the Stop Adani movement, they emphasised the independence of their land rights campaign and drew distinctions between their historic claim for land justice and environmentalist's present-day demands for climate action. The Stop Adani movement acknowledged the linkages between Indigenous dispossession by settler colonialism in Australia and its continuation through coal extraction on Indigenous lands in its narratives of solidarity with the W&J. Messages, banners and session themes at Stop Adani events and activist-focussed conferences such as Beyond Coal and Gas and Power Shift reflected this solidarity with historic Indigenous dispossessions (see Chapter 8).

The W&J collaborated in environmental campaign activities, most prominently the fossil fuel divestment campaign. The world tour of the banks and the collaboration on the divestment campaign constituted one of the key pillars of the W&J–Stop Adani relationship. Such divestment campaigns, based on collaborations between environmentalists and Indigenous groups fighting fossil fuel extraction on their lands, characterise a model of activism associated with the global strategy of halting 'unburnable fuels'.

The W&J extended relationships with environmentalists on a global scale through their involvement with the environmental legal network Earth Justice to appeal to the United Nations about Australia's violation of international Indigenous rights. The W&J's international appeals for justice were strongly aligned with global climate activism's new approach. It was framed within the broader context of climate justice and the historic significance of Indigenous struggles against fossil fuel projects, and it asserted the need for Indigenous sovereignty.

Alongside Indigenous resistances from North America against mega-fossil fuel projects such as the Keystone XL Pipeline and the Dakota Access Pipeline, the W&J's campaign emerged as a leading climate justice and Indigenous rights movement (Lyons 2016). Given historical tensions between Indigenous groups and environmentalists from a non-alignment of visions and Indigenous groups finding their worldviews marginalised within environmentalism, the relationships between the W&J and environmentalists, the independence of the W&J's campaign, and their global reach add new dimensions to Australia's Green–Black relations. Overall, these relationships with environmentalists reflect new patterns in Green–Black politics made possible by environmentalism's new approach.

Environmental activists and environmental justice researchers should consider going beyond seeing such relations as tactical alliances aimed merely to facilitate 'leav(ing) coal in the ground'. While the narratives of climate movements now acknowledge the larger struggles of Indigenous people with colonialism and capitalism as noted in the case of Stop Adani, they also need to systematically consider how Indigenous people's participation in collective resistances can be instrumental in determining a different future and social trajectory (Latulippe and Klenk 2020). Further research on anti-fossil fuel solidarities with Indigenous land struggles can reflect on the question:

How can climate justice activism against fossil fuel extraction centre Indigenous visions and futures in its worldview?

ENGO relations with livelihoods struggles in India

In India, Greenpeace played a leading role in transforming the Mahan community's awareness about their rights and forest ownership, which then shaped Mahan's motivation to resist coal mining. The Greenpeace–Mahan relationship lacked the inter-racial dimensions of the Stop-Adani–W&J relationship, which is a distinctive characteristic of settler-colonial societies. As Mahan represents a social mix between various castes as well as Adivasi and non-Adivasi families (see Chapter 5), their relationship with Greenpeace also lacked any specific inter-ethnic dynamics. The Indian relationship was predicated on a socio-economic divide characteristic of postcolonial societies in the global South: between educated, urban and middle-class activists, and rural, subsistence communities who make up the core of livelihood struggles (see Williams and Mawdsley 2006).

The Greenpeace–Mahan Sangharsh Samiti (MSS) relationship involved trust and co-dependency between the urban activists who educated the community about the laws and the local movement members who helped them understand power structures and social complexities of village life and forest-based living. For the people of Mahan, although their relationship with the forest did not change in a material sense given their continuing dependency on its resources, with an understanding of their legal rights came a distinct sense of ownership, and a heightened sense of what was at stake of being lost (see Chapter 5). For the Greenpeace Mahan team, working alongside the MSS meant both learning to make room for the logic and actions of the Mahan community in their campaign strategies and decolonising their own understanding of subaltern activism in the Southern context.

Overall, the nature of the Greenpeace–MSS relationship can be viewed as a community empowerment model rather than a tactical alliance as seen in the Australian case. Beyond learning about forest rights and mechanisms through which to assert them, the Mahan community's relationship with Greenpeace also influenced a perceptible (even if small) change in their social behaviours around issues of caste and gender. After the success in stopping the coal mine, the relationship resulted in discussions for a sustainable economic alternative for Mahan through schemes for collecting and selling tendu leaves, mahua flowers and other non-timber forest produce. Being a newly politicised community, Mahan did not demonstrate the autonomy or self-assertion (beyond the local scale) that long-established people's movements for livelihoods and forest rights in India are capable of reflecting. Their vision broadened through the association with the Greenpeace activists, leading to their joining the National Association of People's Movements (NAPM).

Participation in national events of the anti-coal campaign such as the action at the Essar headquarters in Mumbai and meeting the Minister for Tribal Affairs in New Delhi about the *Gram Sabha* forgery helped MSS members to understand the bigger significance of their struggle. This broadened perspective was reflected in their interpretation of their actions as dissent in a democracy, and celebrating the victory over the Mahan coal mine as Democracy Day, against the broader context of government crackdowns on civil rights.

As an international ENGO, Greenpeace was able to mobilise the campaign resources and funding to run an extended campaign in a remote Indian location for five years. This capability was a prime factor behind the formation of the movement, its actions, and its relations. The crackdown on Greenpeace and the freezing of its bank accounts raised uncomfortable questions about the ability of Greenpeace to continue supporting this remote campaign that had also proven risky. The alliance formally ended with Greenpeace closing its Mahan office. The attacks on Greenpeace raised broader questions about the vulnerability of relationships forged by international groups with communities at the frontline of coal extraction in the global South.

The relationship with MSS proved beneficial for Greenpeace, particularly around how the ENGO's work was valued by the Indian environmental movement. Grassroots movements in India in general have taken a cautious approach towards trusting international NGOS, regarding them as pursuing their own strategic interests while ignoring the long-term needs of communities. However, an NAPM leader changed his perception about Greenpeace after learning about the community-empowerment approach of the international group's campaign in Mahan (see Chapter 5). But Greenpeace's withdrawal once again put such concerns back on the table. Whether international activist groups can find alignment with the long-term sustainable development needs of subsistence communities, and whether their model of activism is conducive for livelihoods communities in the South, has remained an active topic of debate within India's grassroots movements.

Indigenous community members working in alliance with environmentalists are not only fighting against the challenge of coal mining and climate change; they are resisting persistent structural barriers including illegal occupation and suppression of Indigenous authority to exercise jurisdiction over their lands, ongoing genocide, social, economic and health gaps and higher exposure to environmental harms (Latulippe and Klenk 2020). The community in Mahan, living in the last remaining forests fringing the coal mining ravaged landscape of Singrauli, encountered several such barriers. Their historical and continuing struggles are around the structural challenges of colonial and capitalist domination linked to industrialisation (Whyte 2017).

From their perspective, involvement with a global ENGO (quite understandably) raised legitimate hopes that the engagement could also help to tackle some of their systemic concerns. Greenpeace's engagement in Mahan raises broader questions about what could constitute a long-term process of mutually valuable engagement (see Whyte 2013) between global environmental actors and Southern Indigenous actors. Such an engagement would need to regard the latter as not merely facilitators in global activism's strategy of stopping coal but as instrumental in determining different social trajectories for a post-carbon future. The Greenpeace–Mahan alliance raises an important question that future research needs to consider:

How can engagements of international ENGOs and grassroots campaigns to stop coal in the global South sustain beyond halting coal mines and find a shared vision?

9.3 Indigenous land rights and resistances compared

Both the W&J and the people of Mahan were politicised against coal mining through events that exposed their respective State's biases towards mining corporations and their role in manufacturing Indigenous consent. Native title in Australia and forest rights in India, and experiences of the W&J, and Mahan communities with procedures under the respective laws, broadly reflect similarities and differences in the intent and extent of Indigenous land rights, and the State's role in enabling community rights versus favouring mining in a global North democracy with a settler colonial context and global South democracy with a postcolonial context.

While Australian native title has given Aboriginal Australians a voice in mining-related environmental conflicts but limited their say on what happens on their land by excluding the Free Prior and Informed Consent (FPIC), the Indian Forest Rights Act has given autonomy to village councils to self-determine on mining issues on Indigenous lands. But this difference in the extent of rights enabled legally is perhaps undercut by ineffectual implementation and high levels of violation in the Indian case, caused by several factors including poor awareness amongst communities and a lack of political will to decentralise forest management, apart from the States' obvious interest to facilitate mining (Kohli et al. 2012; Lee and Wolf 2018).

Therefore, while native title covers over 32% of the Australian landmass after 30 years of the NTA's enactment, in India, slow and flawed implementation and high rejection rates of community forest rights claims have meant that only 14.5% of the minimum potential forest areas for forest rights have been recognised in 10 years of the FRA (CFR-LA 2016).[1] It can be argued that a structural divide persists in the implementation of Indigenous land rights by the State between Australia and India. This structural divide speaks to a North–South distinction that Williams and Mawdsley (2006) argue can be experienced even in the case of democracies.

We need to understand a key difference between the Australian and Indian cases of Indigenous self-determination through this argument: in India, despite the law enshrining the provision for self-determination and FPIC, it is overwhelmed by legal violations that ultimately serve to deny rights and historic justice to communities. Given this difference in contexts – in settler-colonial Australia, the limitations of native title itself and in postcolonial India extra-legal challenges around implementation and exercising community rights – the fight for self-determination by the W&J community and Mahan people assumed different dimensions. While in both cases climate justice signified being able to say no to mining on their lands, the former's approach had to be through exposing the limits of native title, and the latter through asserting their democratic rights by demanding that the State allows them their legal rights.

Administrative and legal procedures and actions of governments in both Australia and India reflect a larger process within a neoliberal economic framework that prioritises mining over Indigenous rights and essentially compromises free prior and informed Indigenous consent for mining, as Chowdhury and Aga (2020) argue. The forgery of consent at the *Gram Sabha* in Mahan and the signing of the 'fake ILUA' at the Adani-organised W&J people's meeting indicate similar patterns through which State agencies and the local administration can act in both Northern and Southern contexts to manufacture Indigenous consent. The State–corporate nexus also denied Indigenous community rights through other legal violations (in India) and through amendments to the respective acts.[2]

Given the structural divide mentioned earlier, the ability of communities fighting mining to access justice and visibility for their struggles, differ in scale and degree across both places, as do various aspects of bureaucratic, administrative, and procedural injustice experienced by communities. These qualitative differences unfold within an overarching neoliberal framework, which is operative in both places and favours mining over Indigenous self-determination. I highlight some qualitative differences between the struggles of the W&J and Mahan.

Awareness of rights

The W&J's resistance is seen as an outstanding Indigenous movement for self-determination. Mahan, on the other hand, is yet another central Indian community without a history of prior mobilisation or a lack of awareness of their forest rights. The multi-year process undertaken by Greenpeace to raise community awareness on forest rights demonstrates a fundamental challenge for grassroots and Indigenous mobilisations for self-determination in central India. Consequently, it raises the question of their critical need for support from mainstream NGOs and environmental groups.

Autonomy of resistances

The extent of autonomy in the W&J's movement, its strategic collaboration with the environmental movement, and its ability to access the United Nations to appeal for Indigenous self-determination and expose Australia's violations, stand in sharp contrast with the co-dependency with which the Mahan mobilisation was built up through Greenpeace's support. The co-dependency was evident at other stages too, and not just the initial stages of bringing awareness about forest rights: such as through Greenpeace supporting the Mahan community with freedom of information requests (amongst many other tools and strategies) to find out about the forgeries of *Gram Sabhas*. The extent of this co-dependent relationship signifies challenges for India's Indigenous communities in accessing platforms for appeal and justice at par with their global North counterparts. It further qualifies my last point about the critical need of grassroots struggles for support from urban activists to access justice.

Lived presence on the land

A difference between the land rights campaign of the W&J and the MSS has to do with the latter's lived presence at the site of the project. This difference in the global South – that of the lived presence of communities on the land – adds a further criticality to the human rights issue of loss of lands and livelihoods by communities from coal mining or other disruptive industrial developments. In fact, like many subsistence communities, Mahan residents mostly still live without electricity. Subsistence-based communities in the global South fighting coal extraction can be doubly disadvantaged: through a lack of energy access and the environmental injustice of loss and displacement. As noted in Chapter 3, development-induced displacements that affect Adivasis the most have been a constant feature of India's postcolonial industrialisation, both before and after neoliberalisation. It has therefore remained a central focus of the Indian environmentalism of the poor.

The central theme of human rights violations through land dispossession from coal mining as well as other industrial projects in the global South raises a vital question for global environmentalism: whether its aims can systematically align with land justice and livelihood security for communities at the frontline of extraction?

Intersectionality with land justice

For Indigenous peoples and communities, land justice equates to climate justice. Given the persistence of colonial-capitalist extraction globally, Indigenous people are likely to be revictimised by land use changes in transitions to renewables and climate solutions that ignore Indigenous voices while promoting market-driven technologies (Porter et al. 2020). A global surge in mining for minerals like lithium, copper and nickel that are essential for developing clean energy has already raised critical concerns about Indigenous consent for their mining and protecting Indigenous land rights (Sax 2023). Energy transitions need to acknowledge the historic wrongs of dispossession of Indigenous people and the rupturing of land-based relations and their continuation today; and recognise land rights as a way of initiating such an acknowledgement (Whyte 2021).

A new environmentalism of stopping fossil fuels has to be intersectional with land justice, which for communities is tied to survival and human rights. A land justice approach would go beyond singularly advocating keeping coal in the ground. It would work in solidarity with multifarious extractive challenges to Indigenous lands, historically, today, and in the future, that pertain to sourcing and producing energy. It would include questions of land and human rights for communities at the centre of its vision to transition towards clean energy. Figure 9.1 draws on various land and human rights elements revealed through this book's research that need to be considered by this new environmentalism.

To conclude this section, the resistance of the W&J and Mahan centred around native title and forests rights effectively demonstrate how the provision for free, prior and informed consent is being mobilised by Indigenous struggles against coal mining in the North and South, a settler colonial and a postcolonial context, in two

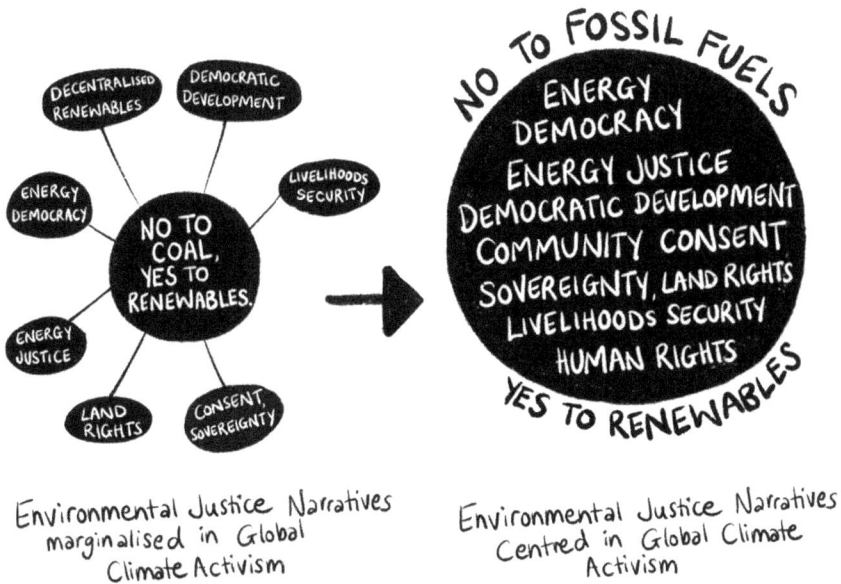

Environmental Justice Narratives marginalised in Global Climate Activism

Environmental Justice Narratives Centred in Global Climate Activism

Figure 9.1 Moving beyond a singular advocacy of 'stop coal and start renewables' by global climate activism. Illustration by Jessica Harwood for Sapna South Asian Climate Solidarity.

democracies. The W&J and Mahan cases show similarities primarily in the tension between the intent of land rights and self-determination, the favouring of mining by the State and the overarching similarities of neoliberal forces at play.

But the scale and structure of the W&J's and Mahan's challenges are different. Mahan faced the prospect of a direct loss of livelihoods from losing forests and lands on which they depend for their economic survival. Solidarity with Indigenous struggles against coal mining essentially needs to keep land justice and human rights of Indigenous peoples and communities at the centre of advocacy and activism, while also recognising the scalar and structural differences between the global North and South even in the case of Indigenous resistances.

9.4 Coal politics and environmental campaigns in Australia and India

The national-level politics of the two anti-coal environmental campaigns in this book offer insights about how the new approach of environmentalism negotiated the political economies of coal, patterns and differences in the social and political dialectic on coal and climate change, and the state of democracy in the respective countries. The campaigns spanned a dynamic period of coal expansion that covered not only neoconservative Liberal Coalition and the right-wing Narendra Modi's Bharatiya Janata Party but also centre-left Labor and Congress-led governments in Australia and India, respectively. The challenges and milestones of these environmental campaigns reveal the paradox of the State under neoliberalism, the extent of a State–corporate nexus in the coal sector and the scope and limits of exercising democracy by CSOs in both countries. The following subsections expand on these themes of comparison between the Stop Adani and Greenpeace India campaigns and reflect on what this comparison signifies for a global outlook for environmentalism.

Transformation through neoliberalism and the minerals boom

The parallel processes of neoliberalisation of the Indian economy and the Australian minerals boom have acted as dominant forces in reshaping the campaign approaches of ENGOs in the respective countries. The arc of Greenpeace's environmental campaigns in India reflects the changing political economy of development under neoliberalism in India, with what Doherty and Doyle (2006) call the 'moral' ground of its advocacy shifting from exposing the role of international corporations in causing environmental pollution, to directly questioning environmental and social risks from domestically owned industrial projects (Talukdar 2019)

Greenpeace India's climate and energy campaign advocated for a reduction of coal usage and a pathway for India to increase renewables, with an emphasis on decentralised and small-scale renewable energy sources, to meet India's diverse energy needs (Bhagat 2018). Its campaign countering coal mining had to necessarily steer an indirect and multi-step course, which first began with the 'Ban the Bulb' campaign that advocated widely acceptable energy efficiency measures. Greenpeace also established the 'moral case' for India to take climate action by highlighting the disproportionate impacts of climate change on the Indian poor, making an argument for intra-generational equity and social justice in the same vein as the environmentalism of the poor (Ananthapadmanabhan et al. 2007).

On the other hand, under the twin effects of the Australian government's lack of action on climate change and its increasing coal usage, the Australian environmental movement strategically transformed into a national anti-coal movement. Due to a range of reasons – impacts on health from pollution, water and effects on farmlands, and concern about climate change – discontent against coal was growing in coal mining regions such as the Hunter Valley in New South Wales during the coal boom that started since the early 2000s.

The failure of the international climate talks at Copenhagen in 2009 also coincided with the failure of the Kevin Rudd Labor government to pass climate legislation due to the divisive nature of Australian climate politics. This caused a disappointment amongst large ENGOs and professional climate networks that relied on lobbying and advocacy approaches, leading to a systematic turn by organisations at all levels – the national, state and local – towards strategies for a direct disruption of coal mining and exports. In the process, the environment movement was transformed from a largely formalised network of professional ENGOs to a diverse network where grassroots and local anti-coal groups played a significant role.

The difference between a national anti-coal climate movement in Australia versus the lack of mass mobilisation on climate change or explicitly linking the issues of coal extraction and climate change in mass-scale movements in India reflects different imperatives, that are defined by material realities, and social and political contexts from which these movements arise. It also puts the spotlight on who is the primary environmental actor in activisms a global North or Minority World context as in Australia, and a global South or Majority World context as in India.

Negotiating the political economy of coal

The two anti-coal movements encountered patterns of power and the prioritisation of coal companies over the public interest by the State. In India, Greenpeace negotiated the controversy and power play over the allocation of the Mahan coal block within the highest levels of the Indian government. The timeline of the anti-coal movement coincided with an unprecedented increase in India's coal and thermal power generation. This prioritisation of coal expansion created divisions between the federal environment ministry and those of coal, power and finance on the issue of allocating coal mines in forests. The Mahan coal block was at the centre of 'Coalgate', the largest government corruption scandal till that date. Coalgate exposed the challenges in transparency in resource management within India's political culture of crony capitalism constituted of a narrow alliance of business and political elites.

Greenpeace's campaign exposed favouritism in the allocation of the coal block, the power of coal companies over governments including through personal favours, ecological costs and the lack of public good in a controversial coal mining project. Greenpeace's role in exposing the corruption and mobilising a grassroots resistance against the coal mine led to legal attacks by the corporation and eventually a crackdown on its campaigns by the Indian government.

In Australia, the Stop Adani movement grew out of and strengthened an existing national anti-coal mobilisation. After massive port developments along the Great Barrier Reef were either cancelled or their proposals drastically reduced due to a withdrawal of investors because of both the structural decline of coal and the divestment campaign, the anti-coal national movement consolidated its fight against mega coal mines in the Galilee Basin in Central Queensland, and specifically Australia's largest proposed coal mine, the Carmichael coal mine.

Australia's coal sector has received special treatment from governments for decades. During the minerals boom, the scale of coal projects and the extent of investment from foreign mining corporations into coal mining played a role in deepening coal's power over Australian governments. Central Queensland became Australia's largest coal-producing region, and the state of Queensland weakened environmental regulations and fast-tracked project approvals to facilitate massive projects.

The first of the Galilee projects received final approvals between 2012 and 2014. These mega mines were anticipated to begin production well after the Paris

Agreement had been signed. The coal mines were also slated to begin production at a time when coal's structural decline was making it financially risky to open new coal mines. The national Stop Adani movement thereby exposed coal's power over Australian politics. The Carmichael coal mine finally commenced at a drastically reduced scale and after a delay of five years. It was propped up by political support regardless of its proven economic unviability.

Despite similarities in the entrenched power of coal over the State in both India and Australia, the case of the 'Coalgate' scandal in India stands out as an extreme case of corruption and undemocratic resource governance, and points to the different scales of challenge in achieving transparency and accountability in resource governance between Northern and Southern coal democracies.

National campaigns framed around climate versus forests

Both the Stop Adani and Greenpeace campaigns reflect a multi-scalar politics and narrative. In the case of Stop Adani, the national mobilisation was driven by mass concern over climate action, while in Queensland the water allocations for the Carmichael coal mine equally mobilised public outrage. The movement in India was driven by the demand for forest rights on the ground and to save forests from coal mining nationally; the solidarity for Greenpeace's anti-coal campaign by Indian civil society after the government's crackdown was framed under the narrative umbrella of democratic dissent. Between the reports produced by Greenpeace Australia (2012a, 2012b) and Greenpeace India (Fernandes 2012), which mapped out the scale and extent of the problem of new coal developments in the Galilee Basin and in central India respectively, can be seen the significantly different ways in which new coal developments were politicised in Australia and India.

Greenpeace India's report highlights the discrepancy between India putting forward the role of its forests as carbon sinks at international climate forums and independent research showing massive 'diversions' of old forests in central India to increase coal and thermal power production. It highlights the social and ecological risks associated with an expansion of coal mining in thick forests in central India – encroachment into national parks, damage to ecosystems and water, impacts on wildlife, the impact on greenhouse gas (GHG) emissions from the destruction of carbon sinks, and the burning of coal, and most significantly, impacts on forest-dwelling communities. The report points out that given the possibility that many coal mines may never go into production due to coal's structural decline, coal mine allocations by the government effectively resulted in a large-scale handover of forestlands to corporations.

Consequently, the organisation made forest rights for central Indian communities a core campaign issue. Countering the impacts of the political economy of coal on the land, forest, and rights of communities, Greenpeace framed its campaign around countering the destruction of forests in central India from coal mining. While protecting forests' and people's rights served as a proxy for its global climate strategy of keeping coal in the ground, the proxy approach proved relevant for global activism's new vision of working in solidarity with communities fighting coal mining.

The Greenpeace Australia report on the Galilee details the impacts of coal mining in the Galilee Basin on the Great Barrier Reef, underground water sources on which farmers rely, freshwater springs sacred to the W&J, and the local ecology. Most significantly, it provides a calculation of the GHG emissions that would be produced from burning coal extracted from all of the Galilee mines, highlighting the risk these emissions pose to the global climate. 'Scope three' emissions – produced from burning Australian coal in overseas thermal power plants – do not count towards Australia's domestic emissions, allowing governments to continue promoting coal exports without being required to assume responsibility for their impacts on the global climate. Attempting to disrupt Australia's coal exports appeared the only way to challenge Australia's leading role in global pollution from coal burning. This defined the purpose of the Stop Adani movement, and it aimed to Stop the Carmichael coal mine as the first step in that process.

To summarise the difference in the social reality within which the Indian and Australian anti-coal environmental campaigns operated, extensive coal mining in the Southern context, apart from causing widespread ecological destruction, also extensively impacts on human rights of communities living and subsisting on the frontlines. Unlike the Australian campaign where the scientific risks of climate change were a central issue that sparked mass mobilisations in cities, in the Indian case, the focus remained on the implications of the destruction of forests on forest-dwelling communities; with this focus serving as an effective proxy in the Indian context for Greenpeace's global strategy of keeping coal in the ground.

Pro-coal and anti-democratic government actions

Civil society in both countries interpreted the pro-coal actions of governments as anti-democratic measures, although their contexts and narratives differed.

The assertion of democracy as a campaign narrative could be seen very strongly in the Indian case. The crackdown on Greenpeace by the Narendra Modi government became a flashpoint for a broader discussion on the discontents of coal. Against the wider context of crackdowns on civil society groups and people's movements in 2014, the court interpreted the actions of the anti-coal movement as a critical act of dissent in a democracy. As noted in Chapter 4, the movement's actions were seen as necessary for questioning the dominant development paradigm that excluded the perspectives of marginalised communities, despite their special protection under the Constitution, signifying anti-coal activism as a constitutional right.

In Australia, demands by environmental groups that governments stop supporting coal projects that lacked 'the social license to operate' and instead act on climate change that most Australians wanted, essentially implied democratic accountability by the State. ENGOs and think tanks have pointed to how political donations from fossil fuel corporations have affected government decision-making and action on climate change and finally the quality of Australian democracy (see Chapters 6 and 7, also Australian Conservation Foundation n.d.). The 'climate wars' in Australian politics, particularly the removal of Kevin Rudd from the position of prime minister by vested interests against climate action, signified how fossil fuel interests had hijacked Australian democracy.

Countering coal proved to be a litmus test for democracy in both the Australian and Indian cases. But finally, the extent of the attack on democratic freedoms and the functioning of democratic institutions, progressive laws and CSOs in India stands out in comparison to Australia. It highlights the need to pay attention both to the necessity for democracy for social and ecological justice, as well as its precarity, in a populous global South context.

Risks of anti-coal activism

Both the Indian and Australian anti-coal campaigns faced a backlash from the coal mining corporations and governments. In Australia, measures by the federal government to restrict campaigning activities against coal in general and the campaign against the Carmichael coal mine in particular included attempting to remove the tax-deductibility status of ENGOs to disrupt their funding and de-allocating government funding for state-based conservation groups and environmental legal networks. Calling the environmental legal challenges against the Carmichael coal mine 'green sabotage', the Tony Abbott government attempted to repeal section 487 of the EPBC Act, which allows for judicial reviews of ministerial decisions and approvals for mining.

In Queensland, the police increased its surveillance on activist activities that tried to directly disrupt mining. The government introduced legislation that criminalised protests by banning the use of 'locking devices'. Adani Australia commenced an aggressive 'attack dog' strategy that threatened activists with legal action. Its retaliation to the W&J set a precedent as the first time that an Australian traditional owner had been made bankrupt by a mining company (Gregoire 2019). The legal persecution of a grassroots activist in Queensland who the company accused of alleged conspiracy sparked concerns amongst Australian journalists about the consequences of reporting sensitive stories about the Carmichael project (Smee 2020).

In India, Essar Power aimed to bankrupt Greenpeace through an A$100 million strategic lawsuit against public participation (SLAPP suit). Local movement members from the MSS faced constant threats and disruptions from local company agents. However, the most significant and consistent attack on Greenpeace's operations came from the Indian government under Narendra Modi in 2014 through the freezing of its bank accounts and the cancelation of its licence to operate in India.

This government attack also targeted international developmental, environmental and human rights NGOs operating in India under the pretext of the misuse of their foreign funding registrations. Greenpeace was singled out in a separate dossier prepared by India's domestic surveillance agency, the Intelligence

Bureau, for its anti-coal campaign in Mahan; the dossier alleged that Greenpeace was acting at the behest of foreign governments who wanted to stall India's development. Greenpeace survived the attacks and successfully challenged the government's actions in the courts. However, by January 2019 it had to wind down many of its campaigns in India following a further freezing of its bank accounts (Talukdar 2019).

Although the link between coal and climate change is well understood by urban-based middle class activists in India, ENGOs do not usually target coal on the grounds of climate change, as noted in Chapter 4. Given the postcolonial developmental anxieties of the Indian government, such advocacy is considered risky, as was confirmed when the Indian government targeted Greenpeace (Talukdar 2018).

Both the Indian and Australian cases demonstrate certain patterns in governments and corporations attempting to delegitimise and intimidate anti-coal activism. But the persistent attacks on Greenpeace because of its anti-coal campaign in Mahan, and severely disrupting its ability to operate in India, are an outstanding example of the risks involved in countering coal mining in a Southern postcolonial context. They also indicate the differences between these Northern and Southern coal democracies.

Tools and forms of mobilisations

The Stop Adani movement had a multi-pronged campaign approach that included a divestment campaign, mass-scale national mobilisations, electoral and local grassroots mobilisations, as well as peaceful direct disruption tactics at the project site and along the transport corridor. It used a variety of new digital organising platforms and mobilisation tactics, similar to movements against fossil fuel projects in North America. These platforms and technologies indicate the availability of a critical mass of urban supporters who could be mobilised on the issue. It indicates that Stop Adani, as a Northern environmental movement with a national and a global reach, was able to mobilise the resources and access the latest campaigning technologies.

Although Greenpeace in India did use digital mobilisation tactics to target its urban supporters, such as when a Greenpeace activist spent a month on a tree in the central Indian forests to raise awareness about the risk to forests from coal mining, given the dominance of the grassroots MSS mobilisation in its anti-coal campaign, Greenpeace's scope for using the latest campaign and organising technologies for urban outreach was significantly limited.

The majority of tactics and mobilisations in the case of the Indian campaign related to the community at Mahan and organising the local resistance there. On-the-ground organising required Greenpeace's Mahan team to be present at the site for most months in the year. Digital technologies could not be accessed in the area due to the lack of both electronic devices amongst most community members and digital connectivity. Communication and campaign promotions mostly depended on one-on-one interactions and words of mouth. Mobility for MSS members was either entirely on foot as in the case of women, or on bicycles, or occasionally on motorcycles. The Greenpeace Mahan team had the only car available for campaign work in the community. Some of the local movement's most prominent actions, such as women protecting trees from being felled in the forest at the peak of the campaign (see Chapter 5), resonate with iconic community struggles to protect forests such as Chipko (discussed in Chapter 2), and reflect the timelessness of what Scott (1985) calls the 'weapons of the weak' that environmentalism of the poor deploys.

Essentially, the difference in the forms and tools of mobilisation between the Indian and Australian cases indicates a characteristic Northern versus Southern demographic difference in terms of the main actors in environmental campaigns.

To summarise Section 9.4, the political economy of coal shows clearly demonstrable patterns of power, a State–corporate nexus and their effects on democracy, in both Australia and India. Consequently, countering coal proved to be a litmus test for democracy in both cases. But outstanding corruption in resource governance and the extent of attack on the rights and activities of civil society in India point to stark differences between the global North and South, even in the case of democracies. The two national campaigns also bring out the reality of a demographic difference between the main actors in environmental campaigns between the global North and South. The imperatives for activism of dominant environmental actors in the global South as well as the political stance of postcolonial governments towards climate change are essentially different from those in the global North and the politics of climate denialism; this influences a different political approach of ENGO campaigns around coal mining.

9.5 Discussion – possibilities and challenges in building a North–South intersectional outlook for environmentalism

In Sections 9.1 to 9.4 and the two following tables, I have suggested certain norms and possibilities for a global outlook for climate justice activism and research that can bridge disparities across the global North and South. In Section 9.1, while delineating the varieties of climate justice that emerged from the Australian and Indian cases, I have underlined the importance of the role of the globally oriented interlocutor, essentially a climate justice activist or researcher who can hold all the threads of claims and contestations from the global South and tie them to the global narrative frames of climate justice. In Section 9.2, while discussing the nature of environmentalists' relations with other actors in the Australian and Indian resistances, I have emphasised the need for climate justice activists to go beyond tactical relations towards putting sustainable futures for Indigenous communities at the centre of their vision.

In Section 9.3, while discussing structural differences that set the global North apart from the South even in the self-determination and land and forest rights struggles of Indigenous communities, I have proposed an intersectional model of solidarity with land justice issues for environmentalism. In Section 9.4, I delineated differences in the processes of economic neoliberalisation and the coal rush, various social, economic and political contexts that led to specific framings

Table 9.1 Various environmental actors and varieties of climate justice in the Australian and Indian cases

Environmental actor	Dominant imperative for countering coal	What does climate justice signify?	Claims, demands, approach	Relationship with environmental groups	Learning for climate justice research and activism
Stop Adani supporters	Climate change	A safe climate future	Abstract carbon made tangible through coal	Volunteers, supporters (including financial donors) to Stop Adani; Predominantly urban-based, often tertiary educated.	Stop Adani's dominant climate change frame is because of their urban supporters.
Farmers in the Galilee Basin	Hardship during 2018 drought and weak government response; Free and unlimited water allocation for Adani during drought; Governments' climate-denialism heightening concern for secure livelihoods.	These experiences shaped their stand against coal as a call for climate action; Climate justice signified protecting water sources and security for agriculture.	Stop free and unlimited water allocation for coal; Protect agriculture by taking climate action by stopping coal.	Environmentalists are meeting farmers where they are (at); Unlikely for green–farmer alliances in the developed world to question structural inequality or resist capitalism as compared to the developing world where these themes can be central to mining conflicts; This green–farmer alliance nevertheless created new political possibilities for climate action.	Future research can explore how green–farmer relations can play a sustained role in energy transition and sustainable futures.

(*Continued*)

Table 9.1 (Continued)

Environmental actor	Dominant imperative for countering coal	What does climate justice signify?	Claims, demands, approach	Relationship with environmental groups	Learning for climate justice research and activism
Wangan and Jagalingou	Their experience with the repressive native title system that favoured mining corporations and did not allow Indigenous native title groups to exercise FPIC.	Owing to Indigenous dispossession in settler colonial Australia, climate justice signified sovereignty, including through the legal system; They articulated an Indigenous climate justice based on the demand to redress historic injustices; The W&J also articulated climate injustice as an extension of their grievances against coal mining in the Galilee Basin, reflecting a dialectical process of meaning-making.	Links with anti-fossil fuel Indigenous struggles in North America strengthened their claim for sovereignty over Country; 'No Means No' was a land rights campaign to expose the failure of Native Title, but it acknowledged the climate change problem; Climate injustice through fossil fuel projects on Indigenous lands formed the basis for their international activism; The W&J campaign emerged as a leading Indigenous climate justice movement.	The W&J emphasised the independence of their campaign and drew distinctions between their historic claim for land justice and environmentalist's present-day demands for climate action; Stop Adani acknowledged linkages between Indigenous dispossession by settler colonialism and its continuation through coal extraction on Indigenous lands in Australia in its narratives; A new dimension to Indigenous–green relations in Australia predicated on environmentalists meeting traditional owners 'at the crossroad'.	Further research on anti-fossil fuel solidarities with Indigenous land struggles can reflect on the question *How can climate justice activism against –fossil fuel extraction centre Indigenous visions and futures in its worldview?*

Mahan community	An understanding of their legal rights brought a sense of what was at stake and the motivation to resist coal mining; the State's bias towards the Corporation and withholding their rights motivated them to fight.	Concerns over climate change not directly articulated at Mahan; Anti-coal movement narratives from the global South are often re-interpreted as climate justice narratives by other globally oriented actors even when the grassroots movement itself might not directly express climate change as a concern; Against a fraught political economic context, climate justice for forest communities impacted by coal mining signifies both being able to save their forests and livelihoods and to protect their legal rights under the FRA.	Against a context of government attacks on civil society groups and democratic institutions movement claims collectivised around a call for democracy and forest rights; Owing to a lack of connection with the scientific language of climate change rural communities articulate a different (from urban middle-class activists) narrative of environmental resistance that is situated in their immediate injustices.	Greenpeace transformed Mahan's awareness about their rights over forests; The Greenpeace–MSS relationship represented a socio-economic divide characteristic of postcolonial global South societies: between urban, educated, middle-class activists, and rural, subsistence communities; The nature of the Greenpeace–MSS relationship can be viewed as a community empowerment model; For Greenpeace, working with MSS meant learning to make room for the latter's logics and actions in their campaign strategy and decolonising their own understanding of subaltern activism; As an international ENGO, Greenpeace was able to mobilise campaign resources and funding to run an extended campaign in a remote location for five years, which was a prime factor behind the formation of the movement, its actions, and its relations; The government attacks on Greenpeace raised uncomfortable questions about the ENGO's ability to continue supporting the Mahan movement.	The vulnerability of Mahan's newfound rights poses a challenge for theorising climate justice in a transitional Southern context; Persistent structural divisions in the global South raise questions of agency, access to various platforms, and visibility for communities, that need to be considered in climate justice research; The role of the interlocutor – to hold all the interconnected threads of discourses emerging from the global South – is important to resist the transplanting of Northern perspectives on to Southern contexts; The central theme of human rights violations through land dispossession from coal mining in the global South raises a vital question for global environmentalism: *whether its aims can align with land justice and livelihoods security for impacted communities.*

Table 9.2 Similarities and differences between Australian and Indian politics and resistance of coal

Issue	Patterns and differences	Significance for a global outlook
ENGOs		
Political economic context of the ENGO campaigns	Both campaigns spanned a dynamic period of coal expansion that not only covered neoconservative Liberal Coalition and the right-wing Narendra Modi's Bharatiya Janata Party but also centre-left Labor and Congress-led governments in Australia and India, respectively; both movements encountered patterns of power and the prioritisation of coal companies over the public interest by the State under neoconservative and right-wing as well as centre-left governments.	The challenges and milestones of these environmental campaigns reveal the paradox of the State under neoliberalism, the extent of a State–corporate nexus in the coal sector, and the scope and limits of exercising democracy by CSOs in both countries; The parallel processes of neoliberalisation of the Indian economy and the Australian minerals boom have acted as dominant forces in reshaping the campaign approaches of ENGOs in the respective countries.
Basis on which the dominant environmental actor countered coal	Although the Stop Adani and Greenpeace campaigns reflect a multi-scalar politics and narrative, they used significantly different ways in countering coal; The issue of the Adani coal mine in Australia was dominantly framed as a climate change problem; the issue around the Mahan coal mine in India was dominantly framed around the forest rights of communities and assertion of democratic rights.	This indicates social and political differences, who is the dominant environmental actor in the movement, and a difference between the Northern and Southern contexts of coal and climate change politics; Coal mining extensively impacts on the human rights of communities on the frontlines of extraction in the global South and therefore they challenge coal on the basis of its implications for their immediate survival and livelihoods off the land and forests, as opposed to scientific climate change; the multiple and urgent human justice issues in the global South result in climate often being used as a 'proxy' issue by campaigns there to link with

Table 9.2 (Continued)

Issue	Patterns and differences	Significance for a global outlook
		global activism's narrative; The Southern context puts a critical mass of human justice issues on to a global platform of climate and environmental activism, and through its socio-economic complexities it challenges the relatively 'neat' framing of issues from the North.
Resources and tools of activism	Stop Adani was able to access the latest campaigning technologies. Although Greenpeace in India did use digital mobilisation tactics to target its urban supporters, given the dominance of the grassroots MSS mobilisation in its anti-coal campaign, Greenpeace's scope for using the latest campaign and organising technologies was limited; the local movement's prominent actions such as women blocking trees from being felled resonate with iconic struggles for forests such as Chipko and reflect a historicity in methods used by environmentalism of the poor.	The difference in the forms and tools of mobilisation between the Indian and Australian cases indicates a characteristic Northern versus Southern demographic difference in terms of the main actors in environmental campaigns.
Indigenous communities		
Green-Indigenous relations	The autonomy in the W&J's movement, its strategic collaboration with the environmental movement, its ability to access the United Nations to appeal for Indigenous self-determination and expose Australia's violations contrast with the co-dependency with which the Mahan movement was built up through Greenpeace's support.	This cannot be considered a characteristic North–South difference in Indigenous mobilisations, but it does point to persistent structural divisions in global South societies; Owing to which, the question of agency, access to various platforms, and visibility for communities, points to a North–South difference in climate justice struggles even in the case of Indigenous resistances.

(*Continued*)

Table 9.2 (Continued)

Issue	Patterns and differences	Significance for a global outlook
Indigenous resistances	Mahan faced the prospect of a direct loss of livelihoods from losing forest and lands on which they depend for economic survival, making the human rights challenge even more acute for global South communities; Subsistence-based communities in the global South fighting coal extraction can be doubly disadvantaged through a lack of energy access and the environmental injustice of loss and displacement; this once again speaks to a structural divide between the global North and South which extends to Indigenous communities.	Activists and researchers should not transplant Northern environmental discourses onto the South; This hampers the emergence of a contextualised understanding such as of the relation between land, livelihoods and environmental justice; Solidarity with Indigenous struggles against coal mining needs to keep land justice and human rights of Indigenous peoples and communities at the centre of advocacy while also recognising structural differences between the global North and South; The new environmentalism needs to extend solidarity to multifarious extractive challenges to Indigenous lands, historically, today, and in the future, that pertain to sourcing and producing energy.
Governments		
Responding to ENGO campaigns	Countering coal proved to be a litmus test for democracy in both the Australian and Indian cases; Both governments retaliated to environmentalists' anti-coal campaigns. However, the scale of retaliation differed, with the Indian government's crackdown on Greenpeace standing out as an extreme measure.	Although countering coal proved to be a litmus test for democracy in both a global North (Australia) and global South (Indian) case, the difference in the state of civil society and protest effectively indicates a difference in the state of democracy between Australia and India;

Table 9.2 (Continued)

Issue	Patterns and differences	Significance for a global outlook
Responding to Indigenous resistances	The W&J and Mahan cases show similarities primarily in the tension between the intent of land rights and self-determination, favouring of mining by the State, and neoliberal forces at play.	This similarity points to the challenges Indigenous resistances face in a democracy and revel the role of neoliberalism in jeopardising Indigenous self-determination.
Indigenous consent for mining	Administrative and legal procedures and actions of governments in both countries reflect a larger process within a neoliberal economic framework that prioritises mining over Indigenous rights and compromises FPIC for mining; Both governments colluded with mining corporations to manufacture Indigenous consent and amended and diluted Indigenous rights in favour of mining; But the extent to which the FRA in India was violated stands out as extreme; A structural divide persists in implementation of Indigenous land rights by the State between India and Australia.	This structural divide in the implementation and state of Indigenous rights speaks to a North–South distinction that can be experienced even in the case of democracies; This difference in the state of Indigenous rights by extension also indicates a difference in the state of democracy in between Australia and India; There is a need to pay attention both to the necessity for democracy for social and ecological justice, as well as its precarity, in a global South developmental context.
State–corporate nexus and pro-coal actions	Both governments demonstrated special treatments towards coal-mining corporations, and a strong support for coal. Civil society groups deemed such actions as undemocratic. Once again, the scale of government corruption in the management of coal blocks in India stands out as an exceptional case of 'crony-capitalism'.	This difference once again indicates a difference in the state of democracy between Australia and India, between accountability and transparency in the governance of natural resources between Northern and Southern coal democracies.

(*Continued*)

Table 9.2 (Continued)

Issue	Patterns and differences	Significance for a global outlook
Climate politics and its connection to coal	Unlike Australia's climate-denialist politics, India positions itself as a supporter of climate action; The Indian government's support for climate action does not generate the same political imperative for mobilisation on the climate issue as in Australia. But the government's strong support for coal mining makes its pro-climate approach paradoxical; and ENGOs do not usually target coal on the grounds of climate change owing to the postcolonial developmental anxieties of the Indian government, which makes such advocacy risky.	The imperative for activism of dominant environmental actors in the global South as well as the political stance of postcolonial governments towards climate change are essentially different from those in the global North and the politics of climate denialism; this influences a different political approach of ENGO campaigns around coal mining.

of the national anti-coal campaigns, tools and tactics used by the movements, government responses to anti-coal activism, and risks to activism and democracy from coal's entrenched power, in Australia and India. Pointing to the outstanding case of corruption in resource governance and the extent of attack on the rights and activities of civil society in India, I suggested the need to pay attention both to the necessity for democracy for social and ecological justice, as well as its precarity, in a populous global South context.

Drawing on the previous sections and weaving them with insights from my own experience in environmental campaigning in India and Australia, here I reflect on some crucial elements that determine a global outlook for a climate justice activism focussed on keeping coal in the ground, discuss the challenges around each of these elements, and make recommendations for activists and researchers.

Recognise past, present and future of land injustice from energy production and supply

During my fieldwork in Mahan, I learnt that many of the families had already been twice displaced by large dams built in the 1980s in the region, to settle in Mahan where the risk of coal mining induced displacement visited them in 2012. Both Mahan and the W&J have experienced prior displacement and removal, one from colonisation and the other from relatively recent energy related industrial developments. By acknowledging colonisation of lands and its continuation through extractive fossil fuel mining, the new approach of environmentalism has taken the past along with the present of land injustice towards Indigenous peoples and global South communities into account. It can also consider future scenarios of Indigenous land injustices from energy systems.

First nations people are at once finding themselves on the frontlines of climate impacts, changed land relations for critical mineral extraction, and setting up of renewable energy projects. This scenario begs the question whose climate and energy justice are we talking about? In India for example some massive solar renewable projects are repeating systemic issues caused by coal projects, such as land dispossession, livelihood disruption, and inability for local communities to access electricity from the project (Roy and Schaffartzik 2021). The surge for large-scale 'clean' and renewable energy systems has already begun creating winners and losers as certain land-based communities are being sacrificed. The relations of ENGOs with other actors in environmental resistances who have a direct stake in the land and natural resources being extracted is likely to assume further criticality going ahead.

As the new environmentalism is predicated on solidarity with communities on the frontlines of extractive energy production, it can broaden and deepen the grounds of its advocacy to focus on systemic and structural causes behind human rights violations in the political economy of energy systems. Environmental activism will need to look beyond today's singular narrative of 'Stop Coal' towards a deeper 'Just and Clean Energy' approach. Sustained solidarities with Indigenous self-determination struggles that foreground just and secure Indigenous futures as goals can begin to address cyclical and systemic land issues related to energy production that visit communities. Through this it can ensure that the post-carbon and renewables-powered society it envisions does not replicate the systemic human rights concerns of fossil fuel regimes.

Recognise land and forest rights as crucial human rights for communities on the frontlines of extractive climate and energy solutions

In Australia, the W&J asserted their rejection of mining on Country as a human right, making the case for the inseparability of the issue of land rights from human rights for Indigenous people. In another example, the youth-based climate coalition Youth Verdict has legally challenged the ecological destruction from the Galilee Basin coal mines on the basis of a breach of human rights, particularly the right to life and the cultural rights of Aboriginal and Torres Strait Islanders People (Daly and Douvartzidis 2020). In Mahan the risk of losing lands and forests where they live and on which their livelihoods depend resulted in a human rights issue for the community. Accounts of the W&J and Mahan community in this book make a clear case for a just future for Indigenous peoples and global South communities through secure land and forest rights, livelihoods, and forest commons.

The principle of FPIC is enshrined as a human right in the United Nations Declaration of the Rights of Indigenous Peoples. The W&J and Mahan cases make a clear case for ensuring that communities can self-determine about industrial developments on their lands through Indigenous legislations that allow FPIC, and the sound implementation of these legislations so that communities can exercise their legal right to self-determination. Mahan demonstrates the critical need for Adivasi and forest-dependent communities to be able to retain, secure and strengthen decision-making powers and governance over forests and their lands particularly through the mechanism of *Gram Sabhas*. As even renewable energy and other so called 'climate solutions' can revictimise Indigenous peoples through land use changes, anti-coal climate justice activists need to approach a just transition by advocating for secure land and forest rights and tenures for Indigenous peoples and global South communities as their essential human rights.

Advocating for rights to forests, land and self-determination as essential human rights for Indigenous peoples and communities becomes increasingly necessary given that Indigenous and subsistence-based communities fighting coal extraction are likely to experience human rights violations on top of the slow violence of climate change.

Advocate for strengthening democracy for a just transition

The Mahan movement exposed how democracy is compromised in Schedule 5 Indigenous majority areas in central India through violation of Adivasi rights. It raises the need to approach just transition in central India through securing grassroots democracy, through the proper implementation and preventing violations or weakening of the FRA. As community-based forest governance is recognised as essential for protecting biodiversity and consequently safeguarding from climate impacts, strengthening grassroots democracy by providing autonomy for local governance and rights to resources is essential both for climate justice for forest-dependent communities and mitigating climate impacts (IPBES 2019, Agarwal and Dash 2022).

These linkages point to the need for international solidarity movements with grassroots Indigenous uprisings against coal in a global South context like India to advocate for strengthening the functioning of local and grassroots democratic systems particularly in its mineral-rich regions. The Indian case study highlights both the necessity and the precarity of democracy on the ground in coal mining regions. This makes it difficult to theorise climate justice in India's transitional context and makes it essential for researchers and activists to grapple with this contradiction.

But the crackdown on Greenpeace by the Narendra Modi-led Indian government for campaigning against coal extraction's ecological and social injustices indicates a contentious political context of energy transition in India, and points to the need to pay attention to democracy even further up from the grassroots level. Operating with anxieties around energy security India continues to increase its coal-fired capacity even as it rapidly increases its renewable capacity; India's sensitivity to global commentary on its coal policy is likely to make the climate and energy campaigns of international ENGOs operating in India vulnerable to government scrutiny and interference.

But the widespread targeting of the civil society sector under the Narendra Modi government since 2014, especially organisations and projects defending the human rights and livelihoods of communities, points to a further level of challenge to the state of democracy in India today, as well as the indispensability of democracy in having the necessary debates around the climate and environment and justice and rights of communities.

The Australian case also demonstrated the need for strengthening democratic accountability of governments towards the public interest and supporting Indigenous self-determination as opposed to the profits of private coal mining corporations. But the extent to which the democratic right to dissent was challenged in the Indian case makes it necessary for international climate campaigns to grapple with the precariousness of global North–South anti-coal alliances and the challenges for global ENGOs to operate within a southern economic context of anxieties of the State around energy security. As well as to make democracy an essential plank of advocacy for international climate justice campaigns – for a just energy transition and for social and ecological justice of communities. Not just to move away from coal but to socially and politically design clean and renewable energy through consent and the democratic representation of the rights, livelihoods, forests and lands, and futures of communities.

Recognise the role of the interlocutor from the global South

Political ecology is an effective framework for understanding and addressing the underlying social and political drivers of environmental change and developing ethical solutions. Which is why it was so relevant for this book – for consistently delineating the social and political make-up of the ground and the context from which environmentalisms arose in Australia and India, and to advocate solutions that are relevant for the context in which the environmentalisms are operating. This was especially important for India, with a persistent tendency of global climate change advocacy and climate justice research to transplant Northern frames on to Southern contexts.

Particularly Indigenous and peasant populations in the South fighting coal extraction are likely to face three kinds of injustices: not only are they vulnerable to climate impacts and the effects of coal mining on their lands, but they also lack access to electricity, the very commodity that coal produces (Talukdar 2017). The priorities of such communities can significantly differ from those that are fighting coal extraction in the global North. But political ecology as a framework has been rightly criticised for failing to grapple with multiple scales and for an overemphasis on the local, which can pose challenges with 'scaling up' local issues and making them relevant in a global (amongst other scales) context.

Through this book I have bridged the dissonance between the local and the global scale of climate justice campaigns and contestations, as well as drawn parallels across the transnational scale. To stitch together dissimilarities in the politics and narratives of environmental movements and to draw out their common ground has been a fundamental aim and a constant endeavour in my work in environmental campaigning over two decades. I have referred to this role as that of an interlocutor in other sections of this chapter. I see the role of the interlocutor as fundamental in tying the various scales of politics and resistance on the environment and climate together, both in climate justice activism and climate justice research.

To add to Naomi Klein's (2016) description of the role of the climate activist to 'overcome the various disconnections and connect our various movements', I specifically propose the role of the interlocutor from the global South, who, by virtue of their connection with local communities and grassroots resistances in a global South setting, as well as with the campaigns of international climate organisations and international climate justice research agendas, can continuously translate the similarities and differences between the ground and the international approach, and suggest ways of bridging the gap between these two through approaches of solidarity and narratives of climate justice that strengthen and bolster stories of communities.

In effect, the interlocutor from the global South slowly and steadily creates space and understanding for the complex and interconnected factors of gender, socio-economics, environmental impacts, caste, class, dynamics of rural or urban societies, and other intersectional factors as the case may be, amongst international activists and research scholars in climate justice.

In the Introduction to this book, I mentioned my work in building understanding amongst Australian climate activists about the socio-political context behind community struggles against coal in India. I have broadened and deepened this work through regular workshop dialogues between South Asian and Australian climate justice activists and researchers. The importance of this slow and sometimes imperceptible work to build the global North community's understanding of the global South contexts needs to be recognised and supported in activist and research spaces.

9.6 Contributions to political ecology and environmental justice research

Here I discuss how the case study chapters have built on existing political ecological texts and addressed known gaps in the field of study. I suggest areas for further study that can build on findings from this book. I also summarise this book's contribution to environmental justice research.

Contributions to Political Ecology

In Chapters 1 and 2, I discussed *Varieties of Environmentalism: Essays North and South* (Guha and Martinez-Alier 1997) whose methodological approach – that of conceptualising different environmental resistances by delineating their social, political, and economic contexts – underpins the approach of this book. The idea of this text was to offer a scholarly response to the hegemony of a certain kind of Northern environmentalism by delineating several, particularly Southern and subaltern, environmentalisms. This book has built on it by bringing the concept of various environmentalisms into today's climate justice activism in the global North and South.

This book has built on texts on Indian environmentalism (discussed in Chapter 2) by bringing the global climate change issue into dialogue with grassroots and historic disaffections. The case studies have introduced the dynamics of multi-scalar politics and multiple non-State actors at a site of environmental conflict. For First World Political Ecology, the Australian case has brought a current anti-coal resistance in a settler colonial society into dialogue with colonisation and industrialisation and their social, political and ecological effects by tracing the State's historic relation with various anti-coal actors.

The Indian (3, 4 and 5) and Australian (6, 7 and 8) chapters are organised around a central analysis of the State and governance regimes across multiple scales – the international, national, regional and local – both historically and during comparable time frames of neoliberal development and the minerals boom respectively. In Chapters 3 and 6, this approach has helped to anchor the analysis of the question 'how environmentalisms have transformed from their previous versions' in geographic areas that have been transformed by neoliberal resource extraction through State-support while community rights have been neglected. This line of analysis has helped to discuss national political trends – the politics surrounding land rights and the decade of climate inaction in Australia between 2007 and 2017 and dilution of the progressive LARR and FRA legislations and crackdowns on CSOs in India since 2014 – that had a bearing on the narratives and politics of the resistances down to the very local level.

In Chapters 4 and 7, tracing of the historical arc of coal-led growth and the State's central role has established fundamental differences between the neoliberal Australian versus developmental Indian States. It has highlighted similarities in the processes of favouring coal developments, especially against today's realities of worsening climate change and decreasing economic viability of coal. Through a central focus on the State, Chapters 5 and 8 have qualified how cumulative State actions since the colonial period have shaped the context around which the current site-based conflicts are playing out, making evident the historicity and the marginality of local actors and their claims.

This book has addressed a known gap in political ecological literature which has 'kept circling State theory without fully engaging with it' (Loftus 2018, p. 140) by engaging various conceptualisations of the State. The case study chapters have analysed settler colonial development particularly in Queensland and the effects of the neoliberalising political economy and resource curse on Australian politics. And conceptualisations of the postcolonial developmental State and its transformation under neoliberalism in India.

But these concepts have failed to adequately justify all State behaviours: the extent of corruption in the Coalgate coal block allocation scam and violent withholding of community rights in central India require a more intricate inspection of crony capitalism and the inherent contradiction in the welfare and developmental roles of postcolonial States. The extent of the resource curse on Australian politics also needs a more intricate inspection of the grip of the Rupert Murdoch owned NewsCorp news media on Australian climate politics. Comparative studies across relatively similar contexts (North–North or South–South) will be able to help delineate patterns and nuanced differences in the behaviours of captured States.

Several earlier political ecological texts have studied ecological changes at sites through interactions of political and economic processes along the vertical scale of the local, regional, national and international (see Blaikie and Brookfield 1987). Rangan and Kull's (2009) argument about the need for research that focusses on how scale is being used to politicise ecological change is directly relevant for climate activism's new strategy to politicise climate change through anchoring it in local coal-extraction while also scaling up grassroots anti-coal resistances as climate justice movements.

The scalar politics of States and the counter-scaling politics of the anti-coal movements have emerged as analytical themes through the interpretation of research results in both cases. The treatment of the Galilee Basin and Singrauli as sacrifice zones indicates the State's assertion of coal's significance along the vertical scalar geographies of the region, State and nation. The mobilisations in both countries attempted to reconfigure this paradigm by building alliances across these territorial boundaries. The notion of Hunter Valley as Carbon Valley and the Carmichael coal mine as a carbon bomb indicate a simultaneous scalar movement towards grounding climate change and globalising coal extraction. Although climate change did not play a similar mobilising role in India, Mahan's anti-coal struggle for forest rights was scaled up as a critical assertion of democracy through nationwide civil society support.

Further, as political ecology traces the long arc of industrialisation and its eco-social effects, it has allowed this research to establish continuity by tracing the colonial-industrial origins of today's climate crisis, and to make Indigenous people's historical disenfranchisement by colonial capitalism relevant in their present struggles against coal extraction. This framework has therefore proven effective for the goal of this book to understand how climate activism can achieve common ground both across and within geographies with struggles of historically marginalised peoples.

Finally, while governments continue to support fossil fuel projects even with advancing climate change, resistances 'from the ground up' to fossil fuel projects, consisting of communities who are more often than not left out of policy considerations, are asserting their voices in debates on energy transition and energy justice. The North–South comparative framework of research on environmentalism in this book, that acknowledges the difference between cases being compared from industrialised versus industrialising socio-economic contexts (amongst others in a North-South comparison) can be effective for in-depth comparisons of grassroots uprisings facing similar challenges of fossil fuel extraction and its effects on their lands and natural resources.

Contributions to environmental justice

Building up a systematic body of such comparative ethnographic research can complement the research scholarship of global projects such as the Global Atlas of Environment Justice (https://ejatlas.org), a global database of environmental conflicts and resistances that serves as a valuable tool for researchers and activists, by providing an in-depth understanding of critical similarities and differences in grassroots mobilisations on similar issues.

Increasing scales of violence against environmental activists and Indigenous land defenders against coal mining and other fossil fuel projects point to the need for research to document human rights violations in environmental conflicts and to help assess protection measures (Feng et al. 2020). The documentation of the violation of forest rights at Mahan, the risks and intimidations to the local MSS movement, and the government crackdown on Greenpeace contribute to environmental justice research on human rights violations in environmental conflicts.

Summarising other areas for further research discussed earlier

Areas of future research previously suggested in Sections 9.1 to 9.4 and Tables 9.1 and 9.2 of this chapter include exploring how green–farmer relations can play a sustained role in energy transition and sustainable futures and reflecting on the question

> *How can climate justice activism against fossil fuel extraction centre Indigenous visions and futures in its worldview?*

I hope that how I have approached political ecology in this book can make a meaningful contribution for future comparative research on climate justice activism along the lines of recommended above.

Conclusion

This book has analysed the politics of coal and anti-coal resistances in two coal-led economies, one in a global North context in Australia and one in a global South context in India, where globally oriented climate and environmental groups have worked in alliance with communities to counter coal mining.

It has examined how the anti-coal activism of the Stop Adani movement has affected the political debate on climate and energy in Australia, how its activism has targeted the political, economic and infrastructural structures of the coal economy through a multi-pronged campaign including divestment, national mass mobilisations, grassroots and electorate-level local actions, and peaceful direct disruptions of the coal project. It has investigated the relational politics of the environmental movement with local farmers against coal mining, and its relationship with the campaign for land rights of the W&J traditional owners. It has discussed what narrative and politics their collective resistance has generated in Australian politics and society, and what this collective movement means for environmental activism in Australia.

The book has also examined how the anti-coal activism of Greenpeace India has affected the political debate on climate and energy in India, and how its activisms have symbolically (rather than substantially) targeted the political, economic, and infrastructural structures of the coal economy through a multi-level campaign that included exposing State–corporate collusion in destroying India's old growth forests, urban mobilisations, and a significant community mobilisation at the proposed mine site. Based on an ethnographic approach, the research phase of this study investigated the relationship between the ENGO and the local community at Mahan, and how through the process of their interactions they negotiated a shared significance and narrative for their collective action. Through an examination of the civil society debate about the value of anti-coal activism in India, the case study has analysed the significance of the anti-coal movement for environmentalism in India and its democratic debate.

This book provides a timely comparison of environmental activist campaigns of stopping coal extraction. Solidarity with communities at the frontline of coal-extraction, and a disruption of the coal-economy at multiple-scales, together constitute a new template for this environmental activism. Targeting of coal extraction has also brought the national political economies of energy-development under the purview of environmental activism. This has brought environmental activism into direct conflict with issues of energy security and the national interest. The anti-coal activisms in both India and Australia have challenged coal's power at various scales from the local to the global, and in the process provoked a backlash from governments that have deemed these campaigns as risks to the national economic interest, regardless of coal's structural decline.

Drawing on and elaborating the themes identified through a discussion on works of literature in Chapter 2, and based on findings from the two case studies in Chapters 3–8, this final chapter has thematically discussed the parallels and distinctions between the two contexts, and between the movements, their politics, narratives and relationships. It has made key recommendations for climate activists and researchers toward making campaigns and agendas of global climate justice activism for an energy transition intersectional from a global North–South perspective.

This chapter has delineated the varieties of climate justice emerging from the politics and resistances of various environmental actors in the two case studies, highlighting the various imperatives including outside of the risk of climate change, and pointing to how climate justice is often applied as an overarching narrative frame to struggles resisting the destruction of land, forests and Country from coal mining by non-environmentalist actors. The new relational politics of environmental activism also points to the need to go beyond the urgent purpose of stopping coal extraction in alliance with these communities, towards collectively envisioning alternative futures beyond fossil fuels that are ecologically and socially just.

Under pressure from global climate activism, particularly the campaign for divestment from fossil fuels, and withdrawal of investors from coal extraction and thermal power projects, governments in countries with coal-driven political economies could likely demonstrate increasing anxieties around coal-driven energy security. The collective resistances of environmentalists and local community actors signify a necessary politics to shift the discussion on energy transition by directly intervening to facilitate the end of coal. As communities fighting coal extraction on the front line are usually the ones left out of policy decision-making, the purpose of their alliance with environmentalists can extend beyond the immediate necessity of stopping coal extraction, towards co-generating post-carbon economic alternatives.

There is a strong resonance between the anti-coal activisms in Australia and India, and the shared meanings they generate with their non-environmentalist movement allies, based on an emphasis on human rights and land rights of Indigenous people. However, a structural divide can persist between the global North and South even in case of Indigenous societies, and their means, ability, and access to various platforms for effective resistance. However, overall, this convergence constitutes a fundamental common ground between Northern and Southern disparities in environmental activisms' new campaigns, and deepening and strengthening this emphasis holds distinct possibilities for its politics and advocacy on energy transition.

Through an emphasis on human and land rights issues, global activism can offer a systemic critique of the fossil fuel regime in both Northern and Southern contexts that is underpinned by justice as well as the impacts of colonialism and capitalism on Indigenous communities. Through advocating for strengthening democracy especially at the grassroots level in the global South, climate advocacy can act not just in solidarity with anti-coal struggles but in anticipation of the various extractive challenges that even so-called clean energy and climate solutions can bring to communities. By looking at the past, present and future of land justice for communities, climate activism can move away from a singular focus of 'Stopping Coal' and emphasise the systemic challenges that large-scale energy systems can pose towards communities and their lands. Last but not the least, through paying attention to the crucial yet often invisible role that interlocutors from the global South play, global climate activism can learn to create space for building a fully contextualised understanding of the intersecting challenges that communities in global South locations encounter, and act in meaningful solidarity.

This chapter also highlights contributions of this book towards the field of political ecology particularly around its focus on the movement of the anti-coal resistances across scales, and the central role of the State in exacerbating or ameliorating environmental injustices through substantially engaging with various conceptualisations of the State in Australia and India. Finally, this chapter makes recommendations for areas for further research in comparative environmentalism and global climate justice research to build upon this book's findings.

Notes

1 A 2021 circular by the Indian government to state governments gives the latter the responsibility to review and facilitate the proper of implementation of forest rights, stating that 'despite a considerable lapse of time since it (FRA) came into force, the process of recognition of forest rights is yet to be completed' (Nandi 2021).

2 Compromises to the provisions of the FRA and rights of communities have also risen from the shortening of approval timelines for environmental clearances, and relaxing the requirement of public consent for coal mining projects under the FRA (CSE 2016). To boost coal mining, India loosened forest clearance processes in 2020 that compromised

the autonomy of *Gram Sabhas* (CFR-LA 2020). The democratic provisions and intent of the FRA to redress historic injustices have been further diluted by the Indian Narendra Modi government in 2022 through the notified new Forest Conservation Rules, allowing people's forests to be diverted for projects even without a No Objection Certificate from *Gram Sabhas*.

References

Adve, N. 2020, 'South Asian coalition links climate with social struggles', *Ecologist*, 26 February, viewed 20 September 2020, <https://theecologist.org/2020/feb/26/south-asian-coalition-links-climate-social-struggles>.

Agarwal, S. & Dash, T. 2022, Policy Brief: Securing Climate Justice for India's *Forest-Dependent Communities*, viewed 20 November 2022, < https://rightsandresources.org/blog/effective-implementation-of-the-forest-rights-act-can-help-india-secure-a-just-and-sustainable-pathway-to-climate-change-mitigation-and-adaptation-new-study-says/>.

Ananthapadmanabhan, G., Srinivas, K. & Gopal, V. 2007, Hiding Behind the Poor, Greenpeace India, Bangalore, viewed 20 March 2020, <www.greenpeace.org/india/Glo bal/india/report/2007/11/hiding-behind-the-poor.pdf>.

Arashiro, Z. 2017, 'Mining, social contestation and the reclaiming of voice in Australia's democracy', *Social Identities*, vol. 23, no. 6, pp. 661–673.

Australian Conservation Foundation. n.d., *Fossil Fuel Money Distorting Democracy*, ACF, viewed 20 September 2020, <https://d3n8a8pro7vhmx.cloudfront.net/auscon/pages/17065/attachments/original/1605244960/Fossil_fuel_money_distorting_democracy.pdf?1605244960>.

Bhagat, A. 2018, *Breaking Renewable Energy Myths*, Greenpeace India, 5 November, viewed 20 September 2019, <www.greenpeace.org/india/en/story/2177/breaking-renewa ble-energy-myths/>.

Blaikie, P. & Brookfield, H. 1987, *Land Degradation and Society*, Methuen, London.

CFR-LA. 2016, *Promise and Performance, Ten Years of the Forest Rights Act in India*, Citizen's Report on the Promise and Performance of Scheduled Tribes and Other Traditional Forest Dwellers (Recognition of Forest Rights) Act 2006.

Community Forest Rights-Learning and Advocacy (CFR-LA). 2020, *Community Forest Rights and the Pandemic: Gram Sabhas Lead the Way*, October, viewed 20 September 2022, <https://rightsandresources.org/wp-content/uploads/2020/10/CFR-and-the-Pand emic_GS-Lead-the-Way-Vol.2_Oct.2020.pdf>.

Chowdhury, C. & Aga. A. 2020, 'Manufacturing consent: mining, bureaucratic sabotage and the Forest Rights Act in India', *Capitalism Nature Socialism*, vol. 31, 2020, no. 2, pp. 70–90.

Connor, L. G., Freeman, S. & Higginbotham, N. 2009, 'Not just a coalmine: shifting grounds of community opposition to coal mining in Southeastern Australia', *Ethnos*, vol. 74, no. 4, pp. 490–513.

Centre for Science and Environment. 2016, Report Card: *Environmental Governance Under NDA Government*, 22 June, viewed 20 September 2019, <www.downtoearth.org.in/cover age/governance/report-card-environmental-governance-under-nda-government-54359>.

Daly, S. & Douvartzidis, L. 2020, 'The convergence of human rights law and environmental and climate change litigation in Australia', *Johnson Winter and Slattery*, June 2020, viewed 20 October 2020, < https://jws.com.au/insights/articles/2020-articles/the-conv ergence-of-human-rights-law-and-environment>.

Doherty, B. & Doyle, T. 2006, 'Beyond borders: Transnational politics, social movements, an modern environmentalisms', *Environmental Politics,* vol. 15, no. 5, pp. 697–712.

Feng, J., Mildenberger, M. & Stokes, L. C. 2020, 'Inhumane environments: global violence against environmental justice activists as human rights violations', in M. Stohl & A. Brysk (eds), *A Research Agenda for Human Rights*, Edward Elgar Publishing.

Fernandes, A. 2012, How Coal Mining *Is* Trashing Tigerland, Greenpeace India Society, Bangalore, viewed 14 August 2019, <www.greenpeace.org/india/en/publication/984/how-coal-mining-is-trashing-tigerland/>.

Forsyth, T. 2014, 'Climate justice is not just ice', *Geoforum*, 54, vol. 54, pp. 230–232.

Greenpeace Australia. 2012a, Boom Goes the Reef: Australia's *Coal Export Boom* and the *I*ndustrialisation of the Great Barrier Reef, March 2015, viewed 20 June 2020, <www.greenpeace.org.au/news/boom-goes-the-reef/>.

Greenpeace Australia. 2012b, Cooking the Climate Wrecking the Reef: The Global Impacts of Coal Exports from Australia's Galilee Basin, December 2015.

Gregoire, P. 2019, 'In defence of country: an interview with Wangan and Jagalingou Council's Adrian Burragubba', *Sydney Criminal Lawyers*, 14 September, viewed 20 October 2020, <www.sydneycriminallawyers.com.au/blog/in-defence-of-country-an-interview-with-wangan-and-jagalingou-councils-adrian-burragubba/>.

Guha, R. & Alier, J. M. 1997, *Varieties of Environmentalism: Essays North and South*, Routledge.

IPBES (Intergovernmental Science-Policy Platform on Biodiversity and Ecosystem Services). 2019. Summary for Policymakers of the IPBES Global Assessment Report on Biodiversity and Ecosystem Services, United Nations, viewed 20 November 2023, <https://ipbes.net/sites/default/files/inline/files/ipbes_global_assessment_report_summary_for_policymakers.pdf>.

Klein, N. 2014, *This Changes Everything: Capitalism vs. the Climate*, Simon and Schuster, New York.

Klein, N. 2016, 'Let them drown: the violence of othering in a warming world', *London Review of Books*, vol. 38, no. 11, 2 June.

Kohli, K., Kothari, A. & Pillai, P. 2012, *Countering Coal?*, Discussion paper by Kalpavriksh and Greenpeace India, New Delhi/Pune and Bangalore, viewed 15 September 2019, <www.greenpeace.org/india/en/publication/989/countering-coal-community-forest-rights-and-coal-mining-regions-of-india/>.

Latulippe, N. & Klenk, N. 2020, 'Making room and moving over: knowledge co-production, Indigenous knowledge sovereignty and the politics of global environmental change decision-making', *Current Opinion in Environmental Sustainability,* vol. 42, pp. 7–14.

Lawhon, M. 2013, 'Situated, network environmentalism: a case for environmental theory from the south', *Geography Compass*, vol. 7, no. 2, 128–138.

Lee, J. I. & Wolf, S. A. 2018, Critical assessment of implementation of the forest rights act in India, *Land Use Policy,* 79, 834–844.

Loftus, A. 2018, 'Political ecology II: whither the state?', *Progress in Human Geography*, vol. 4, no. 1, pp. 139–149.

Lyons, K. 2016, 'Australia's coal politics are undermining democratic Indigenous rights', *The Conversation*, 26 October, viewed 20 October 2020, <https://theconversation.com/australias-coal-politics-are-undermining-democratic-and-indigenous-rights-66994>.

Murphy, K. 2017, 'Indigenous people victims of 'green' fight against Adani mine, says Marcia Langton', *The Guardian*, 7 June, viewed 20 October 2020, <www.theguardian.com/australia-news/2017/jun/07/indigenous-people-victims-of-green-fight-against-adani-mine-says-marcia-langton>.

Nandi, J. 2021. 'Review implementation of forest rights: Union Ministries to state governments', Hindustan Times, 7 July, viewed 20 November 2022, <www.hindustanti mes.com/environment/review-implementation-of-forest-rights-union-ministries-to-state-govts-101625562374847.html>.

O'Neill, K. 2012. 'The comparative study of environmental movements', in P. F. Steinberg & S. D. VanDeever (eds), *Comparative Environmental Politics: Theory, Practice and Prospects*, MIT Press, pp. 115–142.

Porter, L., Bosomworth, K., Moloney, S. & Naarm, 2020. 'Decolonising climate change adaptation', Planning Theory and Practice, May.

Purdy, J. B. 2015, 'Environmentalism's racist history', The New Yorker, 13 August, viewed 20 September 2018, <www.newyorker.com/news/news-desk/environmentalisms-racist-history>.

Rangan, H. & Kull, C. A. 2009, 'What makes ecology 'political'?: rethinking scale in political ecology', *Progress in Human Geography*, vol. 33, no. 1, pp. 28–45.

Roy, B. & Schaffartzik, A. 2021, 'Talk renewables, walk coal: the paradox of India's energy transition', *Ecological Economics*, vol. 180, February 2021, no. 106871.

Sax, S. 2023, 'Scramble for clean energy metals confronted by activists calls to respect Indigenous rights', *Mongabay*, 21 April, viewed 20 March 2024, <https://news.monga bay.com/2023/04/scramble-for-clean-energy-minerals-confronted-by-calls-to-respect-indigenous-rights/>.

Scott, J. C. 1985, *Weapons of the Weak: Everyday Forms of Peasant Resistance*, Yale University Press.

Sethi, N. 2019, 'Modi government's move to amend forest rights takes a giant leap backwards', Business Standard, 1 April, viewed 20 March 2020, <www.business-standard. com/article/economy-policy/modi-govt-s-move-to-amend-forest-act-takes-a-giant-leap-backwards-119040101292_1.html>.

Smee, B. 2020, 'Adani legal action sparks fears journalists could be targeted over Carmichael mine leaks', *The Guardian*, 28 September, viewed 20 October 2020, <www.theguardian. com/business/2020/sep/28/adani-legal-action-sparks-fears-journalists-could-be-targeted-over-carmichael-mine-leaks>.

Talukdar, R. 2017, 'Hiding neoliberal coal behind the Indian poor', *Journal of Australian Political Economy*, no. 78, pp. 132–158.

Talukdar, R. 2018, 'Sparking a debate on coal: case study on the Indian government's crackdown on Greenpeace', *Cosmopolitan Civil Societies: An Interdisciplinary Journal*, vol. 10, no. 1, pp. 47–62.

Talukdar, R. 2019, 'Profit before people: why India has silenced Greenpeace', *New Matilda*, 29 March, viewed 20 June 2020, <https://newmatilda.com/2019/03/29/pro fit-before-people-why-india-has-silenced-greenpeace/?fbclid=IwAR2NwczXun5q1Ug-g3kSxvaOJLInTfwNlnS6Up3Xxts7A7XZF_wbvOsHKYc>.

Whyte, K. 2013, 'On the role of traditional ecological knowledge as a collaborative concept: a philosophical concept', *Ecological Processes*, vol. 2.

Whyte, K. 2017, 'Way beyond the lifeboat: an indigenous allegory of climate justice', in D. Munshi, K. Bhavnani, J. Foran & P. Kurian (eds), *Reimagining Global Climate Justice*, University of California Press.

Whyte, K. 2021, 'Time as kinship', in J. Cohen & S. Foote (eds), *The Cambridge Companion to Environmental Humanities*, Cambridge University Press.

Williams, G. & Mawdsley, E. 2006, 'Postcolonial environmental justice: government and governance in India', *Geoforum*, vol. 37, pp. 660–670.

Epilogue

As I write the postscript in the simmering South Asian summer of 2024, looking at the period in Australian and Indian politics on coal that I researched – between 2011 and 2018 – from hindsight makes apparent two contrasting realities. The first is around the political rhetoric on coal, and how it has changed, and yet not changed, in both countries. Australia has seen a change of government from a conservative and often climate-denialist Liberal Coalition to a centre-left Labor in 2022. In India, the Bharatiya Janata Party has begun a third successive term in government under Prime Minister Narendra Modi, albeit without the sweeping majority that this right-wing political party enjoyed after the 2019 Indian elections.

Both countries have committed to reduce coal usage on their own terms and are also continuing to open new coal mines. Curiously and yet predictably, the massive build-up of renewable energy projects in both countries – India has the world's largest privately operated solar park in Gujarat and Australia is gearing to becoming a renewable energy production and export superpower – has not slowed down the plans of either country to mine more coal. Outside of the essential performative politics on the world stage – for India as a developing economy asserting sovereignty for its development and for Australia as a trade-dependent resource-rich industrialised country continuing to repeat the 'no-regrets' approach – the forces of neoliberalism, privatised energy production and global capital are shaping the narratives around coal and renewables in both countries, with governments as mouthpieces. Which part of the political spectrum a government comes from – conservative or not, centre-left or right – has very less to do with this predictable economic approach. The difference, if any, lies merely in their choice of words and phrases.

However, where the contrast between Australia and India appears striking is at the ground level – at the sites of the Mahan and Carmichael coal mines – and the dialectic between people and the State and Corporation. Take into consideration the developments at Carmichael and Mahan, particularly in the last two years. Although the Carmichael coal mine is under operation, it is a mere fraction of the grand design that the Adani group and Australian governments once envisioned. Also, a protest by the Wangan and Jagalingou continues near the mine site on their

DOI: 10.4324/9781003410416-10

traditional lands. The world is constantly reminded that the Carmichael coal mine is operating without Indigenous consent.

Meanwhile in Mahan, people who had won respite from coal mining in 2015 are now having to worry about the possibility of underground coal mining that will eventually degrade and destroy the forests that serve as their economic, social and cultural basis. Two coal blocks in the Mahan forests have been allotted for private coal mining following coal block auctions in early 2024; the larger of them that holds 956 million tonnes of coal belongs to Adani Enterprises. If we consider the prospect of communities and activists to express dissent against coal mining, the difference between the Australian and Indian democracies appears significant. In fact, how Adani got allocated the coal block in the Mahan forest effectively repeats the old story of the Indian government's corrupted coal-block allocations, as a series of investigative articles by the independent Reporters Collective has shown. In terms of governing India's coal resources, democracy and accountability have not won the day.

How democracy and accountability operate on the ground level today has critical implications for climate justice. A slide in democracy, particularly grassroots democracy, has a fundamental bearing on the rights and justice of communities that either continue to face the impacts of coal projects or can be exposed to new forms of extraction for renewable energy projects, carbon-offset-related afforestation programmes, or any other projects that operate on their lands and in their forests without consent or consultation.

But as this book pays tribute to the indefatigable spirit of environmental justice resistances in Australia and India, it will be pertinent to conclude by acknowledging some movements that have embarked on a long journey for climate justice, regardless of whether they use that term explicitly or not, and by noting the resilience and spirit of these movements in the face of various odds. Over a decade long, the Indigenous resistance to save the old growth Hasdeo forests from the onslaught of coal mining in the central Indian state of Chhattisgarh has once again exposed an old conflict at the heart of coal mining in India – whose development is coal mining serving? In Australia, as the extraction of gas gains renewed attention from governments and corporations as a 'less carbon-intensive' export commodity compared to coal, and as mining for critical minerals unfolds on a large scale, Indigenous communities are facing and challenging new extractive risks to their traditional lands.

Such movements can continue to serve as guiding cases for effective solidarity across a global North and global South context, at a time when the world is transitioning its energy sources. Climate justice activists will have to remember to recognise the similarities and discern the differences in challenges experienced on the ground, in a rapidly industrialising versus an industrialised democracy such as India and Australia, while working towards building solidarity. .

Index

For Product Safety Concerns and Information please contact our EU
representative GPSR@taylorandfrancis.com
Taylor & Francis Verlag GmbH, Kaufingerstraße 24, 80331 München, Germany